Elemental Analysis of Airborne Particles

Advances in Environmental, Industrial and Process Control Technologies

A series edited by **Tuan Vo-Dinh,** Oak Ridge National Laboratory, Oak Ridge, Tennessee, USA

Volume 1
Elemental Analysis of Airborne Particles
Edited by Sheldon Landsberger and Marsha Creatchman

In preparation

Gas and Particle Phase Measurements of Atmospheric Organic Compounds
Edited by Douglas A. Lane

This book is part of a series. The publisher will accept continuation orders which may be cancelled at any time and which provide for automatic billing and shipping for each title in the series upon publication. Please write for details.

Elemental Analysis of Airborne Particles

Edited by

Sheldon Landsberger

and

Marsha Creatchman

University of Texas at Austin, USA

Gordon and Breach Science Publishers

Australia Canada China France Germany India
Japan Luxembourg Malaysia The Netherlands
Russia Singapore Switzerland

 Printed in Malaysia

Amsteldijk 166
1st Floor
1079 LH Amsterdam
The Netherlands

British Library Cataloguing in Publication Data

Elemental analysis of airborne particles. – (Advances in environmental, industrial and process control technologies ; v. 1 – ISSN 1028-3587)
1. Aerosols 2. Air – Pollution 3. Air – Analysis
I. Landsberger, Sheldon II. Creatchman, Marsha
541.3'45'15

ISBN 90-5699-627-4

CONTENTS

INTRODUCTION TO THE SERIES

The aim of this book series is to provide an international and multidisciplinary forum for new advances in the areas of environmental, industrial and process control technologies. In the last few years, important advances have been made in improving existing monitoring methods and/or developing new techniques and instrumentation. Many of these techniques have been described in specialized technical journals or reported at scientific symposia but have not been integrated in a comprehensive book series. The need for such a series is critical because of the increasing demands of the regulatory environment in such areas as hazardous waste remediation, monitoring of indoor air pollution, and industrial hygiene control. Process control issues also address the current emphasis on improving industrial productivity and product quality for cost minimization and global competitiveness.

This series focuses on techniques and methods that have potential for real-life, practical applications in environmental, industrial and process control. Topics included are technology-specific (e.g., lasers for remote sensing, ion-trap mass spectrometry, field-screening spectroscopy methods, spot tests for pharmaceuticals, remote chemical sensors, biosensors, and process control techniques in petroleum industries), problem-specific (e.g., polychlorinated biphenyls, dioxins, and bioremediation of organic wastes), or of programmatic interest (e.g., methods for risk assessment in the nuclear energy industries, and environmental monitoring in Eastern Europe, Asia and so on).

Tuan Vo-Dinh

PREFACE

Over the past few decades it has become increasingly clear that along with the continuing rise in the standard of living in developed and developing countries, comes the price of pollution. Perhaps no single type of contamination other than air pollution can make the public acutely aware of the environmental and health-related consequences resulting from modern-day industrial technology. The study of chemical constituents in the atmosphere is driven by several factors: effects on biological systems (human, animal and plant), materials damage, transport, deposition and radiative processes, radon and other airborne radioactivity, basic physio-chemical sciences, government regulations, environmental global change and atmospheric aesthetics. There have been hundreds of studies involving the analysis of gases and airborne particles in urban, rural and remote regions of the world employing state-of-the-art air sampling devices, common and cutting-edge analytical measurements, and various simple and sophisticated mathematical models.

This book was the outcome of many years of personal frustration at not having a collection of detailed works specifically devoted to the most popular methods in elemental analysis of airborne particles that could serve graduate and undergraduate students, colleagues, and myself. While many review articles have appeared on various chemical methods, few have focused on their application for the trace element analysis of air samples. The intent of this publication is to offer investigators in this field a comprehensive book on several of the most common laboratory analytical methods currently used in trace element analysis. This is not intended to be a book on air sampling, detailed principles of analytical methods, or treatment of detection limits. The reader is encouraged to investigate the many general reviews on these individual matters. In the distant and recent past there have been several excellent works on air sampling, air pollution engineering and specific analytical methodologies (Copper and Alley, 1986; Flagan and Seinfeld, 1988; Gratt, 1996; Harrison and Perry, 1977; Hewitt, 1991; Kouimtzis and Samara, 1995; Leithe, 1971; Liu, 1976; Lodge, 1989; Ness, 1991; Patrick, 1993; Perry and Young, 1977; Sell, 1992; Shields, 1993; and Willeke and Baron, 1988). Dzubay (1977) was one of the first to edit a book on x-ray fluorescence of environmental samples, with several chapters devoted to airborne particle measurements.

The analytical methods chosen for this book include: atomic absorption spectrometry; inductively coupled plasma-atomic emission spectrometry; inductively coupled plasma-mass spectrometry; ion chromatography; x-ray fluorescence; accelerator based ion beam techniques (including particle induced x-ray emission, particle-induced gamma-ray analysis, particle elastic scattering and Rutherford backscattering); and neutron activation analysis. Specific sections on quality assurance and quality control (QA/QC), and source-receptor modeling have also been included since they are critical additions in such an endeavor. Specific references to QA/QC and source-receptor modeling are also included in the individual chapters. Other well-known methods such as polarography, selective ion electrodes, anodic stripping voltammetry, and emission spectrographic analysis were not included. Total reflection x-ray fluorescence (TRXRF) was also not considered. Although it is a powerful method, it is not as yet used extensively for elemental analysis of airborne particles.

In summary, as needs continue to grow for elemental analysis, environmental interpretation and mathematical modeling of airborne particles, it is hoped that this modest contribution will serve both the research and monitoring communities.

REFERENCES

Cooper, C. D. and F. C. Alley, *Air Pollution Control: A Design Approach*, Waveland Press (1986).

Dzubay, T. G., *X-Ray Fluorescence Analysis of Environmental Samples*, Ann Arbor Science (1977).

Flagan, R. C. and J. H. Seinfeld, *Fundamentals of Air Pollution Engineering*, Prentice Hall (1988).

Gratt, L. B., *Air Toxic Risk Assessment and Management: Public Health Risk from Normal Operations*, Van Nostrand Reinhold (1996).

Harrison, R. M. and R. Perry, *Handbook of Air Pollution Analysis*, Chapman and Hall (1977).

Hewitt, C. N., *Instrumental Analysis*, Elsevier Applied Science (1991).

Kouimtzis, Th. and C. Samara, *Airborne Particulate Matter*, in the Handbook of Environmental Chemistry, Springer, Volume 4 Part D, pp. 175–200 (1995).

Leithe, W., *The Analysis of Air Pollutants*, Ann Arbor Science Publishers (1971).

Liu, B. Y. H., *Fine Particles-Aerosol Generation, Measurement, Sampling, and Analysis*, Academic Press (1976).

Lodge, J. P., *Methods of Air Sampling*, Lewis Publishers (1989).

Ness, S. A., *Air Monitoring for Toxic Exposures: An Integrated Approach*, Van Nostrand Reinhold (1991).

Patrick, D. R., *Toxic Air Pollution Handbook*, Van Nostrand Reinhold (1993).

Perry, R. and R. J. Young, *Handbook of Air Pollution*, Chapman and Hall (1977).

Sell, N. J., *Industrial Pollution Control*, Van Nostrand Reinhold (1992).

Shields, J., *Air Emissions Baselines and Environmental Auditing*, Van Nostrand Reinhold (1993).

Willeke, K. and P. A. Baron, *Aerosol Measurement-Principles, Techniques and Applications*, Van Nostrand Reinhold (1988).

1 TRACE ELEMENT ANALYSIS OF AIRBORNE PARTICLES BY ATOMIC ABSORPTION SPECTROSCOPY, INDUCTIVELY COUPLED PLASMA-ATOMIC EMISSION SPECTROSCOPY, AND INDUCTIVELY COUPLED PLASMA-MASS SPECTROMETRY

PETER M. GROHSE

Research Triangle Institute, Research Triangle Park, North Carolina, 27709, USA

INTRODUCTION

Advances in trace element measurements by atomic spectroscopy over the past 25 years have paralleled a dramatic increased concern with environmental issues, including human exposure to trace toxic elements through inhalation. The introduction of various atomic absorption (AA) techniques has provided laboratories performing routine environmental analyses, with the capability of determining picogram per cubic meter (pg/m^3) to nanogram per cubic meter (ng/m^3) concentrations of trace elements when very large air volumes are collected. One limitation of available instrumentation was the lack of a multiple element measurement. When plasma emission instrumentation such as inductively coupled plasma-atomic emission spectrometry (ICP-AES) and direct current plasma-AES (DCP-AES) became commercially available, multielement determinations for up to 65 elements were possible. Initially, the equipment was fairly costly.

When sequential plasma emission instruments became available in the early 1980s, multielement instrumentation in turn, became more affordable for the small analytical laboratory. However, sensitivity was inferior to that of the most sensitive AA techniques until improved nebulizers and torch designs were introduced. In the mid 1980s, ICP-mass spectrometry (ICP-MS) emerged as an ultratrace multielement technique providing detection limits two to three orders of magnitude lower than the most sensitive ICP-AES or atomic absorption spectroscopy (AAS) configurations. Instrument costs for ICP-MS have been high, but lower cost versions are becoming increasingly available making sensitive instrumentation affordable for small laboratories.

Tedious sample preparations such as hotplate usage have been largely supplemented in the past ten years with microwave-assisted procedures. Most of these systems employ closed vessel methodology that is rapid, efficient, and minimizes contamination from laboratory air. Automated microwave systems have become commercially available in the past five years, but this equipment is of limited value for solid samples (such as filters).

In this chapter, commonly used sample digestion/extraction procedures and their subsequent measurement approaches are covered. Advantages and disadvantages of each are discussed, as well as their applicability to air particle samples from various environments; (i.e., ambient, indoor, industrial workplace, or combustion).

SAMPLE PREPARATION

The major challenge to the atomic spectroscopic techniques is the requirement that some solubilized form of the analyte element must be presented to the instrument for the measurement. Three disadvantages are immediately apparent:

1. The preparation involves sample destruction.
2. There is an increased potential for contamination.
3. The sample is necessarily diluted.

However, sample digestion can be advantageous. For example, a filter digest is obviously more representative of the sample than direct solid sampling techniques that examine a small portion of the filter, e.g., solid samples for graphite furnace atomic absorption (GFAA). In addition, one can expect interferences to be less pronounced in a diluted sample. Most commonly used dissolution techniques involve the selection and placement of a representative aliquot of a sample in a suitable container with appropriate leaching/digestion acid or acid mixture added to it. The sample is then usually agitated and/or heated using a number of possible mechanisms: mechanical or manual shaking, ultrasonication, classical hot plate heating, or the more recent microwave-assisted approach. For highly insoluble matter, fluxes may be employed at a high temperature as an intermediate step prior to acid dissolution.

Sample preparation activities are influenced by several laboratory and sample conditions:

- required detection/quantity limit capabilities
- potential laboratory contamination
- volatility of the analyte(s)
- solubility of the analyte/matrix
- stability of the analyte in the digestate
- filter/sampling medium

No discussion of trace analysis is complete without mentioning contamination issues. The analyst should be aware of the reasonable level of contamination control for the given matrix. For example, open beaker digestions in a conventional fume hood might be suitable for filter samples of industrial stack emissions, but are totally inappropriate for ultra low-level indoor or ambient air samples where it would be impossible to differentiate trace analyte concentrations from high and inconsistent blanks. Conversely, it is excessive to use subboiled acids and a Class 100 clean room environment for the combustion of samples from industrial sources. It is important for a laboratory intending to provide the spectrum of low- to high-level determinations to dedicate clean laboratories for indoor air/ambient air samples and assign separate, less rigorous facilities for high-level particle samples. In addition, it may be useful to place in perspective the required facilities for the given analyte. For elements such as Al, Ca, Fe, Mg, Na, and Si and, to a lesser degree, trace elements, Cr, Mn, Ni, Pb, and Zn, the above considerations are crucial to the quality of the results. These species are ubiquitous in the environment and contamination control is difficult. If the target analyte is, Be or U, extraordinary measures for cleanliness may not be required even for ultratrace measurements. It is up to the analytical chemist to be aware of these issues and determine the extent of potential contamination in a given laboratory environment.

Contamination in trace metal determinations originates from a variety of sources. These include, at a minimum:

- laboratory personnel (hands, hair, etc.)
- laboratory environment (dust, rust, etc.), including ventilation hood (metal surfaces, rust, unfiltered air, etc.)

- laboratory water
- digestion solvents (acids, etc.)
- inadequately cleaned glassware/plasticware
- filters used for the removal of particulate matter following digestion
- instrument sample introduction systems (i.e., exposure of open autosampler tubes to room air)

Personnel should always wear laboratory coats and metal-free (preferably powderless) gloves during sample preparation and measurement. Since many shampoos contain high levels of metals such as zinc and selenium, long hair needs to be tied back or laboratory head cover worn. For the ultimate low-level determination, laboratory clothing covering the entire body (including the feet) is appropriate.

The normal laboratory environment contains unacceptably high levels of a number of the metallic elements noted previously. When this same air is pulled through a conventional hood where open digestions are being conducted, the potential for contamination is high. Outside dust is also drawn into the hood as makeup air. At least three approaches involving increasing levels of contamination control are possible. At a minimum, the chemist can digest or extract samples in a closed system, such as a laboratory microwave oven. Expose contents of microwave vessels only for very short periods of time to allow for transfer to volumetric plasticware/glassware. An additional measure would involve the use of a Class 100 hood, constructed entirely of plastic. It is supplied by air that has passed through high-efficiency particulate air (HEPA) filters. A final measure would involve the use of an entire Class 100 clean room supplied by filtered air with the use of nonmetallic surfaces and plumbing fixtures.

Use water for final rinsing of glassware and preparation of samples that at a minimum meets ASTM Type II specifications (ASTM, 1991). In addition use Type I water 18 mΩ or better (ASTM, 1991) for ultratrace applications, and nonmetallic water fixtures. Acids used for digestion, extraction, or acidification should be trace metals grade or better. At least two grades are specifically available for trace metal determinations. The stated purity levels need to be carefully evaluated by the laboratory. Other reagents used in sample preparation also require the highest possible purity consistent with the application.

Designate a facility that is separate from the analytical area for glassware cleaning. Sample containers and digestion glassware cause positive and negative errors due to surface effects from adsorption, desorption, and leaching. Improper removal of environmental contaminants in the laboratory or during the manufacturing process also causes errors. Initially clean glassware with laboratory grade detergent (no chromic acid cleaners) and water followed by a four-hour soak in 1:2:9 HNO_3:HCl:H_2O mixture (US EPA, 1991). Glassware is then rinsed with tap water and ASTM Type I water (18 mΩ) and allowed to either drain or, preferably be stored while filled with a dilute acid such as 1% HNO_3 until use. Immediately prior to use, the glassware is rinsed again thoroughly with Type I water. Other cleaning procedures, such as soaking glassware in 50% HNO_3 for 24 hours followed by Type I water rinsing, are also acceptable (US EPA, 1980; 1995d).

Before using plastic disposable containers, it is frequently useful to select several at random that been have soaked overnight in 5% HNO_3 and analyze the acid for the analyte(s) of interest. If the analyte is not present at detectable levels, it is sometimes advisable to use the glassware or plasticware directly to avoid potential for the introduction of contamination during cleaning procedures for certain analytes, such as Cr, Cu, Fe, Mg, Ni, and Zn. Store cleaned glassware in a sealed, nonmetallic, dust-free area (ziplock bags, Class 100 hoods, etc.).

Plastic vessels are usually preferred for the storage of standards and acid digests. Polyethylene, polypropylene, polymethylpentene, and Teflon® are all usually suitable. It is necessary to have appropriate solution storage periods because there is potential for losses due to adsorption to container walls and precipitation (Moore and Meredith, 1977). Consequently, calibration or quality control check standards are normally prepared daily. Sample digests containing very low parts-per-trillion concentrations to minimize sources for errors require analysis as soon as possible. The laboratory must monitor the stability of the analyte in the storage container of choice at the given concentration and tailor its quality control measures accordingly.

Filter Extraction and Digestion

Dissolution strategies employed for trace metal measurements of air particles is driven by many factors (as previously described). The measurement technique frequently imposes constraints on acids used for dissolution. For example, digestion media other than nitric acid are not recommended for GFAA and ICP-MS determinations. Chlorine-containing acids are particularly inappropriate for GFAA and ICP-MS for many elements. In GFAA, analyte chlorides formed in the furnace are prone to volatilization prior to the atomization step, which leads to low recoveries. In ICP-MS, chlorine and sulfur species form numerous polyatomic mass spectral interferences.

Although nitric acid alone may be appropriate for dissolution of many elements, certain ones, such as Ag, Sb, Si, Sn, Ti, and others, may require halogenated acids (HF, HCl, $HClO_4$) for solubilization and stabilization, once in solution. Removal of dissolved gaseous NO_x species from the final digest for samples to be analyzed by hydride generation atomic absorption (e.g., for As, Se, and Sb) is desired. In addition, the volatility of the analyte must be considered. If a refractory element (e.g., Ti) and a volatile species (e.g., As) are both target elements in the same sample, clearly an open beaker high-temperature wet ashing technique or dry ashing technique would most likely lead to losses of the volatile species. Therefore a good choice is a closed Teflon® vessel technique (microwave or other heat source) to solubilize and retain elements of variable volatility.

The filter material used for sample collection also influences the preparation method. Target analytes attached to particulate matter embedded in the glass or quartz fiber of a high-volume filter might not be quantitatively recovered unless some combination of hydrofluoric and nitric acids is employed to completely digest the filter medium. Other analytes, such as Cd or Pb, might be satisfactorily extracted from the glass or quartz fiber using a dilute acid solution in conjunction with ultrasonication or water bath (90°C) heating, for example. Nitric acid alone easily solubilizes mixed cellulose ester membranes. The addition of another acid is sometimes required to further solubilize certain difficult elements (Sn, Ti, etc.). Teflon® membranes that are resistant to acid mixtures are subjected to leaching procedures, usually with nitric acid or a combination of nitric and hydrochloric acids, which lead to instrument constraints. However, other acids are sometimes required to solubilize certain metals from the surface of this material. Paper filters require combinations of nitric and perchloric or sulfuric acid to achieve complete digestion and avoid undissolved matter that can clog nebulization systems. Alternatively the digest can be filtered.

In the past 10 years there has been considerable progress in the area of microwave-assisted acid dissolution techniques (Kingston and Jassie, 1988). Microwave techniques are particularly effective when conducted under high-temperature, high-pressure (HTHP) conditions. These methods have been developed for a wide range of applications including digestion of particles from geological and combustion sources. Extraction efficiencies equal to or greater than those for lengthy classical techniques, typically two to eight hours, are usually accomplished in 10–30 mins. These techniques are particularly suitable for ultratrace picograms/nanogram determinations since they are carried out in a closed vessel, thereby avoiding room dust contamination during the heating process.

The digestion and vessel material is a major issue. Most plastics–polyethylene, polystyrene, polypropylene, polysulfone, and polymethyl pentene are suitable for low temperature (<100°C) and ultrasonic extractions. Vessel materials such as Teflon® can withstand higher temperatures and pressures and are ideal for microwave work. Borosilicate glass and quartz are fairly temperature resistant (especially quartz) but clearly inappropriate for high-pressure applications unless used with a metal (stainless steel, etc.) jacket. Domestic microwave units are not recommended since they are not equipped with corrosion-resistant safety interlocks and are more prone to magnetron damage when operated empty. Laboratory microwave equipment is convenient and safe but expensive, costing about 50 times more than equipment designed for domestic applications. Microwave digestion systems are now available from a number of manufacturers.

A more traditional approach to pressurized acid digestions employs steel jacketed Teflon®-lined bombs. Samples are digested in the presence of acid (HNO_3, HCl, HF) and are enclosed in a Teflon® liner which in turn is jacketed in stainless steel. The assembly is usually heated in a muffle furnace approximately 140°C. A weakness of this method is corrosion of the jacket resulting in contamination and poor sample throughput.

If complete dissolution of particulate matter is desired but not accomplished with acids or acid mixtures, fluxes considered high-temperature acids and bases can be employed as an intermediate agent to convert the sample to an acid soluble form. With this technique, a (filter) sample is placed in a crucible, graphite, platinum, etc. A fluxing agent is added (e.g., lithium metaborate), and the crucible is heated at a high temperature until the sample flux mixture is molten. When cool, the "meld" is usually dissolved in dilute acid. Awareness of the acid/base and redox properties of fluxes is necessary. For example, silicate materials are acidic and would benefit from a basic flux such as lithium metaborate. Conversely, samples high in calcium and magnesium might be successfully solubilized using lithium tetraborate as the intermediate agent. Oxidizing fluxes are more common than reducing fluxes and generally assume their behavior only in the presence of air (Alfassi and Chien, 1992; Minczewski *et al.*, 1982). Fluxes cannot be obtained with the low contaminant levels of high purity acids. They are not recommended for volatile elements. Finally, dissolved fluxes need considerable dilution to avoid clogging nebulization systems. A summary and overview of available general preparation techniques that may apply to various air particulate sample matrices is found in Table 1; and advantages and disadvantages of each are shown in Table 2.

Although mercury, a highly volatile element, is not viewed as particulate in nature, Hg species may be adsorbed to the surface of a particle. Mercury as an analyte presents special problems for digestion, primarily due to volatilization and sample storage. Generally speaking, Hg species are lost during any kind of ashing or open beaker evaporative process. Closed vessel, low-temperature (<100°C), or reflux type digestions are recommended. Mercury from ambient air frequently exists as an organic

Table 1. Digestion/extraction methods.

- Hot plate acid digestions (including block heating)
- Acid reflux digestions (Kjeldahl)
- Autoclave (mild heating with acid)
- Ashing techniques/acid dissolution (low and high temperature)
- Fusion techniques/acid dissolution
- Ultrasonic extraction with dilute acids (and other mechanical agitations)
- Closed vessel techniques
  — Microwave-assisted acid digestion
  — Acid bomb
  — Oxygen bomb over acid

Table 2. Summary of major extraction methods.

Method	Hot plate	Ashing	Ultrasonication	Microwave
Time of digestion	Hours	Hours	15–60 min	2–30 min
Throughput	Low to high	Medium	Medium to high	Medium to high
Sample size (g)	<1	Up to 20	<1	<1
Temperature (°C)	150–400	400–1200	<100	100–200
Unattended	No	Yes	Yes	Yes
Operating cost	Low	Low	Low	Low
Capital cost	Low	Mid	Mid	High to very high
Comments	Possible losses, Possible refractory insolubility	Loss of volatile elements	Refractory metals insoluble	Fairly universal

derivative, e.g., methyl mercury chloride, dimethyl mercury. Consequently, it is important that the digestion is sufficiently oxidative in order to break down the compound into Hg^{2+} species. Many techniques employ a mixture of sulfuric and nitric acids in conjunction with potassium permanganate, but this approach is not suitable for multiple metal determinations (US EPA, 1979; 1995a). Frequently, using only nitric acid under pressure with microwave techniques is adequate (Thompson and Grohse, 1996a, b). Mercury when stored, is readily adsorbed on all common container/labware surfaces: glass, plastics, Teflon®, and quartz. Losses of Hg^{2+} are also due to reduction/volatilization, hydrolysis, and coprecipitation (Alfassi and

Chien, 1992). The degree of loss is dependent on storage time and temperature, cleanliness of the vessel surface, pH, etc. Glass containers are preferred to plastics. Mercury vapors from a contaminated environment will penetrate polyethylene storage containers.

Preconcentration Techniques

Once an analyte has been solubilized by a given dissolution technique it is possible to enhance the concentration or isolate the analyte species in order to obtain low detection limits with less costly instrumentation and/or remove potential analytical interferences. A number of techniques are available that provide one to three orders of magnitude enhancements, but the following are the most commonly used:

- Coprecipitation of the analyte onto the surface or into the crystal lattice structure of a host species such as $Fe(OH)_3$ or MnO_2, followed by isolation and dissolution of the resulting precipitate (Alfassi and Chien, 1992).
- Solvent extraction of a chelate of the target element(s) (Alfassi and Chien, 1992).
- Ion exchange of the element–metals are collected on the ion exchange resin. Major elements such as alkali and alkaline earth metals are first removed from the resin with ammonium acetate, for example. Heavy and transition metals are then eluted with dilute acid (Alfassi and Chien, 1992; US EPA, 1995a). Packages for automated ion exchange preconcentration are commercially available.

Although the concentration enhancement can be dramatic, the method(s) are usually labor intensive, potentially add contamination due to the added sample manipulation, and are limited to transition and heavy metals.

Quality Assurance/Quality Control Samples (QA/QC)

Low recoveries of trace element species are a function of the following:

- losses due to volatilization of the analyte during the digestion extraction process
- incomplete solubilization of the analyte during the above activity
- instability of the analyte in the digest/extract due to container wall absorption/adsorption and precipitation of insoluble salts
- low recoveries arising from the measurement process (e.g., matrix suppression of the analyte, which is discussed later)

High recoveries can result from a heterogeneous sample or contamination of the sample. Highly variable results likewise may reflect a heterogeneous sample or sporadic blank contamination from particulates in the laboratory air.

The recovery of a given analyte in the overall method is best determined by carrying a known quantity of the analyte through the entire analytical process, which includes digestion and measurement. Whenever possible, use reference materials containing the analyte matrix and filter material used during sample preparation. Some of the available spiked filter reference materials including filter matrix and analyte are given in Table 3. Materials with representative analytes and matrices are not always available. Laboratory personnel can however, prepare useful control samples. Materials that reflect a combustion process sample can be prepared by accurately spiking milligram levels of a National Institute of Standards and Technology (NIST) flyash standard reference material (SRM) onto an appropriate blank filter. Ambient air samples can come from a wide variety of sources. A soil or sediment SRM can be used to spike a blank filter to simulate a rural area air sample, and urban dust SRMs are similarly utilized. Combinations of the above are employed to represent indoor air samples. The use of matrix QC samples is crucial in the method validation and development process. Some reference solids used for method validation of specific flyashes, sediments, topsoils, etc. are listed in Table 4. At a minimum, use liquid multielement solution spikes of the filter matrix to provide data quality information. Low recoveries are likely due to:

1. filter matrix suppression — this is also seen with the postdigestion and method of standard additions (MSA) spike of a portion of an actual sample digest;
2. analyte instability in the solution — this occurs if the digest is allowed to sit around for a few days;
3. analyte volatility during digestion.

This type of filter will not test the ability of the method to fully solubilize the analyte(s) in a given matrix different from the standard; but spiked filters will test the low-recovery items 2 and 3 above.

Summary of Methodology

Prior to 1989, dissolution of airborne particles was performed with single acid or acid combinations to either leach the target analyte or solubilize the entire filter sample to access the more difficult elements, such as Al and Si (Sneddon, 1983). Leaching methods involved hot-plate heating and ultrasonication. More complete digestion methods included acid bombs with HF and HNO_3, fusions with solid flux materials ($LiBO_4$, etc.), and high-temperature ashing methods for carbonaceous filters. More recently,

Table 3. Some available filter reference materials (liquid spikes).

Reference materials	Source	Filter type	Certified analytes
NIST 2676	NIST[19]	Cellulose ester	Cd, Mn, Pb, Zn
NIST 2677[a]	NIST	Cellulose ester	As, Be
NIST 3087	NIST	Cellulose ester	Ba, Cd, Cr, Fe, Mg, Ni, Pb, Se, Sn
QC-TMFM series (three concentrations)	HPS[20]	Celluose ester	Ag, As, Ba, Be, Cd, Cr, Co, Cu, Fe, Mn, Ni, Pb, Se, Ti, V, Zn (plus an additional 17 (elements)
XICP-TM series, five filters	ChemService, Inc.[21]	Cellulose ester	33 elements

Table 4. Suggested matrices for filter spiking.

	Reference materials	Number of certified elements[a]
SRM 2704	Buffalo River sediment	28
SRM 2709	San Joaquin soil	26
SRM 2710	Montana soil (high)	22
SRM 2711	Montana soil (low)	24
SRM 694	Western phospate rock	13
SRM 1632b	Coal (bituminous)	23
SRM 1633b	Coal flyash	24
SRM 1635	Coal (subbituminous)	14
SRM 1648	Urban particulate matter	14
SRM 2689 2690 2691	Coal flyash	9 ea
SRM 2583	Trace elements in indoor dust	Low-level Pb and other elements (in preparation)
SRM 2585	Trace elements in indoor dust	Pb and other elements (in preparation)
SRM 2586	Trace elements in soil	Nominal 500 μg/g Pb and other elements

[a]These materials contain many other elements with uncertified values.

microwave-assisted acid dissolutions have aided the chemist by providing greater throughput, cleaner digestions, and in some cases, more complete dissolution.

Contamination from laboratory dust, acids, other digestion media, and glassware (plasticware) require careful consideration. Although reference filters containing true sample matrices are not available, there are a variety of QA/QC measures that test the acceptability of the test dissolution and measurement methods.

MEASUREMENT METHODS — AN OVERVIEW

Trace elements in air particles are measured by atomic absorption and emission methods involving a variety of atomization, sample introduction, and detector configurations. Flame atomic absorption (FAA) is the oldest technique (Walsh, 1955). It became available as a commercial instrument in the early 1960s, and is still used today for some applications (e.g., alkali and alkaline earths). The cold vapor atomic absorption technique specific for mercury was introduced in 1968 and remains the most commonly used technique for mercury determinations (Hatch and Ott, 1968). The instrumentation uses an inexpensive sample introduction modification to the FAA system. The most sensitive measurement of Hg is performed using cold vapor atomic fluorescence spectrometry (CVAFS), available commercially and easily fabricated in the laboratory (Bloom and Fitzgerald, 1988; US EPA, 1995b). For low level determinations of elements that form volatile hydrides such as: As, Ge, Sb, Se, Sn, Te, and even Pb, hydride generation atomic absorption (HGAA) provides detection limits (DL's) in the picogram–nanogram range for filters that correspond to low- to submicrograms per liter in solution (Caldwell *et al.*, 1973). GFAA became commercially available around 1970 (L'vov, 1959) and is still utilized for low parts per billion (micrograms per liter) determinations of As, Cd, Pb, Sb, Se, and other elements. Although over 50 elements may be successfully determined with this technique, many laboratories find it preferable to use a multielement technique such as ICP-AES (Wendt and Fassel, 1965; Fassel and Kniseley, 1974). This technique became commercially available in the mid to late 1970s, and is widely used today. However, detection limits for many elements are not as good compared to GFAA. Various torch configurations and sample introduction systems recently introduced have alleviated this shortcoming to some degree (Vandecasteele and Block, 1993). Many laboratories are now turning to ICP-MS for ultratrace, sub-parts-per billion, multielement analyses (Date and Gray, 1981; Gray, 1974; Douglas *et al.*, 1983; Houk *et al.*, 1981). The first commercial instruments appeared in the early 1980s. Detection limits are usually superior to those of GFAA for more than 75 elements. This method is considered routine for many applications. It also provides sensitive radionuclide and isotopic information, not available with other atomic spectroscopic techniques.

The measurement method selected for a given air particle matrix should reflect a consideration of detection limits, interferences, instrument cost, and cost of operation. An approximate comparison of detection capabilities of the major methods and lower concentrations of interest for each air matrix: ambient, indoor, industrial workplace, and industrial stack emission, expressed as solution concentrations is shown in Figure 1. Estimated instrument detection limits for various elements are compared in Table 5. When making approximations, the following factors must be taken into account: air sample volume, concentration of interest (weight/unit volume), total digestion and extraction solution volumes, and expected compromise of instrumental detection limits due to the real sample matrix effects.

Implentmenting enhancement techniques such as ultrasonic nebulization, lowers detection limits an order of magnitude or more for the various ICP-AES and ICP-MS methods. Characteristics for the major atomic spectroscopic methods are summarized in Table 6 and the most sensitive instrumental methods are shown in Table 7.

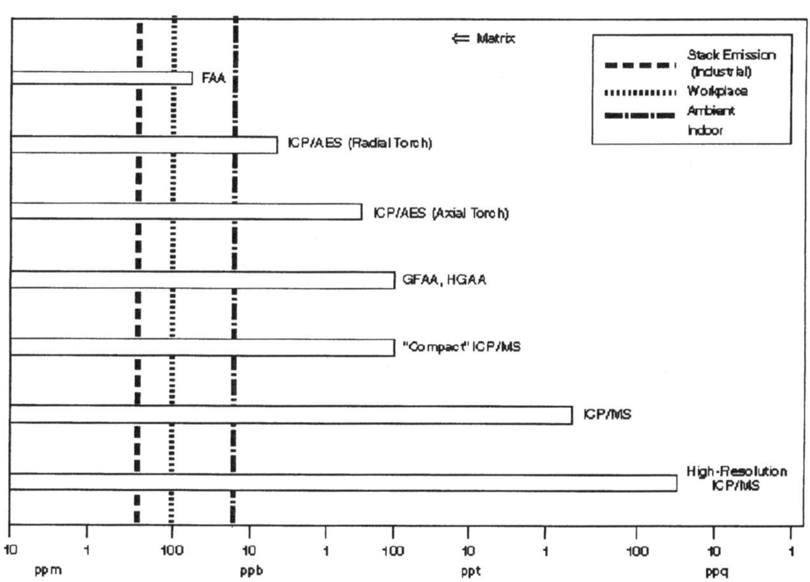

Figure 1. Comparison of measurement methods.

Table 5. Instrument detection limits (μg/L)[a] (estimated).

Elements	Flame AA	Furnace AA[b]	ICP/AES[c] (radial)	ICP/AES[d] (axial)	ICP/MS
Ag	10	0.2	7	0.6	
Al	100	0.5	45	1.9	e
As	f	0.6	53	4	0.002
B	f	f	5	0.2	0.01
Ba	100		2		0.005
Be	5	0.02	0.3		0.002
Ca	80	g	10	0.2	e
Cd	25	0.02	4	0.2	0.001
Co	50	0.3	7	0.8	0.01
Cr	50	0.2	7	0.4	0.01
Cu	20	0.4	6		0.003
Fe	30	g	7	0.5	0.03
K	10	0.4			
Mg	1	g	30		e
Mn	10	0.1	2		0.01
Mo	100	0.4	8		0.002
Na	2	g	30		e
Ni	40	0.6	15		0.005
Pb	100	0.6	45	5	0.002
Sb	f	1.1	32		0.001
Se	f	1.4	75		0.05
Si	f	g	60		e
Tl	100	0.9	40		0.001
U	f	f	f	f	<0.0005
V	0.8	2	8		0.01
Zn	5	g	2		0.002

[a]Appropriate DL — 3 times std. dev. of blank.
[b]20 μL injection.
[c]Using most interference free wavelength.
[d]Not all values available — general 10–20 times improvement over radial torch.
[e]Blank or interference limited.
[f]Not suitable by method — method too insensitive.
[g]Not suitable by method — too sensitive for air particulate.

Table 6. Measurement method comparison.

	ICP/MS	ICP/AES	Flame AAS	GFAAS
Detection limits (approx.)	< 0.001–1 μ g/L	5–100 μ g/L	50–1000 μ g/L	< 1–5 μg/L
Number of elements	> 75	> 70	> 65	> 50
Sample throughput (aprrox.)	2–5 min/sample	2 min/sample (simultaneous) 5–10 elements/min/sample (sequential)	15 s/element/sample	2–4 min/element/sample
Linear dynamic range	10^5–10^8	10^5	10^3	10^2
Precision				
Short-term	1–3%	0.5–3%	0.1–1%	1–5%
Long-term	< 4%	< 5%		
Interferences				
Spectral	Moderate	Moderate/severe	Minimal	Minimal
Matrix	Moderate	Moderate	Moderate/severe	Moderate/severe
Ionization	Mild	Mild	Moderate	Mild
Sample usage	1–2 mL	1–5 mL	5–20 mL	< 0.05 mL
Semiquantitative analysis	Yes	Yes (poor)	No	No
Isotope analysis	Yes	No	No	No
Routine operation	Yes	Yes	Yes	Yes
Skill level required	Moderate	Moderate	Low	Moderate
Unattended (i.e., overnight)	Yes	Yes	No	Yes
Costs				
Operating	High	Moderate	Low	Moderate
Capital	High/very high	High	Low	Moderate

Table 7. Most sensitive methods for selected trace elements (approximate solution detection limit in μg/L).

Element	ICP/MS DL	Alternate method(s) (DL)
Ag	0.01	ICP/AES[a] or GFAA[b] (0.02)
As	0.01	HGAA[c] (0.02)
Be	0.01	ICP/AES or GFAA (0.05)
Bi	0.001	GFAA (0.5)
Cd	0.005	GFAA (0.05)
Co	0.005	GFAA or ICP/AES (0.5)
Cr	0.01	GFAA or ICP/AES (0.5)
Cu	0.01	ICP/AES (0.1)
Hg	<0.001 (CVAFS)	CVAA (0.01)
Mo	0.002	ICP/AES (0.5)
Ni	0.005	ICP/AES (0.5)
Pb	0.001	GFAA (0.5)
Sb	0.001	HGAA (0.04)
Se	0.05	HGAA (0.02)
Tl	<0.001	GFAA (0.05)
U	<0.001	d
V	0.01	GFAA (1)

[a]With axial torch configuration.
[b]Using a 20-μL sample.
[c]Hydride generation using a 50-mL sample.
[d]No other atomic spectroscopy method suitable for trace analysis.

FLAME AND VAPOR GENERATION ATOMIC ABSORPTION SPECTROSCOPY AND ASSOCIATED TECHNIQUES

FAA instrumentation has been commercially available for almost 35 years. Modifications of the basic instrumentation have included in chronological order: cold vapor atomic absorption (CVAA) for mercury, GFAA, and HGAA techniques. CVAA and HGAA accessories remain important for ultratrace determinations in particulate matter, but flame atomization has faded due to the development of more sensitive and rapid multielement techniques.

While FAA is no longer a commonly used analysis of metals, it has several useful applications. This method is the most cost effective analysis for laboratories doing low-budget metal analyses requiring monitoring levels of parts per million. It is rapid in terms of measurements per minute, requires minimal operator training, and is suitable for the determination levels of parts per billion for alkali metals in environmental samples. FAA's elements, optical conditions, atomization conditions, linear range, and detection limits are listed in Table 8.

The method is still applied to air samples representing very high sample volumes, highly contaminated sources, or major elements likely to have high concentrations. In FAA spectroscopy, the sample solution (digest or extract) is introduced into the air or nitrous oxide acetylene flame as an aerosol using one of several nebulization techniques. In the flame, the aerosol droplets are evaporated producing dry particles, which in turn are disassociated and form ground state atoms.

A light beam from a hollow cathode lamp (HCL) (Sullivan and Walsh, 1965), or an electrodeless discharge lamp (EDL) (Dagnall and West, 1968) is directed through the flame into a monochrometer and thereafter onto a photomultiplier (PMT) detection system that measures absorbed light (Figure 2). Absorption depends on the presence of the free unexcited ground state atoms generated in the flame from the sample. The source emits light that is characteristic of the element being measured because the electrodes in these lamps are constructed of the analyte element.

Hollow cathode lamps generally last two years or longer and are characterized by fairly rapid warm-up times. With older single beam instruments, this was an important feature since signal drift was a major consideration during warm-up. With double-beam instruments, drift is compensated for and lamp warm-up time is not a major issue. Current instrumentation electronics are far more stable, but lamp warm-up times still require consideration during instrument setup. Lamps for volatile elements age more

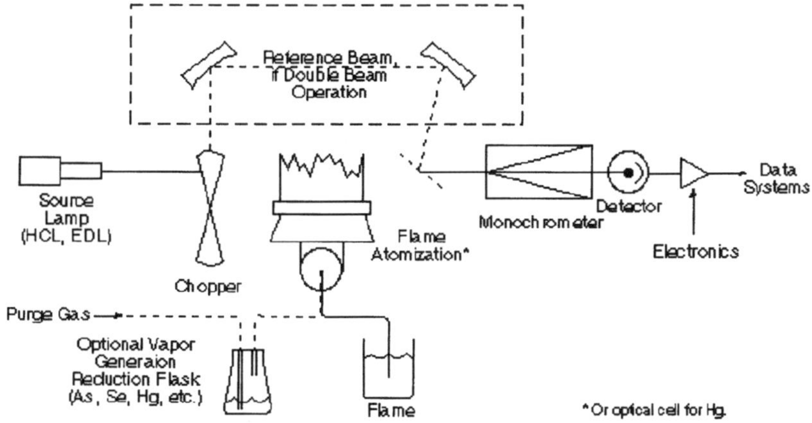

Figure 2. Basic atomic absorption spectrophotometer.

Table 8. Recommended conditions for flame atomic absorption-selected elements (US EPA, 1979; 1986).

Element	Wavelength (nm)	Oxidant/fuel	Range (μg/mL)	IDL (μg/mL)	Inferences, comments, etc.
Ag	328.1	A[a]/Ac[b]	0.1–3	0.01	Light sensitive, stability issues — use ionization suppressant or matrix match.
Al	324.7	N[c]/A (rich)	5–50	0.1	Slight ionization interference — use 0.1% K.
Ba	553.6	A/Ac (rich)	1–20	0.1	Use narrow slit, 0.1% K.
Be	234.9	A/Ac (rich)	0.05–2	0.005	If high Al, use 0.1% fluoride.
Ca	422.7	A/Ac	0.2–7	0.08	Oxyanion complexation — use La, many other interferents.
Cd	228.8	A/Ac (lean)	0.05–2	0.025	
Co	240.7	A/Ac	0.5–5	0.05	Excess transition metals — use matrix matching.
Cr	357.9	N/Ac, A/Ac (rich)	0.5–10 (N/Ac)	0.05	If high alkali, use 0.1% K.
Cu	324.7	A/Ac (lean)	0.2–5	0.02	
Fe	248.3[d]	A/Ac (lean)	0.3–5	0.03	Narrow slit, avoid contamination.
K	766.5	A/Ac (lean)	0.1–2	0.01	Ionization — use 10x Na concentration.
Mg	285.2	A/Ac (lean)	0.02–0.05	0.001	See Ca "comments."
Mn	279.5	A/Ac (sl.lean)	0.1–3	0.01	
Mo	313.3	N/Ac (rich)	1–40	0.1	Ca, Sr, SO_4, Fe are severe — add 1000 μg/mL; Al, Co, Cr, Fe — use N/Ac.

Mo	313.3	N/Ac (rich)	1–40	0.1	Ca, Sr, SO_4, Fe are severe — add 1000 μg/mL; Al, Co, Cr, Fe — use N/Ac.
Na	589.6	A/Ac (lean)	0.03–1	0.002	
Ni	232.0	A/Ac (lean)	0.3–5	0.04	
Pb	283.3/217.0	A/Ac (lean)	1–20	0.1	e
Sb	217.6/231.1	A/Ac (lean)	1–40	0.2	Pb spectral — use 231.1 nm; high Cu, Ni — use N/Ac.[e]
Sn	286.3	N/Ac (rich)	10–300	0.8	e
Tl	276.8	A/Ac (lean)	1–20	0.1	Avoid HCL.[e]
V	318.4	N/Ac (rich)	2–100	0.8	Many interferences — use 0.1% Al.
Zn	213.9	A/Ac (lean)	0.05–1	0.005	Si, Cu, PO_4 — 1.5% Sr removes Cu, PO_4.

[a]A — air.
[b]Ac — acetylene.
[c]N — nitrous oxide.
[d]Several other wavelengths available.
[e]GFAA recommended.

quickly due to rapid cathode sputtering. The EDLs are particularly recommended for the volatile elements such as: As, Bi, Pb, Se, Sb, Sn, and Te. EDL lamp lifetimes are similar to those of HCL's, but are drastically reduced when EDLs are operated below recommended power settings.

The sample introduction system for flame atomization consists of the nebulizer and spray chamber. The nebulizer provides a fine aerosol of the sample solution. The standard type is normally a pneumatic nebulizer, either cross-flow or concentric. The solution is injected with a supersonic blast of air that produces droplets in the range from 1 to 100 μm in diameter. The resulting fog either strikes an impact bead or a flow spoiler to produce small droplets from larger ones. The fog, droplets diameter $< 10\,\mu m$, is mixed with more oxidant, followed by fuel, and carried into the burner for atomization. The overall nebulization efficiency is approximately 10–15%. Alternatively, a flow spoiler can be used in place of the bead to remove large droplets. The spoiler improves nebulization efficiency only a few percentage points, but solutions with high solids may be used. The bead system does provide lower detection limits than the spoiler for many elements. The most commonly used flame is the air acetylene flame with a temperature of about 2300°C. The hotter N_2O/C_2H_2 flame, approximately 2700°C, is used to overcome chemical interferences and dissociate refractory oxides. It requires a different burner.

A monochromator with a spectral bandpass of 0.1–2.0 nm is placed immediately following the atomization cell or flame isolating analyte source lines emitted by the light source. Grating monochromators are usually offered in today's instrumentation. Frequently, two gratings are employed to cover the entire range from 180 to 800 nm. The slit width determines the portion of the spectrum observed by the detector, and is usually kept small to minimize the flame emission reaching the photomultiplier and reject unwanted lines from the source.

Many manufacturers of AA spectrophotometers provide double beam models to compensate for source drift. An analytical and reference beam that bypasses the flame, is produced from the source by means of a mechanical chopper or rotating mirror. One disadvantage of this system is that there is some light loss resulting in reduced intensity due to the added optical components. Also, the double beam system cannot compensate for drift due to changes in the flame conditions.

Single-beam instruments are now quite stable due to improvements in lamp design, power supplies, and other electronic components over the years. Automated baseline correction can compensate for those few elements that require longer lamp warm-up times. The monochromator passes the isolated absorption line on to the PMT detector that is sensitive to radiation in the 1900–8000 λ range. The PMT function is to produce an electrical signal proportional to the light energy that is striking it at any given time. Log amplifiers are used to convert the signal due to absorption into absorbance that is proportional to the concentration (Beer's Law). The processed signal is then available via a recorder, digital readout, or computer-assisted data station.

Chemical Interference

Chemical interferences are caused by the molecular combination in the flame of the analyte atom with other sample species. For example, the analyte can form a thermally stable compound that a flame cannot dissociate because of insufficient energy (too cool). The use of a hotter flame like N_2O/C_2H_2 can be employed. Also a compound, such as EDTA, is added to preferentially complex the cation. Releasing agents, like lanthanum, that preferentially form a lanthanum–anion complex, thereby freeing the metal ion of interest are usually used.

Ionization Interference

Ionization interferences occur when the flame temperature is sufficiently hot to convert the ground state atom to the nonabsorbing ionic species. A cooler flame is likely to increase chemical interferences. Therefore adding an excess of an easily ionized element such as alkali metals is recommended. This material results in the flame containing a large number of electrons that suppress ionization of the analyte of interest. Due to its relative purity and lack of visible emission in the flame, potassium chloride is usually preferred. Lanthanum also acts as an ionization buffer.

Physical Interference

Physical interferences are caused by species in the sample that generally affect nebulization efficiency relative to that of the calibration standards. Examples include high levels of dissolved solids, $>3\%$ or organic solvents in an aqueous solution. Ideally matrix matching of the samples and standards will minimize this effect. However this is not always realistically feasible or possible. The alternative approach is the MSA (discussed later). This is a commonly used technique and involves spiking several sample aliquots, (preferably four or five), of equal volume with equivalent volumes each of the calibration blank and three or four concentrations of calibration standards. The technique also requires a linear calibration response.

Spectral Interference

Spectral interferences are not common because AAS is highly selective and the absorption spectra are relatively simple. There are relatively few cases of spectral interference. One is the Pb/Sb overlap at the 217–217.6 nm region. This is more of a problem when using multielement lamps where the cathode may be constructed of several interfering elements. This effect is minimized by selecting a smaller slit, alternative wavelength, or single element lamp.

Nonspecific Background Absorption

Nonspecific background absorption can occur in flames from a high concentration of dissolved solids. Particles in the flame scatter and absorb radiation resulting in a positive bias. Elements with absorbance lines below 220 nm are affected the most by this interference phenomenon. Some form of background correction is usually recommended. The existence and extent of the background interference is then determined by measuring the signal at several slit width settings. If the background increases at larger slit widths, background attenuation is likely. MSA can neither identify nor correct for background attenuation.

Although not as important in FAA as in furnace atomization, most AA instruments offer some form of an automated background connection system as an option. With one common system, the continuum source background corrector, radiation from the primary source lamp and continuum source lamp (hydrogen, deuterium) are passed alternately through the atomization cell (Koirtyohann and Pickett, 1965). The analyte in the sample absorbs only radiation from the primary source lamp while the attenuation due to background affects both beams equally. The ratio of the beams from both sources is monitored electronically, thereby compensating for the background attenuation. Continuum background sources for flame AA include deuterium and hydrogen lamps for the far ultraviolet (UV) region, and tungsten lamps for the near UV and visible region of the spectrum. Other more efficient correction systems are available, primarily for GFAA instrumentation, and are discussed briefly later in this chapter.

Measurement Process for FAA

The instrument operator's manual usually has adequate detail for routine operation. A typical procedure for flame AA measurement is as follows:

- Warm up the instrument electronics and source lamp (five to thirty minutes) and ignite the flame.

- Select the appropriate wavelength and slit, and optimize the lamp position.
- Optimize the burner position with respect to the light beam.
- While aspirating a midrange standard solution, optimize the nebulizer settings and flame stoichiometry (fuel/oxidant ratio) for maximum sensitivity.
- Zero the instrument with blank solution and aspirate at least three standards, obtaining a least-squares expression where the correlation coefficient exceeds 0.995, preferably 0.999.

Frequently, a measurement requires less than optimum sensitivity due to high analyte concentrations in the sample that exceed the linear portion of the calibration curve. The most convenient means of achieving reduced sensitivity are to rotate the burner head providing a much reduced path length through the optical cell, and/or select a less sensitive wavelength. In addition, slight adjustments in the oxidant/fuel mixture can achieve this effect. Samples can always be diluted to ensure that absorbance is within the range of those of the calibration standards. This also has the effect of reducing potential chemical and matrix interferences. Corrosion resistant nebulizers are available from manufacturers, but sensitivities are slightly poorer when using these.

Calibration

Deviations from Beer's Law can be significant, resulting in a curvature toward the concentration axis. This is caused by stray light, line broadening, and in some cases, absorption at nearby lines. It is advisable to use when possible, the linear portion of the calibration curve for most applications. As a rule, bracket the sample absorbance by concentrations of the standards. Occasionally, there are applications where the nonlinear portion of the curve is used. Instruments manufactured in the past 20 years usually have incorporated microprocessors capable of automatic curve correction with some degree of success, particularly for instances of slight nonlinearity.

Ideally, to minimize various types of interferences, the matrix of the standard should match that of the samples. In FAA this is usually difficult. The analyst must resort to the MSA, dilute the samples five times or more to essentially dilute out the matrix, or use a hotter flame. The MSA technique is perhaps the most reliable and preferable. It avoids significant sample dilution and subsequent degration of detection limits.

Quality Control and FAA

A number of QC operations are employed to address the following issues:

- accuracy of standards preparation
- signal suppression due to sample matrix
- instrument stability and drift
- memory effects from high standards
- sensitivity of the system.

QC samples that are normally used to address these issues include the following:

- Initial calibration blank (ICB): The solution verifies the intercept of the calibration curve.
- Initial calibration verification (ICV): The solution verifies the accuracy of calibration standards. Ideally, the measured value should be within $\pm 5\%$ of expected but should not exceed $\pm 10\%$.
- Continuing calibration blank (CCB): The solution is run after every 10 samples to verify the intercept and serve as a check for memory effects.
- Continuing calibration verification (CCV): The solution serves as a check for sensitivity drift and should not deviate from the expected value by more than $\pm 10\%$. If it does reanalysis is required for samples run since the last successful CCV analysis.
- Duplicate measurements serve as an additional check for signal stability and memory effects.
- Measurement spikes are analyzed to determine whether a significant matrix suppression exists. If so, sample dilution, matrix matching, or MSA may be required.

Maintenance and FAA

Perform the following straightforward operations, usually described in the instrument operator's manual. This will ensure that the maximum instrument performance is obtained:

- burner head cleaning, as needed
- atomize cell window cleaning as evidenced by visible dust
- disassemble and clean the sample introduction system including spray chamber and end cap; and clean and adjust the nebulizer
- clean the optics which usually requires service personnel.

Vapor Generation Techniques

Mercury and elements that form gaseous hydrides are determined using modifications to the FAA system. The techniques require that the vapor of the element or its hydride formed through the chemical reduction of the analyte species is the sample digest. For mercury, the Hg^{2+} species in the solution is reduced with $SnCl_2$ or $NaBH_4$ to the elemental form Hg^0. Because the latter species is already in the atomic ground state form, no high-temperature atomization process is required, hence CVAA. Instead, the resulting Hg^0 species is bubbled or swept out of the solution with a carrier gas air, such as N_2, through an optical cell. It is nominally at room temperature where the atomic absorption process occurs with high sensitivity. Some systems employ a gold amalgamation device to capture the resulting Hg^0 vapor prior to measurement. Rapid heating of the gold amalgamation device (~800°C) vaporizes and sweeps the Hg^0 into the optical cell, producing a much sharper peak.

Hydrides of As, Sb, Se, etc., are likewise chemically formed through a reduction reaction, usually with $NaBH_4$. This is referred to as HGAA. The hydride is swept into a quartz cell heated electrothermally at about 850–950°C or with an air-acetylene to decompose the hydride to form hydrogen and the atomic ground state of the analyte. This is a highly sensitive technique. Solution detection limits, for selected elements, provided by these methods are shown in Table 9.

Calibration standards and QC samples are all carried through the chemical reduction process. Interferences arise primarily from species that inhibit the reduction reaction. For example, Cu and dissolved NO_x species can greatly suppress the HGAA reduction reaction, leading to low values. Any process that alters the oxidation state is to be avoided. For instance, HGAA is successful for As either as As^{3+} or As^{5+}, but responses are different for the two species. Therefore, the chemical form must be the same for both standard and sample. Certain species, such as Cl_2, water vapor, and organic vapor, serve as spectral interferences for CVAA, but they can

Table 9. Nominal detection limits for vapor generation atomic absorption.

Element	Method	Approximate solution detection limits[a] (μg/L)
As	HGAA	0.02
Bi	HGAA	0.02
Hg	CVAA	0.01
Sb	HGAA	0.1
Se	HGAA	0.02
Te	HGAA	0.02

[a]Assuming a 50-mL sample for batch systems.

Table 10. Interferences encountered in HGAA and CVAA.

Element	Technique	Interference
As	HGAA	Strong oxidant anions, e.g., $Cr_2O_7^{2-}$, MnO_4^-, NO. Noble metals, high concentrations of Fe, Co, Ni, Cu, S^{2-}, etc., and other hydride forming elements.[a]
Hg	CVAA	I^-, Se, Te with $SnCl_2$ reducing agent; Ni, Cu, As, Bi with $NaBH_4$ reducing agent. Noble metals (amalgamating Hg in solution). Background absorption due to water vapor, nitrogen oxides, and organic solvents.[a]
Sb	HGAA	Similar to As.
Se	HGAA	Similar to other hydride-forming elements, plus KI (reduces Se^{4+} to Se^0).

[a]Tsalev, 1984.

generally be avoided with water scrubbers and automated background correction. Interferences found in HGAA and CVAA are summarized in Table 10.

Although commercial vapor generation systems are available either as manual or automated systems, they are fairly easy to construct in the laboratory.

Summary of FAA and Associated Methods

FAA is a rapid technique with approximately three to six samples per minute analyzed, and a relatively simple technique requiring minimal operator training. Although it is the least sensitive atomic spectroscopic method for solutions, it is sufficiently sensitive to meet most measurement requirements for certain major and minor elements, such as Ca, Fe, Mg, Na, and Zn. It is often adequately sensitive for other metals including Cu, Mg, and Pb, when very large air volumes are taken or high emission rates are expected to be found in some industrial stack emissions. The precision or relative standard deviation (RSD) of the technique is relatively good, approximately 1–2% RSD. Accuracies are principally a function of the care taken in sample preparation and interference correction, and should generally be in the 85–115% recovery range. It is traditionally a single element method, but some manufacturers offer rapid, sequential capability through automated hollow cathode lamp changing and automated sample changing.

Vapor generation techniques, CVAA for Hg and HGAA for hydride forming elements such as As, Sb, and Se are easily used with some minor modifications made to the instrumentation and the addition of chemical reduction glassware. The methods offer low parts per billion and high parts per trillion solution detection limits.

GRAPHITE FURNACE ATOMIC ABSORPTION

GFAA or electrothermal atomic absorption (ETAA) utilizes electrical energy to cause volatilization of the analyte(s) of interest. The most common approach is to place a droplet of the solution in a graphite tube and heat it by passing electrical current through the tube. These techniques became available commercially in the early 1970s as a complementary analytical tool to FAA, capable of solution detection limits generally below 5 µg/L for most environmentally significant elements. Since then, many improvements have been made in the areas of furnace design, chemical pretreatment of the sample, background correction, automation, and data collection. Although instrumentation designs continue to improve, the technique has reached maturity in the past 10 years. The most interesting recent developments, commercially, are in the area of multielement capability.

There are two fundamental differences between GFAA and FAA. Some matrix components are eliminated in GFAA prior to the analyte atomization step while its atomization occurs in an inert gas atmosphere. The issues involved in this more complex measurement process dictate that some explanation of matrix effects and appropriate corrective actions be presented.

The graphite furnace performs the same function of generating free ground state atoms that absorb radiation from the source lamp as the flame process does with FAA (Figure 3). In most cases with GFAA, the light beam passes through the roughly 0.6 cm diameter graphite tube atomization cell. The quantity of analyte is completely atomized and confined in a small volume within the tube before it is swept away by the purge gas, usually argon. The resulting residence time for the ground state atom in

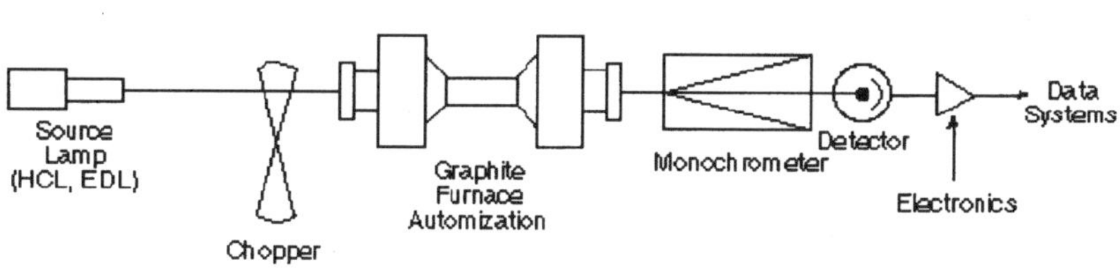

Figure 3. Simplified GFAA spectrophotometer.

the furnace, a few tenths of a second, is up to 100 times that for flame atomization. The analyte vapor is not greatly diluted as with flame atomization, and its efficiency is now theoretically 100% instead of the 2–10% for flame atomization. In addition, the signal is an absorption spike or peak instead of the continuous reading provided in FAA. This results in detection limits that are approximately two to three orders of magnitude lower.

Background interferences are somewhat more imposing for GFAA, and require the analyst to be more involved with the analytical process. Therefore, analyst experience and technique are both definite factors in data quality. GFAA measurements are likely most costly in terms of time and difficulty for laboratories using a number of analytical methods for given sample. Nevertheless, with the commercial introduction of improved furnace designs and background correction systems in the past 15 years, this method is far more universally applicable to difficult sample matrices and provides a detection limit benchmark for other elemental analytical techniques.

The general detection limit capability is summarized in Table 11, and some typical operating conditions are discussed later in the chapter. In the following discussion, topics specific to furnace atomization including: furnace design issues, interferences encountered in GFAA, and means of minimizing them are discussed.

Furnace Atomization

In most recent GFAA instrumentation designs, some form of the mini-Massman type graphite tube provides the atomization process (Massman, 1968). A 5–50-μL sample is dispensed in droplet form or as a spray into a small graphite tube that is heated electrically. The temperature is increased in a stepwise manner to provide smooth drying of the sample, thermal pretreatment to remove the maximum amount of sample matrix while retaining the analyte element, and dissociation and vaporization of the analyte into free atoms. During the drying and pretreatment steps a purge gas, usually argon, is passed through the tube to remove solvent and matrix vapors. During atomization, the gas flow can be shut off for a few seconds to provide the longest possible ground state atom residence time in the optical beam. Alternatively, the gas flow may be increased up to approximately 300 mL/min to lower sensitivity, for higher concentrations. A typical temperature program profile is shown in Figure 4.

As with the flame process in FAA, oxygen, carbon, hydrogen, and other species during furnace atomization can affect the dissociation of a compound into free atoms. Because graphite is not an inert substance at furnace

Table 11. Approximate detection limits of GFAA: selected elements (Perkin).

Element	Furnace design[b]	Pretreatment temperature (°C)	Atomization temperature (°C)	Detection limit[a] (μg/L)	Detection limit (pg)
Ag	P	650	1600	0.03	0.5
Al	P	1700	2400	0.2	4
As	P	1500	2500	1.0	20
Ba	W	1200	2700	0.5	10
Be	P	1500	2500	0.05	1
Bi	P	900	1900	0.5	10
Cd	P	900	1600	0.02	0.3
Co	P	1400	2400	0.1	2
Cr	P	1650	2500	0.05	1
Cu	P	1200	2300	0.05	1
Mn	P	1400	2200	0.05	1
Mo	W	1800	2700	0.2	4
Ni	P	1400	2400	0.5	10
Pb	P	600	1900	0.3	5
Sb	P	1100	2400	0.8	15
Se	P	900	2000	1.5	30
Sn	P	1000	2100	1.0	20
Sr	W	1300	2600	0.1	2
Ti	W	1400	2700	2.5	50
Tl	P	600	1400	0.5	10
V	W	1400	2650	1.0	20

[a]Assuming a 10 μL sample injection, Zeeman correction.
[b]Atomization surface (W = wall; P = platform).

atomization temperatures (1600–2800°C), it can react with these species in a vast number of ways affecting the partial pressure of gaseous species. Therefore, the objective in tube design is to provide a smooth and unreactive surface. Similarly the atmosphere should be inert. Therefore argon is preferred over nitrogen as the purge gas due to the possible formation of CN species.

Traditionally, two types of graphite tubes have been available, one an ordinary high-density, and the other constructed of the same material, but coated with a thin layer of pyrolytic graphite. These tubes are usually provided with grooved ends on the interior surfaces to prevent the sample from creeping toward the cooler ends. Stronger nonspectral interferences are observed with the normal tubes. This is caused by retention of the interfering species, which are released only during atomization. With its

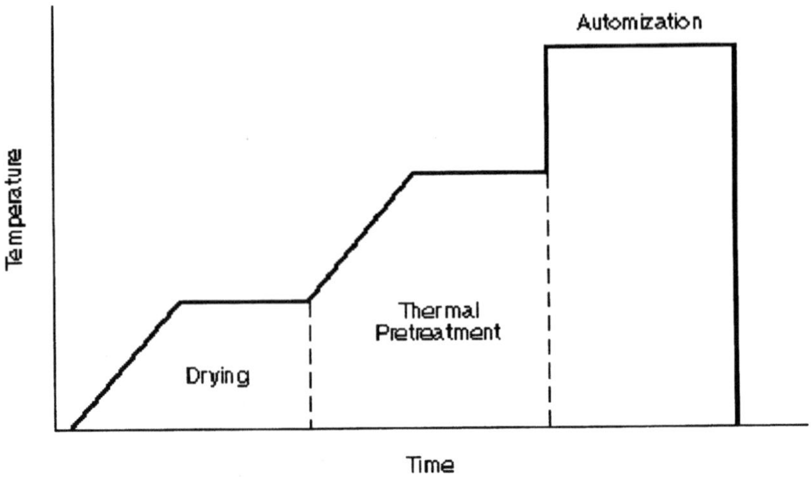

Figure 4. Typical graphite furnace temperature program profile.

more even and dense surface, pyrolytic carbon minimizes this effect. This impermeable surface provides a much longer tube lifetime as well.

The heating characteristics of the graphite furnace are critical to obtaining precise and accurate analytical data. The control of temperature during the heating cycles is of particular importance to the reproducibility of the atomization. Early designs experienced a slow nonlinear rise to the final temperature. In addition, the analyte is volatilized off the wall into an environment that is cooler than the surface of the tube. This rapidly changes its temperature, and expands its volume because the carrier gas is still increasing in temperature. The combination of these effects leads to conditions that are difficult to control, and are the source of numerous interferences. Many current GFAA instruments incorporate a rapid furnace heating cycle using rates of 2000°C/s to guarantee that the furnace is brought to the set temperature as quickly as possible. This allows for a longer period of thermal stability prior to atomization, and in turn lower atomization temperatures.

In the past 15 years, instrument suppliers have incorporated a platform design into the furnace system (L'vov, 1978). In this configuration, the sample is atomized from a small rectangular surface of pyrolytic graphite that exhibits low thermal conductivity in the direction perpendicular to the horizontal plane (Figure 5). Heat transfer from the walls of the tube is minimized, and the sample on the platform is atomized only when the atmosphere within the tube has reached the atomization temperature. This effect allows the system to come to near equilibrium before the analyte

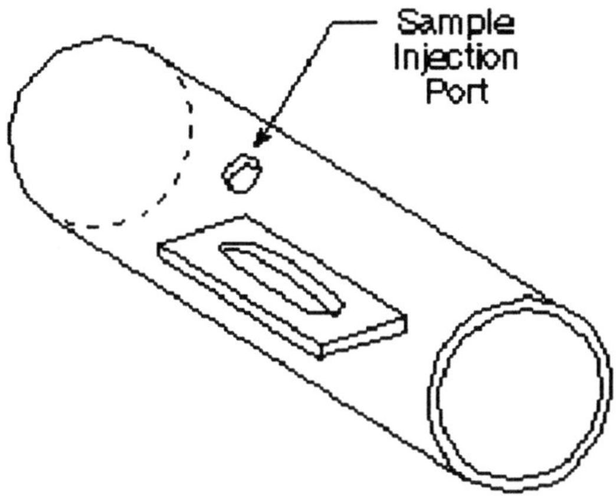

Figure 5. Typical graphite tube/platform assembly.

Table 12. Approximate atomization temperatures (°C) for GFAA (Perkin).

Element	"Off the wall"[a]	Platform
Ag	2700	1500
Al	2700	2300
As	2800	2000
Ba	1900	2300
Cd	2700	1400
Cr	2700	2300
Cu		1900
Mn		1500
Mo		2350
Ni		2300
Pb	2700	1500
Sb	2700	1900
Se		1900
Ti		1600
V		2400

[a]No platform used.

element is atomized. The use of a platform also allows for lower atomization temperatures and therefore longer graphite tube and platform life. Some typical reductions in atomization temperature when using the platform are shown in Table 12.

Fast digital electronics are required to follow furnace atomization peaks. GFAA spectrometers require a response time of less than 20 ms to ensure that absorbance from rapidly atomized atoms is monitored accurately. Many instruments also have the capability to monitor the location of the baseline immediately before the atomization step.

The absorbance peak shape is highly dependent on the matrix of the sample because the rate of vaporization is affected. Unless the matrix effect is virtually eliminated, which is unlikely, peak height measurements is suspect. As a consequence, peak area (integrated absorbance) measurements are recommended for the majority of GFAA determinations. Employing integrated absorbance minimizes the effect of matrix interferences, and the linear range is extended. Peak height measurements are recommended for highly refractory elements whose ground state atoms are slowly and incompletely released from the surface of the tube (i.e., the peak does not reach the baseline within 5 s).

Interferences in GFAA

Interferences in furnace atomization are categorized as spectral and nonspectral. Spectral interferences may be encountered when the following occurs:

- Molecular or atomic lines matrix species absorb source radiation (rare).
- Light scatter and source radiation occur due to particle formation during the atomization cycle.
- Undissociated molecular forms of matrix materials may exhibit broadband absorption spectra.

The latter two types are common in GFAA, and are often referred to as background attenuation or absorption.

Nonspectral interferences directly affect the formation of the ground state analyte atoms and can either enhance or suppress the analyte signal, and are due to the following:

- loss of the analyte to volatilization as a molecular species during the thermal pretreatment step
- reaction of the analyte element with volatilized matrix species to form compounds less easily atomized
- ionization of the analyte element

Background absorption, gaseous molecular species, salt particles, or smoke from the matrix, can be distinguished from the analyte absorption. Although the analyte element can absorb the very narrow spectral line emitted by the source, background attenuation extends over a much broader wavelength region. For instance, the absorption band from chloride salts ranges from below 180 nm to above 360 nm.

The first step in minimizing background absorption is through careful adjustment of spectrometer conditions, including the proper wavelength, slit width, furnace configuration (platform atomization for As and wall atomization for V), purge gas flow, sample droplet size, and length of the signal monitoring cycle. The above settings are performed to minimize background absorption, but are not likely to eliminate it altogether.

The major components of background correction are critical to accurate FAA. Most instruments today employ one of the following automated background correction systems: continuum source background correction, the Zeeman effect background correction system, or the Smith–Hieftje system (Koirtyohann and Pickett, 1965; Prugger and Torge, 1969; Smith and Hieftje, 1983).

Continuum source correction systems are described in the section on flame atomization. The technique is useful when the background absorption is fairly constant in the observed spectral region or slit width. Changing background within the spectral bandpass can cause overcorrection leading to highly inaccurate results. Most instruments manufactured within the past 10 years or more provide an expanded video display of profiles of both the analyte atomization and the background absorption. There are some circumstances where this mode of background correction is adequate for air particulate matter. However, background absorbances in excess of 0.5 will usually exceed the capabilities of this system resulting in overcorrection. A typical atomization profile illustrating this phenomenon is shown in Figure 6. Some matrix difficulties can be minimized in

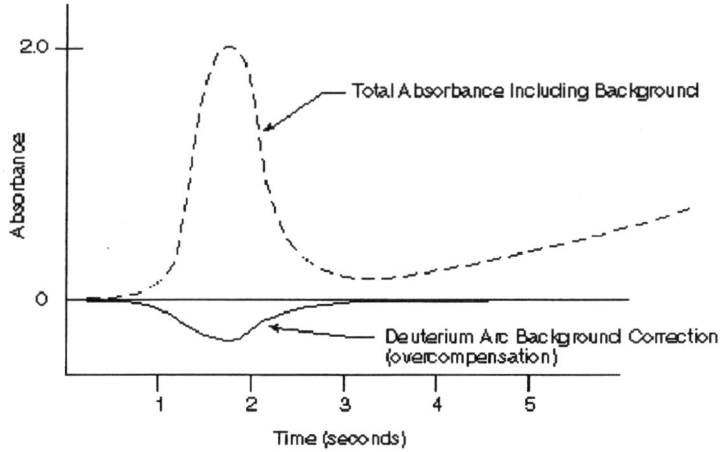

Figure 6. Deuterium arc background overcorrection in the presence of high background.

the sample preparation step, including a more complete acid digestion, or chemical matrix modification in the furnace during the pretreatment step (discussed later). Limitations with the continuum source background correction system are summarized as follows:

- Maximum correctable background is about 0.5 absorbance unit.
- Background absorption must be continuous within the spectral bandwidth.
- The continuum source radiation is absorbed by other atoms in the matrix if they have an absorption line within the spectral bandpass.
- Use two sources to cover the 190–800 nm range — deuterium arc or hydrogen continuum lamps up to about 320 nm and a halogen (e.g., tungsten iodine) lamp from 300 to 800 nm.

The Zeeman effect background correction system is included in many recently manufactured instruments, and provides reliable background correction in difficult matrices. Background absorption up to at least 2.0 absorbance units is permitted. Also, highly structured lines on continuous and noncontinuous background and light scattering are corrected. These systems do not require an additional radiation source, therefore true double beam operation is possible. Consequently, sample preparation steps need not be as rigorous as those required for the continuum correction system. With the Zeeman technique, atomic lines or atomic energy levels are split into three or more components by means of a strong magnetic field (e.g., 1.0 tesla). In many instruments, a pulsed magnetic field is applied to the analyte atoms in the furnace. In the simplest case, the splitting of electron energy levels within the atom produces a π-component at the analyte wavelength, and σ^+ and σ^- components symmetrically shift away from the analyte wavelength. Since it only affects atomic species, the Zeeman effect is used to separate atomic from background absorption. In the transverse design utilizing this effect, the magnetic field is at right angles with the measurement beam, whereas with the longitudinal configuration the magnetic field is parallel to the beam. Transverse designs require a polarizer to reject the π- and pass only the σ-components. In the more recent longitudinal arrangement the nonshifted π-component cannot be received. With the magnetic field on, the signal measured at the analytical wavelength is due to background absorption only; and with the field off, the signal arises from both the atomic and background absorption components. The difference may be attributed to the analyte element alone. The more recent longitudinal design avoids the use of a polarizer to the light beam, allowing more light throughput to the detector with the corresponding potential for improved detection limits. In summary, the Zeeman system is especially applicable for complex matrices, volatile analytes in high backgrounds, and structured narrow-band spectral interferences. A typical atomization profile showing Zeeman background correction is shown in Figure 7.

Some systems employ a correction system developed by Smith and Hieftje (1983) that is simpler than either of the above ones in that it relies on self-absorption of the source lamp. The lamp is pulsed at two different currents to provide the analytical and background signals. Correction is observed at up to 2.0 absorbance units. It is important to measure the extent of the background absorbance to determine whether the system is performing an adequate correction. Most recent instrumentation provides a readout of both signals.

Excessively high background absorption that cannot be fixed with one of the automated background correction systems or through careful manipulation of furnace conditions, are usually changed with thorough chemical pretreatment, especially for most sample matrices involving air particulates. Approaches include:

- modification of sample matrix during the digestion process
- chemical modification of the sample matrix in the furnace during the thermal pretreatment step, formerly called the char or ash step
- use of alternate purge gases, such as oxygen or air in the thermal pretreatment step

Note that chemical pretreatment should be used with background correction.

Ion exchange and solvent extraction are viable techniques used to isolate the analyte from the matrix. The more steps that are implemented during sample preparation, the greater the potential for contamination. Chemical modification in the furnace is the most powerful technique for matrix modification and has received much attention. Recipes are available for each element to reduce matrix background absorption and are shown in Table 13. With chemical pretreatment, the analyte element reacts

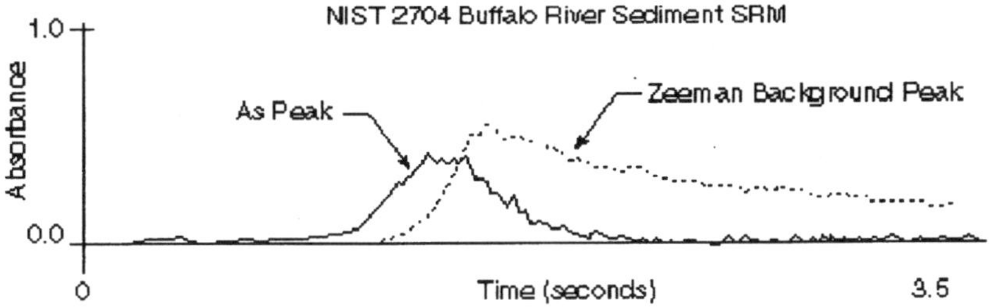

Figure 7. Typical Zeeman background correction profile.

Table 13. Matrix modifiers for selected elements.

Element	Modifier
Al	Mg $(NO_3)_2$
As	Ni[a] or Pd + MG $(NO_3)_2$
Be	Mg $(NO_3)_2$
Cd	$NH_4H_2PO_4$ + Mg $(NO_3)_2$
Co	Mg $(NO_3)_2$
Cr	Mg $(NO_3)_2$
Cu	None or Pd + Mg $(NO_3)_2$
Mn	Mg $(NO_3)_2$
Ni	Mg $(NO_3)_2$
Pb	$NH_4H_2PO_4$ + Mg $(NO_3)_2$
Sb	Ni[a] or Pd + Mg $(NO_3)_2$
Se	Ni + Mg $(NO_3)_2$ or Pd + Mg $(NO_3)_2$
Sn	HNO_3 + $NH_4H_2PO_4$ or Pd + Mg $(NO_3)_2$
Tl	H_2SO_4 or Pd + Mg $(NO_3)_2$
V	None

[a]Ni alone.

with a chemical modifier to form a more stable compound, which in turn allows a higher thermal pretreatment temperature. Alternatively, the modifier reacts with the matrix to increase its volatility during the thermal pretreatment step. In either case, the analyst is seeking to facilitate thermal separation of matrix from analyte. One of the first modifiers used was $Ni(NO_3)_2$ for the stabilization of As and Se. As and Se form a stable compound with $Ni(NO_3)_2$ that allows for pretreatment temperatures up to 1200–1300°C, facilitating greater removal of volatile compounds in the matrix. Other widely used modifiers have included $Mg(NO_3)_2$ and $(NH_4)H_2PO_4$. In recent years, $Pd(NO_3)_2$, in conjunction with $Mg(NO_3)_2$, has been proposed as a more universal modifier. It has proved beneficial to raise pretreatment temperatures for a number of analytes.

For samples containing significant levels of NaCl, background attenuation is experienced for nearly all elements due to the high volatilization of this salt ($>$1000°C). With the addition of ammonium nitrate, it is believed that the NaCl is more readily vaporized through the formation of the more volatile salts NH_4Cl and $NaNO_3$ and excess NH_4NO_3. These modifiers can have significant blank analyte levels and should be checked prior to use. Also, the pretreatment modifier and temperature recommended will vary with matrix, furnace design, and background correction method.

Another technique used to modify the matrix during the thermal pretreatment involves the use of an alternate gas during the pretreatment step. With the use of air or oxygen, the organic content in the sample is ashed rather than charred. This technique is not available for all graphite furnaces but is not likely to be needed in the air particle application.

Under certain circumstances, a spurious signal at atomization is observed that is caused by black body radiation off furnace walls. This phenomenon occurs when working near UV and visible wavelengths with high PMT gain amplification settings. This interference may be minimized by using a narrower slit width, increasing the lamp current, and carefully aligning the furnace in the optical beam, if possible.

Nonspectral interferences affect the analyte signal directly. With this interference, the analyte element exhibits a different sensitivity in the sample matrix than it does in the presence of the calibration or reference matrix. The differences in response is due to:

- volatilization of the analyte during the pretreatment step
- reaction of the analyte element with components in the matrix to form compounds that are less easily atomized
- differences in the degree of ionization caused by sample matrix components

The analyte can also react with the graphite tube inhibiting the atomization process, but this is likely to influence the analyte in the sample and standard matrices equally. Although this is not technically an interference, it is significant and should be avoided because it not only results in sensitivity losses but also in memory effects. This effect is minimized through the use of high purity argon and pyrolytically coated graphite tubes and platforms. Nitrogen should be avoided.

When the reactivity of the analyte with the graphite surface changes over the course of the analysis, the effect qualifies as an interference. An example of this effect is the coexistence of the analyte with an element such as tungsten (W) that reacts with the graphite surface, and forms a refractory carbide that changes the graphite surface permanently. This may reduce the tendency of the analyte to react with the graphite and actually cause an increase in sensitivity. The use of good pyrolytically coated graphite, addition of the coexisting element to the standards, or saturation of the samples and standards in the coexisting element all are helpful techniques in minimizing this interference.

Losses from analyte volatilization during the thermal pretreatment step are the most common manifestation of the nonspectral interference. For example, metal halides are generally more volatile than nitrates. It is a good rule to avoid the use of hydrochloric acid. Nitric acid is almost always preferred.

As shown with some spectral interferences, matrix modification is useful in controlling preatomization losses. The modifier stabilizes the analyte element in addition to enhancing matrix removal at higher thermal pretreatment temperatures. It is important to add a large excess of modifier to both standards and test samples to control the chemical environment during the thermal pretreatment step. Concentrations of the modifier that are beyond the minimum excess may actually increase background attenuation.

Another nonspectral interference involves the formation of compounds of the analyte after release into the vapor phase. This occurs primarily when atomization takes place without a platform and the temperature within the graphite has not stabilized throughout the graphite tube. Vapor-phase interferences occur primarily at the cooler ends of the graphite tube. As mentioned previously, the platform delays atomization until the tube and its atmosphere have reached their final temperature, and are even hotter than the platform surface. Vapor phase interferences can be reduced dramatically by employing a combination of adjustments and techniques that provides for conditions that most closely imitate the stabilized temperature furnace (Slavin *et al.*, 1983). These techniques include: platform atomization, rapid heating for atomization, maximum thermal pretreatment temperature, gas stop during atomization, peak area measurements, and use of matrix modifiers. Although when used properly, these conditions will minimize vapor-phase interferences, the occasional sample may require some version of MSA.

General Measurement Procedures for GFAA

The following is a general guide applicable to air particle measurements. Typical maintenance items are incorporated into the general procedure because they are a integral part of this trial and error process.

1. Select the appropriate lamp EDL or HCL, lamp settings, wavelengths, and slit widths. Allow for adequate lamp warm-up.
2. Select the expected read time of the atomization cycle, noting that readjustments are likely later.
3. Using gloves, check the condition of the graphite atomizer parts as recommended by the manufacturer and carefully install the tube, platform, and contact parts.
4. Check the position of the tube and platform and interaction of the autosampler tip with the platform, by blocking the source beam with a white card and viewing it with a mirror. Clean the furnace windows if necessary.
5. Establish the cooling water and argon purge gas flows. Nitrogen is not recommended due to the formation of CN molecular species.

6. Select recommended sample injection size, chemical modifier, and nominal furnace temperature conditions.
7. Inject a sample with modifier, and observe with a mirror the drying of the droplet. A gradual, smooth, and complete drying process is crucial.
8. Readjust the furnace conditions as necessary to ensure that an acceptable drying program is obtained.
9. Check the thermal pretreatment and atomization conditions. The manufacturer's recommendations are usually adequate but check them by monitoring the cycle for any premature atomization. The atomization peak should be smooth and without doublets.
10. Calibrate the instrument with at least three standards and a blank. Ideally the calibration is linear as with FAA.
11. Inject a sample and observe the atomization, verifying that the background absorption is not excessive. Inject a spiked version of the sample and calculate the recovery. If it is lower the than recommended matrix matching, dilution, or some form of the MSA should be employed. Use simplified MSA procedures if all samples are similar in the matrix.
12. Analyze the usual QC samples at the appropriate intervals.

Summary

GFAA remains the ultratrace measurement approach chosen by most small and medium sized analytical laboratories. Requiring only submiliter sample aliquots, low parts per billion solution detection limits are available for most elements. Instrument design improvements have reduced many of the interferences encountered in the furnace, but the technique is still somewhat dependent on the skills and experience of the analyst with complex sample matrices. The GFAA method is the least productive of the major atomic spectroscopic techniques requiring two to three minute per single element determination. With the introduction of truly effective multielement GFAA instruments, the technique is likely to remain a valuable analytical tool for the environmental laboratory.

INDUCTIVELY COUPLED PLASMA/ATOMIC EMISSION SPECTROMETRY

Introduction to ICP-AES

The feasibility of using ICP as a source of atoms and ions for AES was discovered more than 35 years ago, but the first commercial ICP spectrometer was not introduced until 1975. Since then, the growth in the number of ICP-AES instruments in analytical/environmental laboratories has been enormous. For the past 15 years this technique has been the method of choice for multimetal determinations. It has reached maturity as an analytical technique, attaining acceptance by a number of government regulatory agencies.

ICP-AES provides the capability of the quantitive measurement of up to 65 elements. Generally, ICP-AES provides detection limits that are similar or in some cases superior to FAA limits. Recent instrumental developments have led to commercialization of instruments with detection limits approaching GFAA for many elements. The range of linearity is approximately five orders of magnitude resulting in fewer sample dilutions. As with other atomic spectroscopic techniques, ICP-AES is an indirect method. Photons emitted at widely varying intensities are measured in contrast with for example, mass spectrometric techniques (ICP-MS) that directly measure ions impacting on an electron multiplier detector. Consequently, detection limits vary up to five orders of magnitude, making semiquantitive measurements somewhat inconvenient and considerably less quantitative than semiquantitive methods with ICP-MS. For environmentally targeted elements, this uneven response is significant. For many important elements with wavelengths below approximately 225 nm, namely, As, Hg, Pb, Se, Sb, and Tl, the technique is quite insensitive using conventional nebulization and traditional torch configurations. Instrument detection limits are generally above 30 μg/L. This limitation is not as severe with current nebulizer and torch designs. Conversely, for refractory elements such as Be, Ti, and V, ICP-AES is preferred. This method is more forgiving as far as the solvent system used for digestion of the sample, whereas certain acids, HCl and $HClO_4$, are avoided in GFAA work. Higher acid concentrations, 20% or more, are tolerated. Awareness of matrix suppressions on the analyte signal is necessary. Also, with HF, the sample introduction system should consist of a plastic spray chamber and ceramic torch.

Most instrument manufacturers in the past few years have offered models with torches in an axial configuration. In this configuration, the plasma is pointed directly at the entrance slit and needs no further position optimization. Emission is at the maximum intensity. Detection limits are improved by at least one order of magnitude, and many interferences are similarly enhanced. Desolvation nebulizers provide 30% nebulization efficiency, (5–50-fold improvements in detection limits over conventional nebulizers) but interfering elements are similarly enhanced. These improved designs may provide the ideal compromise in instrumentation for air analysis, and are a reasonable choice for many air particulate applications.

Instrumentation for ICP-AES

The ICP-AES technique typically involves either simultaneous or sequential multielement determination of trace and minor elements in solution. Atomic emission provided by a plasma excitation source is measured by optical spectrometry. Because the plasma operates at over 6000 K, virtually all elements are efficiently atomized and ionized providing a lime rich spectrum. Due to the complexity of this spectrum, a high resolution optical system consisting of multiple detectors or a scanning monochrometer/ detector is employed in order to provide multielement capability. The major components include:

- sample introduction system consisting of a nebulizer and spray chamber
- radio frequency ICP excitation source producing atomic line emission spectra
- spectral dispersion system, i.e. a grating monochromator
- photosensitive device, photomultiplier(s) or diode array to monitor the line intensities
- automated background system to compensate for background intensity
- data processing system.

A simplified ICP-AES system is shown in Figure 8.

Plasmas are hot, partially ionized gases that are energized with high-frequency electromagnetic radio frequency fields or with direct current (DC). The radio frequency (RF) ICP is the most widely used plasma device, somewhat more sensitive than the DC plasma (DCP). Instruments

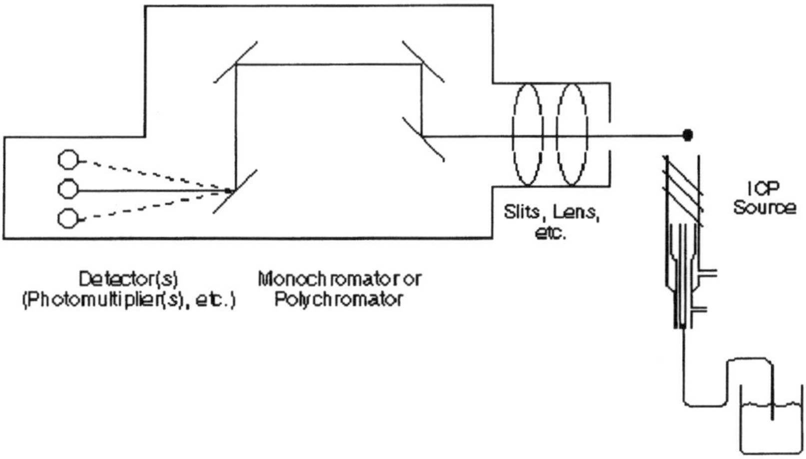

Figure 8. Typical ICP-atomic emission spectrometer.

with DCP's are also commercially available, but to a lesser extent. The DCP is produced by a DC discharge struck between two electrodes. The ICP is preferred because the design allows for better sample penetration into the plasma, and residence times in the plasma are considerably longer than in the DCP. High atomization efficiency, relative absence of molecular species, and lack of significant ionization effects are among the important characteristics of the ICP. The main body of an ICP torch consists of a quartz tube surrounded by an induction coil connected to a high-frequency generator operating usually at 27 or 40 MHz. An inert gas, usually argon acts as a support gas for the plasma and coolant for the quartz tube. Temperatures within the plasma reach 6000–10,000 K.

Sample introduction systems are similar to those found in the FAA systems, in that a nebulizer or spray chamber configuration provides a fine aerosol to the excitation source. Cross-flow nebulizers are the most common. A list of other available devices include: Hildebrand Grid, Meinhard, Babington, and V-groove nebulizers with sample handling capabilities ranging from high dissolved solids to low sample volumes (Vandecasteele and Block, 1993). Spray chambers are similar in design to those found in FAA. Designs involving concentric quartz tubes, such as the Scott or modified Scott (Vandecasteele and Block, 1993), are also frequently employed. Hydrofluoric (HF) acid with quartz systems should be avoided unless they are at very low concentration ($<0.5\%$) for short periods. Ceramic torch tips and plastic spray chambers are used for higher HF concentrations. For the ultimate improvements in detection limits, such as for indoor and ambient air applications, the analyst can resort to the ultrasonic nebulizer and similar desolvation systems such as the Mistral. With these systems, a much denser fine aerosol is delivered to the plasma (ten to thirty times improvement) through sequential heated desolvation and refrigeration, resulting in increased sensitivity. Detection limits are further lowered due to the greatly reduced molecular background of the dry analyte aerosol. The resulting detection limits are lowered by a factor ranging from 5- to 50-fold. Spectral interferences are similarly increased.

Spectrometers are available as sequential scanning instruments or as simultaneous systems. Sequential scanning units either use a moving grating with the detector in a fixed position or the opposite configuration for fixed optics with a moving detector (Echelle systems). Simultaneous systems require dedicated detectors, usually photomultiplier tubes for each analytical wavelength. Photodiode array multichannel detectors have been employed with ICP systems. These devices offer time resolved spectra and measurement of transient events. They provide the capability of analyte peak storage for later display and manipulation. Analytical throughput for simultaneous ICP units is greatly improved because the analytical time for a given sample involves only one read cycle for all elements. Although drift is less likely and read times can be greatly extended without sacrificing analytical throughput, the analyst is limited to those fixed wavelengths (except for diode array detectors). Simultaneous systems are more costly than the more flexible, sequential instruments due to the cost of added detectors.

Most manufacturers provide the option of argon-purged optics or, better still, purged and evacuated optics that result in improved limits of detection (LODs) below 200 nm for elements such as As, B, Hg, I, P, S, Sn, and Ti due to the reduction in molecular background from O_2 and N_2.

Interferences in ICP-AES

Several types of interferences are encountered in ICP-AES. Two types, spectral and physical, are particularly significant. Spectral interferences, usually the most significant in ICP-AES, can take the form of a direct or partial spectral overlap from another element such as Cu interference on the P wavelength line at 214.91 nm, or unresolved molecular band emission, like the hydroxyl band emission at the V-line at 310.23 nm. Another significant type occurs when the wing of the emission band of a very high concentration material overlaps the analyte line. For example, Al on Pb at 220.35 nm or from stray light from a high concentration Al on Sb at 206.83 nm. These examples are depicted in Figure 9.

For the first type where a direct spectral line overlap occurs, compensation is performed by measuring the concentration of the interfering element, multiplying this by the previously determined interfering element correction factor, and subtracting the result, the false analyte concentration, from the total measured analyte concentration. Alternatively another emission line may be selected. Background correction can usually compensate for elevated baselines due to a high concentration of an interfering element. For example, at the 206.83 nm Sb wavelength, Al causes a background elevation that is effectively minimized using background correction techniques. The molecular band type of interference OH, or NO usually calls for an alternate analytical wavelength or altered plasma conditions.

Typical interelement spectral interferences are found in Table 14. Correction factors are not listed because they must be determined for each instrument. Some of these are not significant for certain high resolution systems that employ an Echelle monochrometer (Harrison, 1949). Recommended correction methods are noted. Other alternate wavelengths are found in several reference texts (Boumans, 1987; Phelps, 1992; Winge *et al.*, 1985). An important QC operation during setup of an ICP-AES is the measurement of known high concentrations of potential interferences

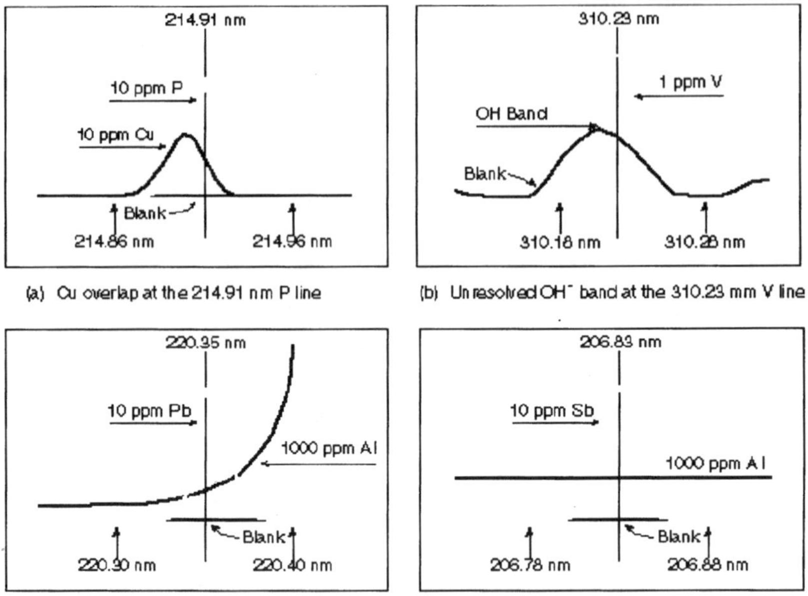

(a) Cu overlap at the 214.91 nm P line

(b) Unresolved OH⁻ band at the 310.23 mm V line

(c) Al wing at the 220.35 nm Pb line

(d) Stray light from high Al concentration at the 206.83 Sb line

Figure 9. Common spectral interferences in ICP-AES.

to validate the various interfering element correction (IEC) factors and background correction positions. In the absence of the analyte of interest, corrections should reduce false analyte values less than two to three times the detection limit.

Background correction (BC) is applied only to elevated baselines that are linear. BC may be needed on both sides of the analytical wavelength if the elevation is not parallel to the baseline. Choose sloped correction points at regions free of other spectral interferences. Large negative readings indicate inappropriate BC positions.

Physical interferences are associated with various sample transport and nebulization effects resulting from large changes in viscosity and surface tension between standards and samples, such as high acidity in sample digests. Whenever possible, matrix matching of standards and test samples should be employed, especially when working near the detection limit. Other techniques that are useful in overcoming these effects include: MSA, internal standardization, or dilution if sample concentrations are not near the detection limit. If samples are all similar in composition, a simplified MSA version can be employed involving MSA for one sample and applying the

Table 14. Spectral interferences for selected analyte wavelengths in the radial (vertical) torch configuration.

Analyte	Wavelength (nm)	Approximate LOD (μg/mL)	Prominent interferences	Suggested correction method
Ag	328.07	0.01	(Fe)[a], V	IEC[b]
	333.29	0.01	Cr	Alt λ[d]
Al	396.15	0.01	Ca	BC
	309.27	0.02	OH band, Mg, (V)	Alt λ, IEC
As	193.70	0.03	Al, (Fe)	IEC
	197.20	0.04	Al	IEC
B	249.77	0.005	Fe	Alt λ
Ba	455.40	0.001	—	—
	493.41	0.001	—	—
Be	234.86	0.0001	Fe	IEC
	313.11	0.0002	(OH)	—
Bi	223.06	0.04	(Cu), (Fe)	—
Ca	393.37	0.0002	(Ca contam.)	—
	317.93	0.006	(Ca contam), Fe	Alt λ
Cd	214.44	0.001	Al, Fe	Alt λ, IEC
	228.80	0.002	As, Co, Ni	Alt λ
Co	238.89	0.003	Fe	Alt λ
	228.62	0.005	Ni	IEC
Cr	206.15	0.003	Al, Zn	BC, Alt λ
	267.71	0.003	Mn, V	Alt λ, IEC
Cu	324.75	0.001	OH band, Fe	Alt λ
	327.4	0.002	V	Bc
Fe	238.20	0.002	NO band, Co	Bc
	239.56	0.002	Co	Alt λ
K	766.49[e]	0.05	Mg	Alt λ
	769.90	0.15	—	—
Li	670.78[e]	0.002	—	—
Mg	279.55	0.0001	(Mn)	—
	279.08	0.020	Mn	IEC or BC
Mn	257.61	0.001	Al, Fe	IEC
	294.92	0.01	Fe, V	IEC
Mo	202.02	0.004	(Fe) Al	BC
	203.84	0.01	Al	BC

Table 14. (continued).

Analyte	Wavelength (nm)	Approximate LOD (μg/mL)	Prominent interferences	Suggested correction method
Na	589.59	0.01	—	—
Ni	221.65	0.004	Si	Alt λ
	231.60	0.01	Co	IEC
P	213.62	0.02	Al, Cu, Fe	Alt λ
	214.91	0.03	Al, Cu	Alt λ, BC
Pb	220.35	0.03	Al	IEC
	283.31	0.09	Fe	IEC
Sb	206.83	0.02	Al, Cr, Fe	IEC, Alt λ, BC
Se	196.03	0.03	Al, (Fe)	BC
Sn	189.99	0.03	Fe	IEC
Sr	407.77	0.0002	—	—
	346.45	0.01	—	—
Ti	334.94	0.001	(OH) (Ca)	—
	323.45	0.002	Ni	Alt λ
Tl	276.79	0.03	Cr, Fe, V	IEC
V	309.31	0.003	Al, (OH)	Alt λ
	311.07	0.005	Mn	IEC
W	207.91	0.02	Al	BC
	218.94	0.03	Fe	IEC
Zn	213.86	0.002	Al, Cu, Ni	IEC, IEC, BC

[a]( ) — only for systems with wide search windows (e.g., not with Echelle).
[b]IEC — interfering element correction.
[c]BC — background correction.
[d]Alt λ — select alternate wavelength in presence of high concentrations of interference.
[e]red-sensitive PM tube recommended.
[f]pirged and evacuated optics to remove O_2.

matrix correction factor(s) to all samples. When using internal standards, it is important that the physical chemical characteristics of the internal standard element(s) match those of the analyte(s). This may be difficult to achieve. Another physical interference, high dissolved solids, causes salt buildup at the tip of the torch and the nebulizer. Water saturated argon and sample dilution is used to overcome this effect.

Chemical interferences are not widely encountered in ICP-AES but occasionally take the form of molecular compound formation and ionization effects. These effects are minimized by optimizing plasma conditions for certain elements and by the usual matrix matching and MSA approaches.

Memory effects result when residues of a previous sample contribute to the signal of the current one. Buildup of a previous sample can occur on the torch tip or improper washout or drainage can occur in the spray chamber. Suitable rinsing between samples minimizes this effect. Normal rinse times range between 30 and 60 s. This time may need to be extended with certain matrices. Samples containing widely varying analyte concentrations more than two orders of magnitude will require a longer washout time.

For air particle measurements, sample digests should not present many difficulties. Some combination of a simplified MSA calibration are usually adequate for those digests that utilize high concentrations, $>5\%$ of nitric or nitric/hydrochloric acid mixtures. Interelement corrections may be necessary for samples containing high Al, Fe, Mn, and Ti.

General Procedures for ICP-AES

Operating procedures are usually described well in the manuals supplied by the manufacturer. The following general steps include some issues related to the digests of air particulate filters and simultaneous and sequential analysis.

1. Before the instrument is started, check sample pump tubing, any O-rings, gaskets, etc. in the sample introduction system for deterioration, and replace as necessary. The entire sample introduction system should be scrupulously clean. Follow manufacturer's recommendations closely. When reassembling the system, ensure that the mist generated by the nebulizer is dense with no spitting. Adjust as necessary.
2. Before initiating power sequences for the spectrometer and the torch, check the status of the ICP argon and RF coil cooling water. Unless the instrument is used less than twice a week, it is recommended using liquid argon instead of gas cylinders due to purity and economy. Use a water chiller instead of the local water supply. It provides a more constant pressure and flow, and the water stays cleaner.
3. Select the rinse and sample uptake time. The rinse time is highly dependent on the sample matrix and will require some experimentation to ensure that the highest expected concentrations are washed out of the sample introduction system prior to the introduction of a blank. Usually 30–60 s rinses are suitable. For the sample uptake, the air bubble in the uptake probe can usually be followed into the spray chamber. Add 10–15 s to ensure adequate uptake time. Alternatively, the analyst can aspirate a high Y solution and time to the appearance and disappearance of the blue bullet in the plasma.

4. After igniting the plasma, adjust the RF power and allow the instrument to warm up for at least 30 min before calibration and sample analysis. During this time, adjust the argon flows to the torch (coolant and auxiliary) and sample uptake rate to ensure smooth operation. Perform optical calibration and torch position optimization during warm-up period. For the torch optimization, select a compromise element, between soft and hard emitting elements, such as Ni or Mn. This process may vary between instruments. The torch can also be optimized by aspirating a 1000 μg/mL Y solution and adjusting the nebulizer pressure so that the blue bullet extends 5–20 mm above the coil.
5. Select the appropriate analytical wavelength. The choice is based on the detection limit requirement and freedom from interferences. Scan the sample(s) at each wavelength to verify the existence, or lack of spectral interferences. An analytical blank, the zero standard, should also be scanned to cover the wavelength range of the sample scan in order to determine whether there are baseline shifts. If there are any spectral overlaps for all suitably intense wavelengths, then determine whether the interference or background correction will adequately subtract the overlap. The mathematical correction will affect the precision of the measurement at low levels and therefore affect the detection limits. Since background correction will lengthen the scan time for each element, choose position(s) that clear the analyte peak and any spectral interferences. Occasionally, correction on both sides of an analyte peak will be necessary.
6. Select the appropriate integration time. Integration times greater than 5–10 s generally do not improve detection limits. Longer times greater than 5–10 s may be necessary for elements with emission lines below 200 nm. If the unit is a simultaneous instrument or direct reader, the longest integration time can usually be selected without sacrificing throughput.
7. Select the number of integrations per scan. If using a simultaneous system, choose an average of at least three readings without sacrificing throughput.
8. Calibrate the instrument using a blank and at least three standards. Perform this procedure when samples are first analyzed using a particular program and then quarterly thereafter. The correlation coefficient (r) of the regression expression should exceed 0.999, or repeat the calibration. For daily operation, reestablish the slope and intercept with a blank and midrange standard. If the result of the blank deviates by more than approximately three times the estimated detection limit, then contamination is suspected. Disassemble the sample introduction system and clean thoroughly. If the re-slope midrange standard deviates by

more than 20% from the original calibration, repeat the calibration using new pump tubing and a thoroughly cleaned sample introduction system. Always use ICP grade spectrally pure single element or multielement source standards.

9. Determine whether there is a matrix suppression. Following a successful slope and intercept recalibration, measure an unspiked and spiked aliquot of a sample digest. If the spike recovery is between 75% and 125% (ideally 85–115%), then proceed without using the MSA. If the spike recovery is unacceptable, perform MSA on only one digest sample and apply a correction factor to the remainder of the sample analyses. This is usually performed automatically through the software. In this case, verify the accuracy of the correction factor at least every 20 samples through the analysis of pre- or postdigestion spikes. If predigestion spikes are recovered adequately, then the analyses of postdigestion spikes are unnecessary.
10. Analyze a solution containing high concentrations, 200–1000 μg/mL, of potentially interfering elements, such as Al, Cr, Fe, Mn, or Ti. ICP quality standards tailored for this purpose are commercially available. This solution is used to check the system for proper interfering element corrections (IECs). If correction factors have not been previously calculated, the following general procedure is performed. Aspirate the high concentration solutions as individual samples. The false single element values obtained are divided by the concentration aspirated of the potential interfering element. For example, a false value of 0.3 μg/mL Pb obtained when aspirating a 1000 μg/mL Al solution will yield a factor of 0.0003. This factor is multiplied by whatever concentration of interferant Al is measured in the actual test sample, and the result is subtracted from the total apparent analyte concentration Pb. This operation is usually automated using data acquisition software.
11. Reanalyze a CCV and CCB standard at least every 10 samples. If the results are not within 10% of the expected value, then reanalyze samples run since the last successful CCV analysis. Similarly, if the CCB analysis deviates from zero by more than three times the estimated detection limit, samples run since the last successful CCB measurement must also be reanalyzed.
12. Analyze CCB and CCV solutions at the completion of the analytical run.

Maintenance of an ICP-AES

There are a number of activities crucial to the proper operation of an ICP-AES instrument. It is particularly important to maintain the sample introduction system in a timely manner because most poor quality data and instrument malfunctions are caused by neglect. Critical operations include the following:

- Clean and lubricate sample pump monthly without allowing lubricant to come in contact with peristaltic pump tubing.
- Replace pump tubing weekly or more frequently, and check daily before startup.
- Clean the nebulizer at least weekly according to manufacturer's directions, and adjust as needed. Check the mist generated daily, a dense mist without spitting should be observed.
- Clean the spray chamber weekly, and if the baffle is removable, adjust.
- Soak the torch in strong acid (HNO_3 HCl or aqua regia) and rinse to remove buildup. If adjustable, carefully follow the manufacturer's directions. Make sure that the aerosol injector tip is centered properly in the torch if it is adjustable.
- Miscellaneous items include: cleaning of dust filters, flushing of coolant water lines periodically, and lubrication of the autosampler.

Summary of ICP-AES

The introduction of ICP-AES has provided the analytical/environmental laboratory with a technique having multielement analytical capabilities and detection limits equal to or better than those of FAA. A wide variety of acids and acid concentrations are tolerated by ICP-AES, provided the appropriate corrections for matrix suppression are used. Although the method has historically been insensitive for elements with emission lines at or below 220 nm (e.g., As, Pb, Se), recent developments utilizing axial torch designs and desolvation nebulizers have provide detection limits closer to those of GFAA. Spectral interferences are numerous and in many cases require some form of mathematical background or interfering element correction. Matrix interferences must also be considered but are usually dealt with using some form of MSA, dilution, matrix matching, or internal standardization. Operating and instrument costs are higher than those for FAA, and maintenance is somewhat more extensive. In general, the technique presents few difficulties in the analysis of air particles.

INDUCTIVELY COUPLED PLASMA/MASS SPECTROMETRY

Introduction to ICP-MS

Although ICP was developed initially as a multielement source for atomic emission, for the past 15 years it has been an excellent ion source for elemental MS. Most emission lines used in ICP-AES are ionic, not atomic.

With improvements in instrumentation, ICP-MS has evolved from a research tool to a reliable means of obtaining sub parts per billion multi-element quantitation of a wide variety of matrices, including geological, environmental, and biological samples. Difficulties normally associated with sample homogenization and digestion of bulk quantities (0.2–5 g or more) of these materials are considered relatively minimal for air particle samples. Similarly, any matrix effects from bulk sample digests containing high acids, etc., may normally require some dilution for ICP-MS analysis and are greatly reduced for air particle samples.

ICP-MS techniques offer the advantage of solution detection limits including those measured by ICP-AES and FAA, for approximately 65 elements on the order of 1–100 parts per trillion excluding Ar, C, F, H, He, N, O, etc. ICP-MS can also be used for a number of metals and nonmetals not previously quantified by atomic spectroscopy, including the radionuclides, rare earths, and some halogens, such as Br and I. With recent advances in ion detection and operating software design, the linear dynamic range of ICP-MS has been extended to eight orders of magnitude. ICP-MS also provides the capability of isotope measurements for isotope tracer studies and source identification (Date and Gray, 1983). Finally, isotope dilution techniques are used to provide extremely accurate measurements (Vandecasteele and Block, 1993).

Compact ICP-MS systems are available which provide GFAA detection limits for most elements and are quite suitable for most air particle applications at a lower cost. Some compromises have been made in the instrument components lens, quadrupole mass spectrometer, and the detector system. High-resolution ICP-MS costing about 1.5–2 times more than ICP-AES instrumentation are well suited for laboratories interested in pg/m^3 concentrations found in semiconductor work or other low-level applications of indoor air measurement. Instead of the quadrupole systems, magnetic sections (MS) are capable of resolving masses that differ by 0.01–0.02 *M/Z* and are highly useful in distinguishing between many singly charged elemental ions and interfering molecular species. In quadrupole systems, these interferences are dealt with either by selecting an alternate isotopic mass, mathematically applying a correction factor with subsequent degradation of detection limit, or in a few cases using a sample preparation technique that minimizes potentially interfering matrix species, such as chloride, carbonate, or sulfate. Recently, cool plasma techniques that eliminate most Ar^+-based interferences have become commercially available. Some special nebulization techniques help to minimize water content, with the corresponding reduction in oxide and hydroxide formation, and other molecular species including organics. Some typical detection limits are shown in Table 15.

Table 15. Approximate ICP/MS detection limits for selected elements (ng/L)[a].

Element	Compact[b]	Conventional[c]	High resolution[d]
As	190	2	<0.05
Ba	30	1	<0.05
Cd	70	1	<0.05
Cr	50	3	<0.05
Cu	50	3	<0.05
Fe	3000	10[e]	10
Hg	>36	13	NA[f]
Mg	240	5	NA
Mn	20	1	<0.05
Mo	40	2	<0.05
Na	2200	≈ 10	NA
Ni	80	5	<0.05
Pb	30	2	NA
Sb	20	1	<0.05
Se	1000	50	NA
U		0.1	<0.05
V	30	5	<0.05
Zn	130	3	NA

[a] Parts-per-trillion (ng/L).
[b] VG Genesis.[17]
[c] VG Plasma Quad.[17]
[d] VG Plasma Trace 2.[17]
[e] With cool plasma.
[f] Not available.

Instrumentation for ICP-MS

Analogous components for ICP-MS and ICP-AES instruments are summarized in Table 16. The basic ICP-MS instrument design is shown in Figure 10. As previously noted, both systems essentially share the same sample introduction systems and ICP sources. The development of the interface between the argon plasma and mass spectrometer has been important to the evolution of ICP-MS. Ions produced at atmospheric pressure in the plasma enter the quadrupole mass spectrometer at operating pressures not exceeding 10^{-4} torr. The solution to this discrepancy in pressure is to establish a region (evacuated by a few torr) bounded by a sampling extraction cone 1 mm in diameter and immediately downstream of the plasma by a skimmer cone 0.5–1 mm. The sampling and skimmer cones are usually constructed of polished Ni or Pt.

Table 16. Instrumental comparison of ICP/MS and ICP/AES.

Function	ICP/AES	ICP/MS
Sample introduction	Nebulization	Nebulization
Source	Photons–inductively coupled plasma (27 or 40 MHz)	Ions–Inductively coupled plasma (27 MHz)
Transfer optics	Entrance slit mirror, lens, etc.	Skimmer, sampling cones, electrostatic lenses
Filter	"Photon" filter (monochrometer)	Mass (ion) filter (quadrupole)
Detector	Light detector (photomultiplier)	Ion (mass) detector (electromultiplier)
Summary	Indirect measurement (photons)	Direct measurement (ions)

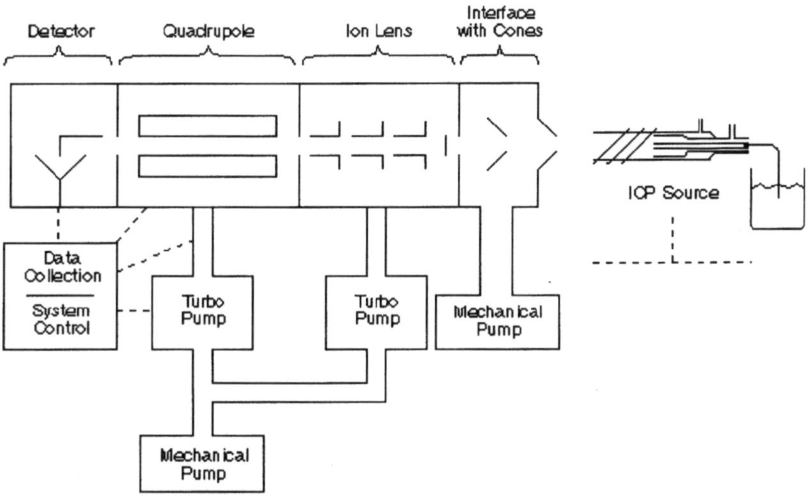

Figure 10. Typical ICP-MS.

After ions are generated in the plasma, they are extracted from the center line plasma flow by the sampling core, which expands into the interface region. The flow of ions is then directed into the ion focusing region, that is maintained at a pressure of approximately 10^{-4} torr, usually by means of turbomolecular pumps backed by rotary pumps. Following the interface, an electrostatic ion lens element, the extraction lens, repels negative ions and directs the positive ones toward the main lens stack and the quadrupole. A photon stop is placed in the path of the ion flow to reduce the light-induced background signal at the detector. The variably charged ion lenses direct the ions around the photon stop in most configurations, and then refocuses the beam where it passes into the mass separation hardware, the quadrupole mass spectrometer. The electric field in the quadrupole oscillates forcing the ions into a wavelike motion. Depending on the quadrupole control, some ions with masses of interest achieve a stable trajectory and pass through the quadrupole region. Most ions are removed in the electric field and collide with the quadrupoles. Those ions with stable trajectories (resonant ions) are sorted according to their mass to charge ratio, *m/z* in the quadrupole. Individual ions are then detected by an ion counting electron multiplier. Usually for ultimate sensitivity and speed, pulse counting is employed. Most instruments also provide an analogue mode for handling higher ion fluxes. Recent improvements in control software allow for a combination of both modes during the same measurement cycle to allow for parts per billion and trillion measurement in the presence of parts per million concentrations of major and minor elements, leading to dynamic ranges of up to eight orders of magnitude.

The features of ICP-MS are derived from several sources. The mass spectrum is greatly simplified compared to the atomic emission spectrum. Instead of thousands of electromagnetic emission lines, there are only a few hundred isotopic masses. Unlike atomic emission where light energy is usually measured at 90°with respect to the torch radial configuration, ICP-MS instrumentation provides detection directly from the torch, mounted axially, pointing more or less directly at the detector. This advantage is not as pronounced when compared to recent ICP-AES configurations where the torch is also mounted axially. Nevertheless, the direct measurement of ions in MS provides a far more uniform response at the detector across the periodic table. This makes the ICP-MS technique particularly suitable for rapid semiquantitative determinations of reasonable accuracy ($\pm 15\%$). Finally, the aperture at the interface between the ICP source and the ion optics is approximately an order of magnitude larger than the entrance slit for the optical system.

Compared to digests or extracts of soils, biota wastes, and geological samples, those of air filter samples are relatively clean. They resemble the more ideal low acid, essentially clean water or trace standards used by instrument manufacturers when making detection limit claims. Compilation of solution detection limits for selected elements, shown earlier in Table 15, were obtained using essentially dilute nitric acid $<0.5\%$ v/v solutions. The three performance levels of commercially available instrumentation are shown to be compact, conventional, and have a high resolution.

The analyst may have difficulty achieving the high level of performance described. First, the method might require a higher acid content in the final digest. The total dissolved solids concentrations in the sample may cause a degradation of the sample introduction and interface cones, thereby affecting precision. High blank values from acids and laboratory water, in addition to inadequate contamination control techniques, further affect detection limits. Imprecision will also impact on mathematical correction of spectral interferences as with ICP-AES. Therefore, detection limits might necessarily be 10 times higher or more for elements frequently requiring interference correction factors, for example As and V in the presence of high chloride. With nickel cones, nickel detection limits are somewhat elevated, but perhaps only tenfold. The determination of rare earths and radionuclides, is particularly well suited to ICP-MS methodology due to the relatively clean mass spectral region of these elements.

Contamination control used with other atomic spectroscopy techniques are not likely to be adequate here. Trace metals grade acids are not sufficiently pure for indoor air sample measurement. Sub-boiled acids of the highest purity and ASTM Type I water are usually required (US EPA, 1986; 1995a, c). Closed vessel digestions conducted in class 100 hoods, or clean rooms are not excessive measures for indoor and ambient air sample analyses. Other techniques might be devised to protect the sample digests from room dust contamination. At a minimum, enclose the sample introduction system. Many commercial systems are configured in this manner. An autosampler placed in an all plastic housing that is continually purged with HEPA filtered air is a useful approach (Thompson and Grohse, 1996b).

Perorm the preparation of daily calibration solutions in ultraclean surroundings; and prepare and store stock solutions, 100 µg/mL or higher in a separate facility. Anticipate contamination when working at low nanogram/picogram levels.

Interferences in ICP-MS

Interferences associated with the sample introduction process for ICP-AES are also encountered in ICP-MS analyses. Measures taken to compensate for these effects are therefore similar. In addition, the use of internal standards in compensating for matrix effects, is more uniformly successful in ICP-MS and usually required. Although the ICP-MS spectrum is relatively simple compared to those for optical and emissions techniques, interferences can significantly affect the accuracy and precision of selected elements. Notable mass spectral interferences include:

- background — arising from the nebulization of water into the argon ICP forming N_2^+, ArO^+, etc.

- molecular — due to reaction with high Cl^-, SO_4^{2-}, or PO_4^{3-} (usually from the acid matrix)
- interelement (isobaric) spectral overlaps between analyte ions
- oxide and hydroxide species originating from sample constituents (analyte and matrix elements)

Typical ICP-MS background species and those arising from the use of specific acids during sample preparation are summarized in Table 17. There are a few serious overlaps in this category: N_2^+ on $^{28}Si^+$, $^{31}P^+$ (NOH^+), ^{32}S (O_2^+), ^{40}Ca (Ar^+), ^{56}Fe (ArO^+), and Ar_2^+ and trace Kr^+ on $^{80}Se^+$. Each of these background species provides multiple isotopic combinations affecting other masses. Although the background spectrum may contain numerous examples of such overlaps, they are in many cases relatively insignificant. The complexity and severity of interferences due to the sample matrix are significantly increased when chloride and sulfate are introduced as the preparation acids. Although chloride is particularly useful in causing the solubilization of many analytes, a wealth of species contributes to the background when all isotopic combinations are taken into account. The serious effect of $^{35}Cl^{16}O^+$ species on $^{51}V^+$ is the most dramatic interference in this category, and the 0.24% $^{50}V^+$ isotope also suffers from a number of interferences. In addition, the intense $^{40}Ar^{35}Cl$

Table 17. Background polyatomic[a] interferences for ICP/MS due to various acid systems.

Solvent system	Interfering species	Analyte	Mass
H_2O/NO_3^-	N_2^+, N_2H^+, NO^+, $NOH^+O_2^+$, O_2H	Si, P, S	28–33
	ArH^+, Ar, ArH^+	K, Ca	39–41
	CO_2^+	Ca	44
	ArN^+, $ArNH^+$, ArO^+ArOH^+	Mn, Fe	54–57
	Ar : Ar	Se	76, 78, 80
Cl^-	ClO^+,$ClOH^+$,	V, Cr, Fe	51–54
	ArCl	As, Se	75, 77
	Cl_2^+	Ge	70, 74
SO_4^{2-}	SO^+, SOH^+	Ti, V	48–51
	S_2^+/SO_2^+	Zn	64
	ArS^+	Ge	72, 74
PO_4^{3-}	PO, POH	Ti	47, 48
	PO_2	Cu	63
	ArP	Ge	71

[a]Including all isotopic combinations.

species significantly impacts on the monoisotopic $^{75}As^{+}$ mass. Therefore, V and As determinations by ICP-MS cannot seriously be considered unless HCl solutions are diluted to less than 0.05%. Nitric acid is preferred for solution stabilization and sample dissolution. Most of the common polyatomic interferences have been identified (Houk, 1986).

Interelement mass spectral overlaps are easily predicted from tables (*Handbook of Chemistry*, 1996) and are correctable. Usually, alternate analytical masses can be used. When, as is the case with monoisotopic elements these are not available, equations are usually provided with instrument software that use the measured interfering element (at another mass) and the natural abundance information to correct for false analyte signal.

Possibly the most serious ICP-MS interference arises from oxide and hydroxide formation. The oxides and hydroxides of the Ca and Ti isotopes overlap with the most abundant isotopes of Co, Cu, Fe, Ni, and a number of lesser ones. The formation of oxide species confounds use of interference correction equations. Since almost all metals form oxides and hydroxides in the plasma, a considerable amount of research has been done on instrumental design to reduce oxide levels. Interference design, plasma gas composition, nebulizer gas flow rates, and forward power all play a role in the degree of oxide formation. Most instrument manufacturers provide specifications for oxide formation for such elements as Ba or Ca. Desolvation nebulizers (ultrasonic, Mistral, etc.) minimize oxide formation by reducing the water solvent load on the plasma. Water-cooled spray chambers are also helpful in achieving reduction in oxide formation.

Another effect grouped with other mass spectral interferences is abundance sensitivity. This refers to the overlaps that may occur when a major element is one mass unit away from an isotope at trace concentrations and is indirectly related to the instrument resolution. Usually instrument manufacturers define a given model's abundance sensitivity in terms of the degree of overlap of Na^{23} on masses 22 and 24.

General Measurement Procedures for ICP-MS

Measurement procedures for ICP-MS differ slightly between manufacturers, but usually follow the general steps listed below:

1. Vacuum pumping procedures are initiated according to the manufacturer's directions. This includes all rotary pumps, interface, and backup pumps to the turbomolecular pumps.
2. Initiate the plasma in a manner similar to that for ICP-AES. Issues related to sample introduction and adjustments are equally of concern here.

3. Analytical conditions including: selection of masses (isotopes), number of times that exposures are counted per analysis, scanning mode, and mass regions to be skipped, are all established at this time or previously stored through the system software. It is often useful to choose additional isotopes to monitor for potential mass spectral interferences. Many instruments also offer the option of scanning the entire region of interest to obtain additional data that is examined later, or rapid peak hopping mode if for example, fewer than three or four analytes are of interest.
4. Make the warm-up period last at least 30 min.
5. Tuning procedures comprise the primary increase in complexity over ICP-AES. Prior to the analytical calibration, a tuning solution is aspirated. Use a solution containing 50–100 µg/L of approximately five elements representing the mass range of interest, and prepared in dilute nitric acid. An example from the EPA methods is a mixture of Be, Co, In, Mg, and Pb (US EPA 1991; 1995a, c). Several other tuning elements are equally useful. This solution is used to tune or check the mass calibration, and possibly the resolution. EPA protocols define good resolution as a peak width of 0.75 amu at 5% peak height for the Mg isotopes (24, 25, 26) at the low end of the spectrum and Pb isotopes (206, 207, and 208) at the high mass end of the spectrum. Mass calibration should be adjusted if any peaks have shifted more than 0.1 amu from center. A second tuning operation involves the calibration of the detector. For those instruments that have a dual mode detection, analog/pulse counting system, a relationship between the modes is established through software protocol that creates a normalized relationship between the two resulting in an extended linear range of up to three orders of magnitude. It is also important to note the peak magnitude for various elements in the tune solution to ensure that instrument sensitivity is acceptable. Daily compare these data to previously recorded peak counts in order to identify unwanted performance trends. Some software configurations establish a response curve across the mass region of interest. Ideally, a smooth, but nonlinear curve should be observed. This is particularly useful for semiquantitative determinations using one calibration element.
6. Calibration standards, QC standards, blanks, and samples are then spiked with internal ones to monitor and compensate for matrix differences between samples and standards. The internal standard spike should contain elements that are not likely found in the samples at measurable levels e.g., Bi, Ho, In, Rh, Sc, Tb, Y, and possibly Li. These elements must represent each significant mass region scanned in the analysis. If only one element is analyzed, e.g., Pb, only one internal standard, nearby Bi, is needed.

7. The analytical calibration is then performed using a minimum of a blank and three calibration standards. Although the range of linearity for ICP-MS instruments goes from 10^5 for older instruments to 10^8 for more recent ones, the analyst is advised not to fully utilize this feature for quantitative determinations. Instead construct the calibration curve over approximately three orders of magnitude for a given analyte. The extended linearity capability is useful for the quantitative determination of consistently higher level parts per million elements, such as Al, Fe, Mg, and Zn in the presence of parts per trillion level elements, such as Be and U.
8. QC samples typically analyzed with ICP-AES are applicable here (ICBs, ICVs, CCBs, and CCVs) using the same data quality objectives and analysis frequency.
9. For older instruments, scan the samples in a less sensitive mode to identify any extremely high concentrations that will damage or reduce the lifetime of the detector. More recent software routines will usually have an automated prescreening mode built into the scan to shut the detector down in the event of a very high ion flux.
10. Initially monitor the analysis for internal standard element recovery. Ideally recoveries should range from 70% to 120%, although some protocols are more relaxed. If they are unacceptably low, due to a severe matrix effect, dilute and reanalyze the sample.

MSA can also be employed, simultaneously with or without internal standards, but is not likely needed. For environmental measurements, the above calibration techniques provide precision and accuracies similar to those found in ICP-AES. For extremely accurate determinations, $\pm 0.5\%$ to 1%, use isotope dilution techniques, as with any mass spectrometric technique. General procedures are found in a number of references (US EPA, 1991; 1995a, c). Isotope ratio determinations are highly useful for the identification of anthropogenic contamination sources, particularly radioactive contamination. The most interference free mass(es) are selected for calibration and all isotopes are then scanned for raw peak counts. Because the relative response of the isotopes is similar, relative abundance of each is easily determined and compared to the known natural abundance.

Hydride generation commercially available as an option for the ICP-MS determination of As and Se, suffer from Cl and Ar interferences when using direct aspiration. Chloride interferences are removed with improved gas–liquid separator designs. While the argon interference for Se cannot be minimized, the Se signal is greatly enhanced. Detection limits of less than five parts per trillion are easily achieved. General purpose flow injection systems are available providing not only for hydride generation analyses but also for automated internal standardization, dilution, and ion-exchange preconcentration of selected elements. Auto samplers are highly recommended.

Maintenance

Maintenance operations are more extensive than for other atomic spectroscopic techniques. In addition to the usual sample introduction maintenance activities, the following ICP-MS specific procedures are required to ensure proper performance:

1. Interface cones require daily inspection and daily to weekly cleaning. Platinum cones provide greater resistance to difficult matrices, e.g., high acid content.
2. Check and/or clean water and air filters weekly.
3. Check multiplier voltages weekly.
4. Clean extraction lens, and check rotary pump oil monthly. Also check oil mist filters.
5. Check and/or clean ion lens stack, and change rotary and turbo pump oils every six months.
6. Any worn O-rings are replaced annually.

Summary for ICP-MS

ICP-MS has become a particularly promising technique for air particle analyses, offering: sub parts per billion solution detection limits, highly accurate isotope dilution measurements, reasonably accurate semiquantitative analyses, and isotopic determinations. No longer exclusively a research tool, ICP-MS is developed to the point of being considered routine and reliable. This is primarily due to advances in the design of ICP-MS interface, and computers/software assisting in the control of operation and data management. While the skill level for ICP-MS is perhaps the greatest among all AAS and ICP techniques, routine operation does not require individuals with research orientation, particularly if computer stored protocols are available for the operator.

Interferences arise from matrix effects normally encountered with ICP-AES as well as a variety of mass spectral interferences due to overlaps from background species in the plasma and solvent system and other elements in the sample. Dilute nitric acid is the solvent system of choice while HCl should be avoided when making measurements in the mass region between 35 and 77. An enhanced awareness of contamination issues. Instrument and operating costs are more significant for ICP-MS

Table 18. Methods for atomic spectroscopic determination of metals in air particulates.

Matrix	Collection medium	Approximate air volume (m^3)	Dissolution method	Solvent	Measurement method[a]	Reference
Ambient	Quartz (glass) fiber filter (and other filter media)	2500	(1) Hot Plate/Boiling	3 M HNO_3	Flame AA (also all others)	CFR, 1978
			(2) Ultrasonic — covered Teflon® beaker	2.6 M HNO_3–0.9 M HCl	ICP/AES, Flame AA (ICP/MS, GFAA)	US EPA, 1980
			(3) Hot ultrasonic — covered Teflon® beaker	1.03 M HNO_3–2.23 M HCl	(ICP/AES, Flame AA (ICP/MS)[a]	US EPA, 1980
			(4) Microwave heating — closed Teflon® vessel	3 M HNO_3	GFAA, ICP/AES, ICP/MS, Flame AA	US EPA, 1995c
			(5) Microwave heating — closed Teflon® vessel	0.89 M HNO_3–2.06 M HCl	ICP/AES, Flame AA (ICP/MS)	
Indoors (residential)	Teflon® membrane	6	Shaker–30 minutes with 50% HNO_3, evaporate; add H_2O_2, evaporate	50% HNO_3, 30% H_2O_2	GFAA, CVAFS ICP/MS	RTI, 1995
Industrial workplace	Polyvinyl chloride	0.5–2	(1) Hot plate or microwave	4 : 1 HNO_3 : $HClO_4$ or HNO_3 : HCl or HNO_3	FAA, GFAA ICP/AES, ICP/MS	NIOSH, 1994[b]
	Teflon® membrane		(2) Hot plate or microwave	HNO_3/H_2O_2	All	NIOSH, 1994
Industrial source (stack)	Quartz (glass) fiber filter	3–4	Hot plate or microwave	HNO_3/HF	All	US EPA, 1995a

[a]HCl causes some spectral interferences for ICP/MS for selected elements and potential volatilization losses during GFAA determinations.
[b]NIOSH recommends consulting other methods for the dissolution of Ag, Sn, W, Ti, etc.

Table 19. Other applicable reference methods for dissolution of air particulates.

Matrix	Agency	Method	Description	References
Solid wastes	EPA	3050	Hotplate extraction with HNO_3, H_2O_2 (HCl optional)	US EPA (1995a)
	EPA	3051	Microwave digestion with HNO_3	US EPA (1995a)
	EPA	3052	Microwave total digestion with HNO_3 (HF, HCl, H_2O_2 optional)	US EPA (1996)
Dusts			Dry ash at 350°C, 30 min; reflux with 25% HNO_3	Jones and McDonald (1983)
			Hotplate digestion with 2 M HNO_3	Tinsley (1983)
			Dry ash at 550°C, 30 min; reflux with 25% HNO_3	Tinsley (1983)
Air Particulates	OSHA		HNO_3, $HClO_4$, heat	OSHA (1991)
			HNO_3, H_2SO_2, heat	CFR
Soils			HNO_3/HF/$HClO_4$, hotplate	Scott and Thomas (1977)
			HNO_3/HF/$HClO_4$	Que Hee (1988)
			50% HNO_3, 2 hr at 80°C	Krueger and Duguay (1989)
			HNO_3, HF, $HClO_4$, hotplate	Veneman *et al.*, (1982)
			HNO_3, $HClO_4$, Kjeldahl	Veneman *et al.*, (1982)

than the other commonly used atomic spectroscopy techniques, and maintenance operations are more extensive.

REFERENCE METHODS

Methods established in the United States for the determination of trace elements in air particulate matter are shown in Table 18. Dissolution procedures are more specific to the media than are the measurement techniques. In many cases, the measurement options listed are virtually identical across a variety of diverse media: water, waste waters, solid wastes, and stationary sources. Regulatory exposure limits for a given analyte matrix usually dictate which of the measurement techniques to implement. Dissolution procedures not only reflect differences in filter media, but attempt to address the expected diversity of particulate composition. In some instances, alternative digestion/extract procedures are provided when acids used in the primary method are incompatible with a given measurement technique, e.g., the GFAA measurement of dilute HCl solutions.

Modifications to a method are acceptable to a regulatory agency when the laboratory has validated it extensively through the use of QC materials matrix spikes, reference materials, etc., or uses another reference procedure from the same agency for a different medium. For example, EPA Method 0060 for stationary source combustion aerosols lists FAA, GFAA, and ICP-AES as options for measurement (US EPA, 1995a). It is usually acceptable to implement an ICP-MS finish to the method since the technique is now widely accepted by EPA for Methods 200.8 — Water and Wastes (US EPA, 1991), 6020 — Solid Wastes (US EPA, 1995a), and 1638 — Ambient Waters (US EPA, 1995d).

The determination of mercury adsorbed on particulate matter is a special case. No atomic spectroscopic method except CVAA or CVAFS is accepted. A variety of reference methods for the dissolution of other media (e.g., solid waste) that are applicable to air particulate matter are shown in Table 19. Other useful methods applied to various particulate matter are also listed.

ACKNOWLEDGMENTS

Acknowledgments are due to Drs. Peter Brown of Fisons Instruments and William Gutknecht of Research Triangle Institute for their comments, suggestions, and insights. Thank you to Ms. Jan Shirley of Research Triangle Institute for her indispensable assistance in the preparation of this manuscript.

REFERENCES

Alfassi, Z.B. and Chien M. Wai, *Preconcentration Techniques for Trace Analysis*, CRC Press, London, 1992.

Annual Book of ASTM Standards, Vol. 11.01, 1991.

Bloom, N. and W.F. Fitzgerald, *Anal. Chim. Acta.*, **208** (1988) 151.

Boumans, P.W.J.M., ed., *Inductively Coupled Plasma Emission Spectroscopy, Pts I and II*, Wiley-Interscience, New York, 1987.

Caldwell, J.S., R.J. Lishka and E.F. McFarren, *JAWMA*, **65** (1973) 731.
Code of Federal Regulations, Title 40, Part 50, Appendix G, pp. 46258–46263, Oct. 5, 1978.
Code of Federal Regulations, Title 40, Part 60, Method 12/55.
Dagnall, R.M. and T.S. West, *Appl. Opt.*, **7** (1968) 1287.
Date, A.R. and A.L. Gray, *Analyst*, **106** (1981) 1255.
Date, A.R. and A.L. Gray, *Int. J. Mass Spectrum Ion Phys.*, **48** (1983) 357.
Douglas, D.J., E.S.K. Quan and R.G. Smith, *Spectrochim. Acta.*, **38B** (1983) 39.
Fassel, V.A. and R.N. Kniseley, *Anal. Chem.*, **46** (1974) 1110A.
Gray, A.L., *Proc. Soc. Anal. Chem.*, **11** (1974) 182.
Handbook of Chemistry and Physics, Section B, Chemical Rubber Co, Cleveland, OH: 1996.
Harrison, G.R., *J. Opt. Soc. Am.*, **39** (1949) 522.
Hatch, W.R. and W.L. Ott, *Anal. Chem.*, **40** (1968) 2085.
Houk, R.S., *Anal. Chem.*, **58** (1986) 97A.
Houk, R.S., V.A. Fassel and H.J. Svec, *Dyn. Mass Spectrum*, **6** (1981) 234.
Jones, K. and A. McDonald, *Environ. Pollut. Series B*, **6** (1983) 133.
Kingston, H.M. and L.B. Jassie, eds., *Introduction to Microwave Sample Preparation: Theory and Practice*, American Chemical Society, Washington, DC, 1988.
Koirtyohann, S.R. and E.E. Pickett, *Anal. Chem.*, **37** (1965) 601.
Krueger, J.A. and K.M. Duguay, *Bull. Environ. Contam. Toxicol.*, **42** (1989) 574.
L'vov, B.V., *Inzh. Fiz. Zh.*, **2** (1959) 44.
L'vov, B.V., *Spectrochim. Acta*, **33B** (1978) 153.
Massmann, H., *Spectrochim. Acta*, **23B** (1968) 215.
Minczewski, J., J. Chwastowska and D. Dybczynski, *Separation and Preconcentration Methods in Inorganic Trace Analysis*, Ellis Horwood, Ltd, Chichester, 1982.
Moore, M.R. and P.A. Meredith, *Clin. Chim. Acta.*, **75** (1977) 167.
National Institute of Occupational Safety and Health (NIOSH). Methods 7082 and 7300, *NIOSH Manual of Analytical Methods*, 4th edn. US Dept. of Health and Human Services (DHHS), NIOSH Publication No. 94-113, August 1994.
Occupational Safety and Health Administration, Division of Physical Measurements and Inorganic Analyses, *OSHA Method ID-121*, US Department of Labor, Salt Lake City, UT, 1991.
Perkin Elmer Corporation, Ridgefield, CT.
Phelps, F.M., *M.I.T. Wavelength Tables, Vol. 2, Wavelengths by Element*, MIT Press, Cambridge, MA, 1992.
Prugger, H. and R. Torge, German Patent 1,964,469 — U.S. Patent 3,676,004, December 1969.
Que Hee, S.S. and J.R. Boyle, *Anal. Chem*, **60** (1988) 1033.
Research Triangle Institute (RTI), Laboratory Operations Protocol, *Analysis of Air Particulate for Lead, Arsenic, Cadmium, and Chromium*, RTI/ACS-AP-209–111, June 15, 1995.
Scott, D. and J.D. Thomas, *Soil Science*, **124** (1977) 319.
Slavin, W., G.R. Carnrick, D.C. Manning and E. Pruszkowska, *At. Spectrosc.*, **4** (1983) 69.
Smith Jr., J.B. and G.M. Hieftje, *Appl. Spectrosc.*, **37** (1983) 419.
Sneddon, *J.,Talanta*, **30** (1983) 631.
Sullivan, J.V. and A. Walsh, *Spectrochim. Acta.*, **21** (1965) 721.
Thompson, L.J. and P.M. Grohse, "The Reliable Low Level Trace Metals Determinations of Biota by ICP-MS" in *The 1996 Pittsburgh Conference on Analytical Chemistry and Applied Spectroscopy*, Chicago, March 3–8, 1996a.
Thompson, L.J. and P.M. Grohse, "Trace Metals Analysis of Biota by ICP-MS" In *The 22nd Annual Conference of the Federation of Analytical Chemistry and Spectroscopy Societies*, St. Louis, October 15–20, 1996b.
Tinsley, D.A., A.R. Baron, R. Critchley and R.J. Williamson, *Int. J. Environ. Anal. Chem.*, **14** (1983) 285.
Tsalev, D.L., *Atomic Absorption Spectrometry in Occupational and Environmental Health Practice*, CRC Press, London, 1984.
U.S. EPA, Environmental Monitoring and Support Laboratory, Method 245.1., In *Methods for Chemical Analysis of Water and Wastes*, EPA-600/4-79-020, 1979.

US EPA, Method EQL-0380-044, *Determination of Lead Concentration in Ambient Particulate Matter by Flameless Atomic Absorption Spectrometry*, US EPA, Environmental Monitoring Systems Laboratory, Research Triangle Park, NC, 1980.

US EPA, Method 7000 Series, *Test Methods for Evaluating Solid Waste*, SW-846, 3rd edn., US EPA, Washington, DC, 1986.

US EPA, Environmental Monitoring and Support Laboratory, Method 200.8, In *Methods for Chemical Analysis of Water and Wastes*, EPA-600/4-79-020, 1991.

US EPA, Office of Solid Waste and Emergency Response, Methods 0060, 3050, 3051, 6020, 7470. In *Test Methods for Evaluating Solid Waste. Physical/Chemical Methods*. 3rd edn. (SW-846) US EPA, Washington, DC, 1986 (January 1995a revision).

US EPA, Office of Water Regulations and Standards, Industrial Technology Division, Office of Water, Method 1631, US EPA Agency, Washington, DC, April 1995b.

US EPA, Office of Research and Development, National Risk Management Laboratory, Center for Environmental Research Information. Draft Method, IO-3.1. In *Compendium of Methods for the Determination of Inorganic Compounds in Ambient Air* (Draft), September 1995c.

US EPA, Method 1638, *Determination of Trace Elements in Ambient Waters by Inductively Coupled Plasma-Mass Spectrometry*, US EPA, 821-R-95-031, 1995d.

US EPA, Office of Solid Waste and Emergency Response, Draft Method 3052. In *Test Methods for Evaluating Solid Waste, Physical/Chemical Methods*, 3rd edn. (SW-846) US EPA. Washington, DC, 1996.

Vandecasteele, C. and C.B. Block, *Modern Methods for Trace Element Determination*, John Wiley and Sons, Chichester, UK, 1993, pp. 241–244.

Veneman, P.L.M., S.M. Bodine, J.R. Murray and J.H. Baker, *Comm. Soil Science Plant Analysis*, **13** (1982) 585.

Walsh, A., *Spectrochim. Acta.*, **7** (1955) 108.

Wendt, R.H. and V.A. Fassel, *Anal. Chem.*, **37** (1965) 920.

Winge, R.K., V.A. Fassel, V.J. Peterson and M.A. Floyd, *Inductively Coupled Plasma-Atomic Emission Spectroscopy: An Atlas of Spectral Information*, Elsevier, N.Y., 1985.

2 X-RAY FLUORESCENCE ANALYSIS OF AMBIENT AIR SAMPLES

J.G. WATSON, J.C. CHOW and C.A. FRAZIER

Desert Research Institute, P.O. Box 60220,
Reno, Nevada, 89506, USA

INTRODUCTION

The energy dispersive X-ray fluorescence (XRF) analytical method can be applied to determine the concentrations of more than 40 different elements in particles on air filters without destroying the sample. The objectives of this chapter are to describe the principles and applicability of XRF to air filter samples, outline practical methods for its application, and identify examples for its use in air quality studies.

Particles with aerodynamic diameters less than ~50 µm (Total Suspended Particulate matter, or TSP) can become suspended into the atmosphere, and those with aerodynamic diameters less than 10 µm (PM-10) can transport over large distances (Lancaster and Nickling, 1993) and enter the human respiratory system (Ferris *et al.*, 1979; Miller *et al.*, 1979). Particles with aerodynamic diameters less than 2.5 µm (PM-2.5) are most effective at scattering light and have a great effect on visibility (Malm, 1979; Watson and Chow, 1994), impairment, and the Earth's radiation balance (Charlson *et al.*, 1992).

The adverse health impacts of suspended particles have caused the United States to adopt the National Ambient Air Quality Standards (NAAQS) for PM-10 of 50 µg/m^3 for an annual average, and 150 µg/m^3 for a 24-hour average (Federal Register, 1987). The United States

Environmental Protection Agency (US EPA) is currently re-evaluating federal standards for suspended particles, and it is possible that new standards will be issued. These will probably retain the annual average concentration of 50 μg/m^3 for PM-10, while adding an annual arithmetic average for particles with aerodynamic diameters less than 2.5 μm (PM-2.5), between 12.5 and 20 μg/m^3, and a 24-hour PM-2.5 average between 18 and 65 μg/m^3. As particle sizes become smaller, XRF elemental determinations become more accurate owing to lower self-absorption of low energy X-rays by large particles.

Particle chemical composition is of special concern for health, since potentially toxic elements such as lead, arsenic, beryllium, cadmium, chromium, copper, selenium, zinc, acidic species, and certain organic compounds can cause both short and long term medical problems. Various chemicals affect the transfer of radiation through the atmosphere in different ways. The application of source receptor models (Chow *et al.*, 1993; Watson *et al.*, 1989; 1990; 1991) also requires chemical characterization of these particles to apportion ambient concentrations to their sources for the development of emissions reduction strategies.

XRF PRINCIPLES AND VARIATIONS

Both XRF (Giaque *et al.*, 1973; Rhodes, 1973) and Particle Induced X-ray Emission (PIXE) (Cahill, 1980; 1993; 1995) spectroscopy are based on the measurement of X-rays produced by the ejection of an inner shell electron from an atom in the sample, thereby creating a vacancy in the inner atomic shell. A higher energy electron drops into the lower energy orbital and releases a fluorescent X-ray to remove excess energy (Jaklevic *et al.*, 1977). The released energy is characteristic of the emitting element, and is used to identify the element. The area of the fluorescent X-ray peak (intensity) in the spectrum is proportional to the number of atoms in the sample. It is used to quantitatively determine a specific element's concentration through a direct comparison with standards (Dzubay and Stevens, 1975). Elements with atomic numbers exceeding 14 (Na) are commonly quantified by XRF. The most common elements measured in aerosol samples are for: Al, Si, P, S, Cl, K, Ca, Ti, V, Cr, Mn, Fe, Co, Ni, Cu, Zn, Ga, As, Se, Br, Rb, Sr, Y, Zr, Mo, Pd, Ag, Cd, In, Sn, Sb, Ba, La, Au, Hg, Tl, Pb, and U.

An X-ray tube is most often used as the excitation source for XRF, while PIXE usually uses protons from a cyclotron or Van de Graff accelerator. The minimum detectable limits for XRF and PIXE are a function of the irradiation time. Longer analysis times yield lower detection limits with an increasing cost per sample. The higher excitation energy of PIXE sometimes results in loss of elasticity of membrane filters, effectively limiting the attainable detection limits. This creates potential handling problems for samples requiring long irradiation times. A disadvantage of XRF and PIXE is that the X-rays from low atomic number elements are emitted at very low energy levels and are strongly absorbed by the filter material and aerosol deposit. Aluminum (Al) is the lightest element that is usually measurable in particle samples. Particles in the coarse fraction are large enough that sizable corrections must be made for self-absorption of X-rays from aluminum and other light elements (Dzubay and Nelson, 1975). The corrections are negligible for the fine particles, however XRF and PIXE are most accurate when applied to thin film samples in which X-ray absorption is minimized or correctable.

XRF methods can be broadly divided into two categories: wavelength dispersive (WDXRF), that utilizes crystal diffraction for observation of fluorescent X-rays, and energy dispersive (EDXRF), that uses a silicon (Si) semiconductor detector. The WDXRF method is characterized by excellent spectral resolution, which minimizes peak overlaps. WDXRF requires high power excitation to overcome low sensitivity resulting in excessive sample heating and potential degradation. Conversely, EDXRF features excellent sensitivity but less spectral resolution, requiring complex spectral deconvolution procedures.

XRF methods can be further categorized as direct/filtered excitation, where the X-ray beam from the tube is optionally filtered and then focused directly on the sample, or secondary target excitation, where the beam is focused on a target of the selected material to produce X-rays of the desired energy. The secondary fluorescent radiation is then used to excite the samples. The direct/filtered approach has the advantage of delivering higher incident radiation flux to the sample for a given X-ray tube power, since about 99% of the incident energy is lost in a secondary fluorescor. The secondary fluorescor approach, however, produces a more nearly monochromatic excitation that reduces unwanted scatter from the filter, thus yielding lower detection limits.

Total Reflection X-ray Fluorescence (TXRF) (Streli, 1995; Streli *et al.*, 1995a, b; Wobrauschek *et al.*, 1991) has recently been applied to measure elements with atomic numbers lower than that of aluminum. In TXRF, X-rays enter a vacuum chamber where they impinge on a W/C multilayer monochrometer. The modified beam is directed to a sample reflector, where the sample is excited and monitored outside the chamber by a solid state detector. Quantitative measurements of B, C, N, O, F, Na, and Mg are possible with the TXRF configuration.

EDXRF is the most easily implemented and commonly used method of elemental analysis for air samples, and only this method is discussed in this chapter. Bertin (1975) provides a comprehensive handbook of approaches to XRF analyses, while recent reviews by Potts and Webb (1992), and Bacon *et al.* (1995) have extensive references to more recent XRF applications and methods.

In EDXRF, hereafter termed XRF, emitted X-rays with energies less than approximately 4 keV affecting Na, Mg, Al, Si, P, S, Cl, and K, can be absorbed in the filter in a thick particle deposit, or even by large particles in which these elements are contained. Very thick filters also scatter much of the excitation radiation or protons, thereby lowering the signal to noise ratio for XRF. For this reason, thin membrane filters with deposits in the range of 10–50 $\mu g/cm^2$ provide the best accuracy and precision for XRF analysis.

XRF is most accurate for particles with geometric diameters less than 10 μm and elements emitting X-rays with high energies (>4 keV). Dzubay and Nelson (1975) have developed mathematical corrections for this absorption that depend on a general knowledge of the particle size distribution on the sample. These corrections are more accurate when size-selective inlets with known transmission characteristics are incorporated into the particle sampling system. XRF analyses are applied to a portion of the filter. Therefore, it is important to acquire a sample with a uniform deposit of known area across the filter so that measured amounts can be scaled to the entire deposit.

COMPARISON WITH OTHER METHODS

XRF is one of several analytical methods that have been applied to elemental analysis of air samples, including instrumental neutron activation analysis (NAA), atomic absorption spectrophotometry (AAS), and inductively coupled plasma–atomic emission spectroscopy (ICP–AES). XRF is a preferred method in many cases because it obtains many elemental concentrations with minimal labor, low detection limits, and without sample destruction. XRF is not applicable to all aerosol samples, and in its most commonly applied form does not accurately quantify elements with atomic numbers lower than 14 (Na).

The detection limits from several different elemental analysis methods are compared in Table 1. XRF detection limits depend on a number of factors including: type and manufacturer of filter media, consistency of filters with respect to thickness and background contamination, analysis counting time and conditions, and the element. XRF detection limits can be substantially lowered by:

- increasing the irradiation energy — the amperage of the radiation tube
- radiation area — the beam spot size to which the filter is exposed

- reducing background radiation — by using thinner filters or monochromatic excitation energies
- increasing analysis times — detection limits decrease by a factor of two each time the analysis time is doubled

The detection limits reported in Table 1 can be lowered substantially by an optimal combination of these methods. Many experiments are needed when an XRF system is configured to determine this optimal combination.

Table 1. Analytical measurement specifications for air filter samples.

Element	Minimum detection limit in ng/m^3 [a]					
	ICP/AES[b,d]	AA/Flame[b,e]	AA/Furnace[b,e]	INAA[b,f]	PIXE[g]	XRF[c]
Be	0.06	2[d]	0.05	NA[h]	NA	NA
Na	NA	0.2[d]	<0.05	2	60	NA
Mg	0.02	0.3	0.004	300	20	NA
Al	20	30	0.01	24	12	5
Si	3	85	0.1	NA	9	3
P	50	100,000	40	NA	8	3
S	10	NA	NA	6,000	8	2
Cl	NA	NA	NA	5	8	5
K	NA	2[d]	0.02	24	5	3
Ca	0.04	1[d]	0.05	94	4	2
Sc	0.06	50	NA	0.001	NA	NA
Ti	0.3	95	NA	65	3	2
V	0.7	52	0.2	0.6	3	1
Cr	2	2	0.01	0.2	2	1
Mn	0.1	1	0.01	0.12	2	0.8
Fe	0.5	4	0.02	4	2	0.7
Co	1	6[d]	0.02	0.02	NA	0.4
Ni	2	5	0.1	NA	1	0.4
Cu	0.3	4	0.02	30	1	0.5
Zn	1	1	0.001	3	1	0.5
Ga	42	52	NA	0.5	1	0.9
As	50	100	0.2	0.2	1	0.8
Se	25	100	0.5	0.06	1	0.6
Br	NA	NA	NA	0.4	1	0.5
Rb	NA	NA	NA	6	2	0.5
Sr	0.03	4	0.2	18	2	0.5
Y	0.1	300	NA	NA	NA	0.6
Zr	0.6	1000	NA	NA	3	0.8

Table 1. (continued).

Element	Minimum detection limit in ng/m³[a]					
	ICP/AES[b,d]	AA/Flame[b,e]	AA/Furnace[b,e]	INAA[b,f]	PIXE[g]	XRF[c]
Mo	5	31	0.02	NA	5	1
Pd	42	10	NA	NA	NA	5
Ag	1	4	0.005	0.12	NA	6
Cd	0.4	1	0.003	4	NA	6
In	63	31	NA	0.006	NA	6
Sn	21	31	0.2	NA	NA	8
Sb	31	31	0.2	0.06	NA	9
I	NA	NA	NA	1	NA	NA
Cs	NA	NA	NA	0.03	NA	NA
Ba	0.05	8[d]	0.04	6	NA	25
La	10	2000	NA	0.05	NA	30
Au	2.1	21	0.1	NA	NA	2
Hg	26	500	21	NA	NA	1
Tl	42	21	0.1	NA	NA	1
Pb	10	10	0.05	NA	3	1
Ce	52	NA	NA	0.06	NA	NA
Sm	52	2000	NA	0.01	NA	NA
Eu	0.08	21	NA	0.006	NA	NA
Hf	16	2000	NA	0.01	NA	NA
Ta	26	2000	NA	0.02	NA	NA
W	31	1000	NA	0.2	NA	NA
Th	63	NA	NA	0.01	NA	NA
U	21	25,000	NA	NA	NA	1

[a]Minimum detection limit is three times the standard deviation of the blank for a filter of 1 mg/cm² areal density.

ICP/AES = Inductively Coupled Plasma with Atomic Emission Spectroscopy.

AA = Atomic Absorption Spectrophotometry.

PIXE = Proton Induced X-ray Emissions Analysis.

XRF = X-ray Fluorescence Analysis.

INAA = Instrumental Neutron Activation Analysis.

[b]Concentration is based on the extraction of 1/2 of a 47 mm filter in 15 ml of deionized–distilled water, with a nominal flow rate of 20 l/min for 24-hour samples.

[c]Concentration is based on 13.8 cm² deposit area for a 47 mm filter substrate, with a nominal flow rate of 20 l/min for 24-hour samples with 100 s radiation time.

[d]Harman (1989).

[e]Fernandez (1989).

[f]Olmez (1989).

[g]Eldred (1993).

[h]Not available.

SAMPLE PREPARATION

Aerosol samples are most often acquired by drawing ambient air through filter material using a pump. The particles remain deposited on the filter, while all gases pass through it. After sampling, filters can be submitted to XRF and other analyses for the desired chemical compounds. However, the intention to perform XRF analyses on filters cannot be separated from the methods used to obtain the sample. Sampling for chemical analysis requires stringent attention to filter media, the sampler used to obtain the filter deposits, and sample handling and storage. When this chemical analysis is intended for source apportionment receptor modeling, the use of sequential sampling systems, particle and gas sampling systems with denuders, and portable samplers may be needed. When chemical analysis of samples is anticipated, the first consideration should be which analysis methods will be applied, and how samples can be obtained to meet the needs of those analyses. Chow (1995) describes how to match sampling equipment, filter media, and analytical techniques such as XRF in the design of an aerosol measurement program.

Filters

Particle sampling filters consist of a tightly woven fibrous mat or plastic membrane that has been penetrated by microscopic pores. Chow (1995) identifies several filter types and manufacturers that have been successfully used in different sampling systems when elemental analysis is desired. The choice of filter media results from a compromise among the following filter attributes:

- mechanical stability
- chemical stability
- particle or gas sampling efficiency
- flow resistance
- loading capacity
- blank values
- artifact formation
- compatibility with analysis methods
- cost and availability

Lippmann (1995) provides a comprehensive list of commercially available filter types, their sampling efficiencies, and manufacturers. XRF provides its best performance when thin membrane filters with homogeneous deposits on their surface are used.

Teflon® membrane filters consist of a porous Teflon® sheet that is either stretched across a plastic ring or supported by a loosely woven Teflon® mat. These filters meet requirements in all categories except flow resistance and carbon blank levels. Because of their low porosity, it is not usually possible to attain the flow rates needed by the size selective inlets in high-volume sampling, but those required for low- and medium-volume inlets can be obtained. The filters usually have very low blank levels for ions and elements. The deposit of particles on the filter surface makes these substrates especially amenable to XRF and PIXE analyses. Gelman 1.0, 2.0, and 3.0 μm pore size Teflon® filters, made of poly tetrafluoroethylene (PTFE) Teflon® stretched across a polymethylpentane ring, are the most commonly used Teflon® membrane filters. Gelman Zefluor filters consist of PTFE Teflon® mounted on a woven PTFE mat. The Zefluor filters are less desirable because their larger mass density decreases XRF and PIXE sensitivity, and the similar appearance of both sides causes them to be mounted upside down with the particles being drawn through the mat rather than onto the surface of the membrane.

Etched polycarbonate membrane filters are constructed from a thin polycarbonate sheet through which pores of uniform diameter have been produced by radioactive particle penetration and chemical etching. These filters have low sampling efficiencies ($>80\%$) , even for small pore sizes (Buzzard and Bell, 1980; Liu and Lee, 1976). This low efficiency is used as an advantage when making size-specific measurements. Polycarbonate membrane filters have low elemental blank levels. The filters hold an electrostatic charge that causes them to be attracted to nearby objects unless substantial effort is invested in discharging them with a small radioactive source (Engelbrecht *et al.*, 1980). The Nuclepore 8.0 and 0.4 μm filters are most commonly used in ambient aerosol sampling. While the 0.2 μm pore size filter provides a higher sampling effectiveness, its higher flow resistance requires excessive vacuum for a reasonable flow rate. Though membrane filters are usually very clean, filter batches may be contaminated, and a sample from each batch of filters (1 out of 50–100 filters) should be submitted to XRF analysis for use in a field study. Recent studies have found elevated levels of lead, calcium, and nitrate in batches of blank Teflon® membrane filters.

Most XRF units cannot accommodate filters larger than approximately 47 mm in diameter. If larger filters are used, a section must be removed for placement in the X-ray chamber. This requires additional labor, has a potential for contamination, and lowers ambient detection limits since much of the deposit is spread over sections of the filter not submitted to analysis. Typical filter sizes for XRF are 47, 37, and 25 mm. Smaller filters require substantial collimation of the X-ray beam, with subsequent loss of sensitivity owing to a smaller beam spot. Smaller filters also increase the probability that the excitation beam will contact the filter cassette, thereby increasing the background of scattered radiation.

Sampling Methods

Filter substrates must be protected from contamination prior to, during, and after sampling. This is accomplished by placing the filter in a holder. These holders must:

1. mate to the sampler and flow system without leaks
2. be composed of inert materials that do not absorb acidic gases
3. allow a uniformly distributed deposit to be collected
4. have a low pressure drop across the empty holder
5. accommodate the sizes of commonly available air sampling filters, e.g., 37 or 47 mm
6. be durable and reasonably priced

Most filter holders are configured as in-line or open-faced. In-line holders often concentrate the particles in the center of the substrate, and this will bias the results if analyses are performed on portions of the filter. Fujita and Collins (1989) show differences as high as 600% between chemical measurements in the middle and at the edges of filters sampled with in-line filter holders. Open-faced filter holders are a better choice for ambient aerosol sampling systems.

Lippmann (1995) displays specifications and photographs for a large variety of commercially available filter holders. The holders most commonly used for ambient aerosol sampling are: Gelman stainless steel, Nuclepore polycarbonate plastic, Savillex PFA Teflon®, and Sierra-Andersen polyethylene rings.

The Gelman stainless steel filter holder accommodates 28 and 47 mm filters. Though it has an open-faced adapter, it is sometimes used with an in-line coupler. As noted above, this in-line holder results in a spot in the center of the filter. These filter holders cost several hundred dollars apiece, filters are generally loaded into them in the field rather than in a laboratory environment, and they cannot be used to sample nitric acid.

The Nuclepore polycarbonate plastic filter holders accommodate 25, 37, and 47 mm diameter filters. Each one is inexpensive making it possible to purchase many of them for laboratory loading and unloading.

The Savillex PFA Teflon® 47 mm filter holder is made of injection molded PFA Teflon®, that was previously noted as having the least inclination to absorb nitric acid. These filter holders have a tapered extender section (called a receptacle) that can be mated to a sampler plenum with an O-ring in a retainer ring. The cost of these filter holder combinations is a few tens of dollars, making it feasible to purchase many for laboratory loading.

The Sierra-Andersen polyethylene filter holders are two piece rings that accommodate 37 mm diameter filters. They are used exclusively with the Sierra-Andersen 240 series of virtual impactor samplers. They are inexpensive (less than 10 dollars each) and can be loaded in the laboratory for shipment to field sites.

The Sierra-Andersen low-volume dichotomous sampler has been used extensively for obtaining 37 mm Teflon® membrane filter samples. This sampler is commercially available as a PM-10 reference sampler, and uses a virtual impactor to separate the PM-2.5 and coarse particle size fractions. Flow rates are controlled by a differential pressure regulator. Ten percent of the fine particles are sampled on the coarse particle filters, and corrections must be made (Evans and Ryan, 1983) to the coarse particle measurements to compensate.

The stacked filter unit (SFU), used in the National Park Service Visibility Network (Cahill *et al.*, 1986), applies the etched polycarbonate membrane filter size separation method described above with a nominal sample duration of 72 h. Flow rates are manually set and recorded before and after sampling. This sampler operated reliably at remote locations for over a decade, and a large data base of elemental concentrations from PIXE analyses has been derived from its samples.

The Sequential Filter Sampler (SFS) was originally designed in the late 1970s for use in the Sulfate Regional Experiment (SURE, Mueller *et al.*, 1983), and the Portland Aerosol Characterization Study (PACS, Watson, 1979). It has been applied in over a dozen subsequent studies. A version of this sampler has since received PM-10 reference method status. The SFS consists of an aluminum plenum to which a SA-254 PM-10 or a Bendix 240 PM-2.5 inlet is attached. Up to 12 sampling ports within the plenum are controlled by solenoid valves that divert flow from one channel to the next by means of a programmable timer. These ports accept filters that have been pre-loaded into open-faced 47 mm Nuclepore filter holders. The sample flow can be divided for simultaneous collection on two or more filter media. The differential pressure volumetric flow controller has been adapted to split the flow between filters and maintain a constant flow rate despite filter loading.

The Interagency Monitoring of Protected Visual Environments (IMPROVE) sampler was developed to replace the SFU at United States National Parks and Monuments in support of regional visibility assessment. This sampler consists of different modules controlled by a common sample switching system. Each module can be tailored to a specific type of measurement. The standard configuration acquires one PM-10 sample for gravimetric and elemental analyses and three PM-2.5 samples, one on Teflon® membrane for gravimetric and elemental analyses, and one on Teflon® membrane for ion analysis, and one on quartz fiber for carbon analysis. Up to four sequential samples can be taken without operator intervention by using a timer to control solenoid valves. Flow rates are controlled by critical orifice.

Other samplers are currently under development to support compliance monitoring for a future PM-2.5 standard. One of the main criteria for these samplers is that they accommodate membrane filters suitable for elemental analysis by methods such as XRF.

Sample Handling and Storage

Because particles are deposited on the surface of XRF filters, and the levels being measured are so low, special precautions must be taken to prevent the loss of particles and contamination of the sample. Dust can deposit on filters prior to and following sampling. This bias can be evaluated by placing dynamic blank filters, through which no air is drawn, in the sampler along with the sampled filters and submitting them to the same chemical analyses. It can be minimized by using samplers that do not expose the filters directly to the atmosphere and by minimizing the passive time in the sampler before and after sampling.

Particles can fall off filters when samples have large deposits and receive rough handling during movement from the field site to the laboratory (Dzubay and Barbour, 1983). Shorter sample durations and lower flow rates may be required to prevent overloading in very polluted environments, especially those in which fugitive dust is a large contributor. Careful handling during transport will also minimize the loss of particles from the filter surface. Various chemicals can be inadvertently introduced to the filter during handling. This can be minimized by testing a few blank filters from each manufacturing lot to ensure that handling prior to receipt has not caused contamination. Subsequent contamination can be reduced by loading and unloading filters in a laboratory setting, keeping them in containers before and after sampling, and eliminating contact of the filters with bare hands.

A process flow diagram of the filter handling process that can be followed for ambient sampling and analysis is shown in Figure 1. Since data from the XRF analyses must be eventually unified with data from the field sampling device and gravimetric analysis, samples should be labeled with unique ID codes that are associated with each data set. Relational data base programs can then be used to merge these data for calculating mass

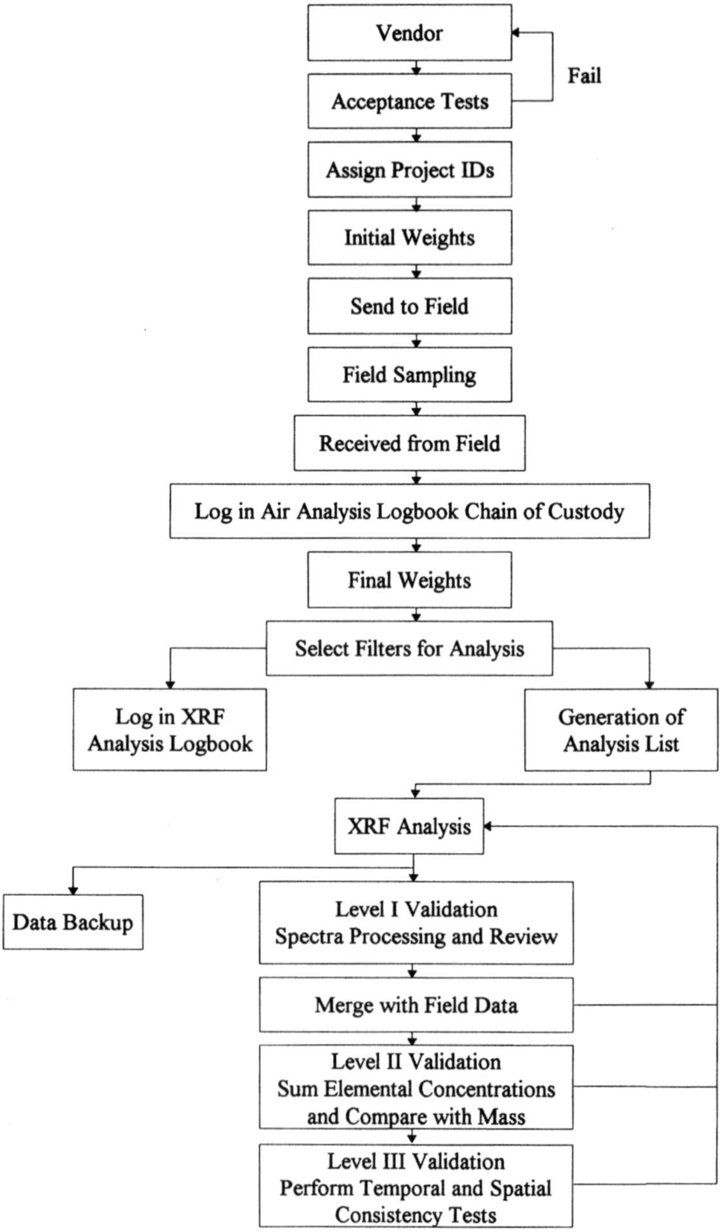

Figure 1. Flow diagram of XRF sample processing and validation.

concentrations and performing validation tests. Three levels of data validation are also displayed in Figure 1. Level 1 verifies that the XRF procedure has been properly applied, and involves checking standards, replicates, and spectrum shapes for each sample. Level 2 occurs after XRF elemental concentrations are merged with mass concentrations. If the sum of the elemental concentrations exceeds the mass concentration, an error has occurred. Level 3 happens after data from several sampling sites and times have been acquired so that spatial and temporal distributions can be examined. If large differences between successive sample or elemental concentrations at nearby sites cannot be explained by an environmental cause, there may be a measurement anomaly that needs to be investigated. XRF is an excellent validation tool because a significant portion of PM-10 mass (typically 20–50%) can be accounted for by the measured elements. Erroneous mass measurements that might have resulted in non-attainment of air quality standards have been discovered when elemental analyses are available for comparison with mass concentrations.

ANALYSIS EQUIPMENT AND PROCEDURES

Watson (1979) describes the equipment and procedures for applying EDXRF with an ORTEC TEFA 6110 system while Dzubay and Stevens (1975) describe its application with a custom built system from Lawrence Berkeley Laboratories (LBL). These units cannot currently be obtained commercially. This chapter describes procedures applicable to the Kevex analyzer because it is available commercially and has the following desirable items:

- multiple X-ray targets in a single tube
- secondary fluorescors
- exchangeable or adjustable collimators
- programmable sample changing mechanism
- desktop computer hardware with interfaces for control and processing software
- a large sample chamber

These features should also be sought in hardware obtained from other manufacturers for the purpose of air sample analysis.

Physical Configuration

The configuration for sample irradiation and detection is seen in Figure 2. The source of X-rays in the Kevex analyzer is a side window X-ray tube

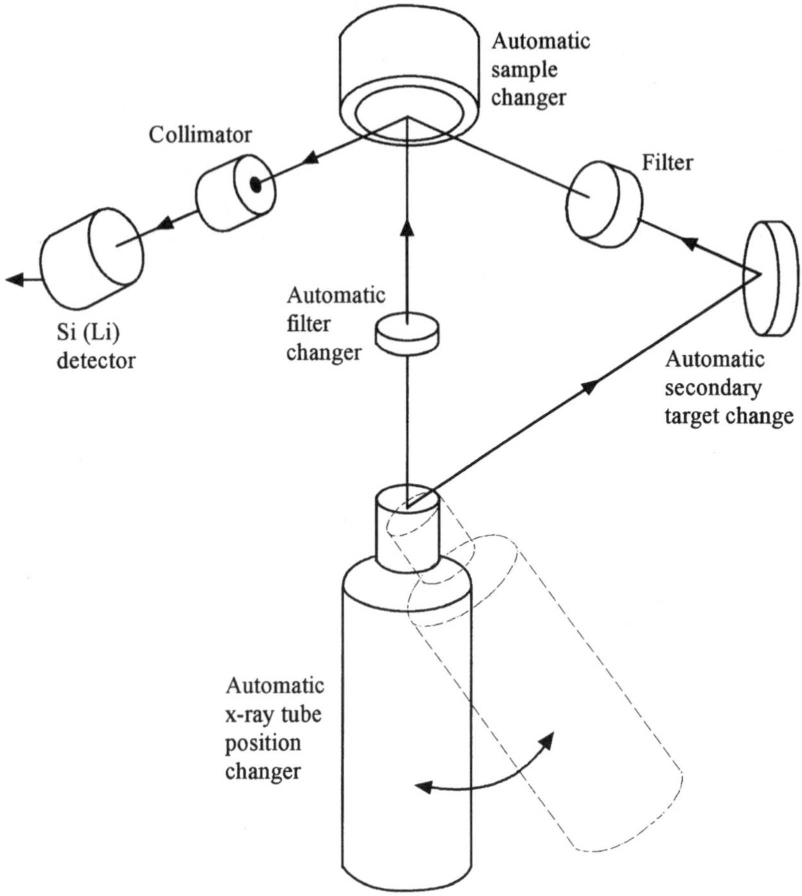

Figure 2. Schematic of sample excitation with secondary fluorescors and filters.

with a rhodium (Rh) anode. These X-rays may be focused directly on the sample or modified using filters and collimators. They may also be centered on one of six secondary targets that in turn emit X-rays used to excite a sample. X-rays from a secondary target or the tube are absorbed by the sample, exciting electrons to high level orbitals. As the electrons return to their ground state, photons are emitted that are characteristic of the quantum level jumps made by the electron. The energy of the emitted photons are therefore characteristic of the elements contained in the sample. The fluoresced photons are detected in a solid state lithium-drifted silicon detector. Each photon that enters the detector generates an electrical charge whose magnitude is proportional to the photon's energy. The electrical signals from the detector are sorted into energy channels, counted, and displayed. A sample spectrum consists of characteristic peaks superimposed on a background caused by the scatter of X-rays from the tube into the detector. Spectra are collected for a specified length of time and stored on disk for later processing.

Five different analysis conditions applied during a single analysis run to maximize sensitivity to the full range of elements reported are shown in Table 2. These conditions were developed after extensive experimentation to maximize the sensitivity for the elements listed that correspond to those of interest in ambient aerosol samples. Each of the analysis conditions, that correspond to different secondary targets; primary X-ray filters, X-ray tube voltage and current, and energy detection range, is designed for a specific group of elements. The elements analyzed by more than one condition are used for quality assurance (QA) purposes. The different protocols, corresponding to increasing irradiation periods, provide improvements in detection limits at the expense of longer analysis periods and are displayed in Table 3.

The configuration of polycarbonate cassettes that are used to mount membrane filters in the Kevex sample carousel is shown in Figure 3. These are not available commercially, and must be commissioned from a precision plastics manufacturer. They consist of a small frame with a recessed lip and compression ring to hold the filters face down.

Interferences and Uncertainties

The ideal conditions are designed specifically for thin films. X-ray spectra are subject to distortion if unusually heavy deposits are analyzed. This is due to internal absorption of both incident and emitted X-rays within the samples. Optimum filter loading is approximately 150 μg/cm^2 (1 mg/filter for 37 mm filters, and 2 mg/filter for 47 mm filters) to comply with these assumptions. Make adjustments in filter deposit area, sampling time, and flow to ensure that deposits are within this range. At low concentrations counting statistics and signal noise will dominate the calculations of elemental concentrations. Adjustments in filter deposit area, and sampling time and flow should be made so that deposits are at least 15 μg/cm^2.

The X-ray beam is focused on an area approximately 2–4 cm^2 in the center of the filter. The results are extrapolated to the entire deposit area of the filter during data processing. Therefore, the center of the filter must be representative of the entire deposit. In addition, each of the five excitation conditions uses a slightly different analysis area, and non-uniformity of the deposit will affect inter-condition comparisons of overlapping elements.

Table 2. Optimal excitation conditions of the Kevex X-ray fluorescence analyzer for ambient aerosol samples.

Parameter	Condition Number				
	1	2	3	4	5
Tube voltage	60 kV	35 kV	30 kV	30 kV	8 kV
Tube current	0.4 mA	3.3 mA	3.3 mA	3.3 mA	1.0 mA
Direct excitation (rhodium anode)					
Filter	Mo	Rh	None	None	Whatman 41
Thickness	0.10 mm	0.13 mm	None	None	3 layers
Secondary fluorescor excitation	None	None	Ge	Ti	None
Filter	None	None	Whatman 41	Mylar	None
Thickness	None	None	1 layer	3.8 mm	None
Analysis time (longer times give lower detection limits)					
Protocol A	100 s	400 s	400 s	100 s	100 s
Protocol B	200 s	800 s	800 s	200 s	200 s
Protocol C	800 s	3200 s	3200 s	800 s	800 s
Protocol D	1600 s	6400 s	6400 s	1600 s	1600 s
Energy range analyzed	0–40 keV	0–20 keV	0–10 keV	0–10 keV	0–10 keV
Elements analyzed	Pd, Ag, Cd, In, Sn, Sb, Ba, La	Fe, Co, Ni, Cu, Zn, Ga, As, Se, Br, Rb, Sr, Y, Zr, Mo, Au, Hg, Tl, Pb, U	K, Ca, Ti, V, Cr, Mn, Fe, Co, Ni, Cu, Zn	Al, Si, P, S, Cl, K, Ca	Na, Mg, Al, Si, P, S

Absorption of both incident and emitted X-rays occurs in the presence of large particles. This particularly affects the light elements analyzed under condition 5 (Na through S). In addition, large particles may be enriched in certain elements and may bias results in the same matter as inhomogeneous deposits.

Table 3. X-ray fluorescence air filter analysis interference-free minimum detectable limits[a] using different analysis protocols.

Filter[c]		Quartz-fiber filter[b]	Teflon-membrane			
Element	Condition number[d]	ProtocolQA – A ng/cm[2e]	Protocol A ng/cm[2d]	Protocol B ng/cm²	Protocol C ng/cm²	Protocol D ng/cm²
Al	5	NA[f]	10	7.2	3.6	2.5
Si	5	NA	6.3	4.4	2.2	1.4
P	5	NA	5.6	4.0	2.0	1.4
S	5	40[g]	5.0	3.5	1.8	1.2
Cl	4	30	10	7.4	3.7	2.6
K	4	40	6.1	4.3	2.2	1.5
Ca	4	100	4.5	3.2	1.6	1.1
Ti	3	50	2.9	2.1	1.0	0.73
V	3	20	2.5	1.7	0.87	0.62
Cr	3	8	1.9	1.4	0.67	0.48
Mn	3	7	1.6	1.1	0.56	0.40
Fe	3	15	1.5	1.1	0.54	0.38
Co	3	5	0.88	0.62	0.31	0.22
Ni	3	4	0.89	0.63	0.31	0.22
Cu	3	4	1.1	0.76	0.38	0.27
Zn	3	6	1.1	0.76	0.38	0.27
Ga	2	8	1.9	1.4	0.68	0.48
As	2	9	1.6	1.1	0.56	0.39
Se	2	5	1.2	0.86	0.43	0.31
Br	2	5	1.0	0.72	0.36	0.25
Rb	2	5	1.0	0.68	0.34	0.24
Sr	2	8	1.1	0.78	0.39	0.28
Y	2	8	1.3	0.92	0.46	0.33
Zr	2	10	1.7	1.2	0.59	0.42
Mo	4	20	2.7	1.9	0.95	0.67
Pd	1	20	11	7.6	3.8	2.7
Ag	1	20	12	8.6	4.3	3.0
Cd	1	25	12	8.6	4.3	3.0
In	1	30	13	9.5	4.8	3.4
Sn	1	40	17	12	6.2	4.4
Sb	1	50	18	13	6.4	4.5
Ba	1	170	52	37	18	13
La	1	190	62	44	22	16
Au	2	NA	3.1	2.2	1.1	0.77
Hg	2	20	2.6	1.8	0.91	0.65

Table 3 (continued).

Tl	2	NA	2.5	1.8	0.88	0.62
Pb	2	14	3.0	2.2	1.1	0.76
U	2	NA	2.3	1.7	0.83	0.59

[a]MDL defined as three times the standard deviation of the blank for a filter of 1 mg/cm^2 areal density.
[b]Analysis times are 100 s for Conditions 1 and 4, and 400 s for Conditions 2 and 3. Actual MDLs for quartz filters vary from batch to batch due to elemental contamination variability.
[c]Analysis times are 100 s for Conditions 1, 4 and 5, and 400 s for Conditions 2 and 3 for Protocol A; 200 s for Conditions 1, 4 and 5 and 800 s for Conditions 2 and 3 for Protocol B; 800 s for Conditions 1,4 and 5 and 3200 s for Conditions 2 and 3 for Protocol C; and 1600 s for Conditions 1, 4 and 5 and 6400 s for Conditions 2 and 3 for Protocol D.
[d]Condition 1 is direct mode excitation with a primary excitation filter of 0.15 mm thick Mo. Tube voltage is 50 kV and tube current is 0.6 mA. Condition 2 is direct mode excitation with a primary excitation filter of 0.13 mm thick Rh. Tube voltage is 35 kV and tube voltage is 2.0 mA. Condition 3 uses Ge secondary target excitation with the secondary excitation filtered by a Whatman 41 filter. Tube voltage is 30 kV and tube current is 3.3 mA. Condition 4 uses Ti secondary target excitation with the secondary excitation filtered by 3.8 mm thick mylar film. Tube voltage is 30 kV and tube current is 3.3 mA. Condition 5 uses direct mode excitation with a primary excitation filter consisting of 3 layers of Whatman 41 filters. Tube voltage is 8 kV and tube current is 0.6 mA. Multi-channel analyzer energy range is 0–40 keV for Condition 1, 0–20 keV for Condition 2, and 0–10 keV for Conditions 3, 4, and 5.
[e]Typical exposed area is 406 cm^2 for standard high-volume filters; 6.4 cm^2 for 37 mm ringed Teflon-membrane filters; and 13.8 cm^2 for 47 mm ringed Teflon-membrane filters.
[f]Information not available.
[g]For Condition 4.

Increased X-ray scattering is caused by greater filter thickness. This in turn increases the spectral background and causes problems in determining accurate background subtraction and quantification of elements at low concentrations. Additional uncertainties are introduced if filter thickness varies considerably within a lot, again making accurate background subtractions difficult. Uncertainties in results from analyzing the damaged portion of a filter are obvious. However, the X-ray signal is heavily dependent upon distance of the filter from the tube and the detector. If the filter is damaged and causes sagging or puckering these distances will be changed, and analysis results will also be affected.

Calibration

Three types of XRF standards are used for calibration, performance testing, and auditing: vacuum-deposited thin-film elements and compounds (Micromatter), polymer films (Dzubay *et al.*, 1981), and National Institute of Standards and Technology (NIST) thin glass films. The vacuum deposits cover the largest number of elements and are used as calibration standards. The polymer film and NIST standards are employed as quality

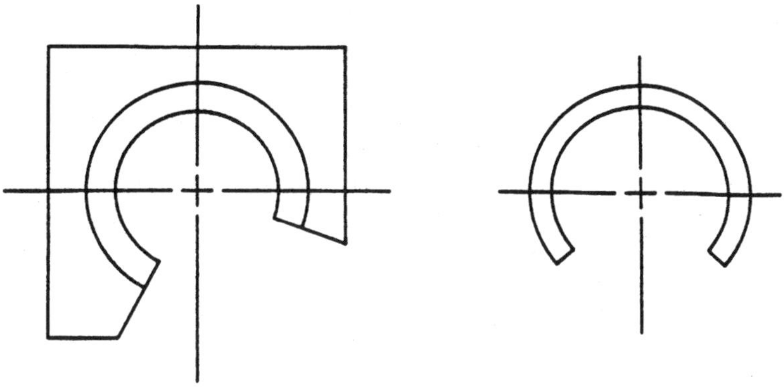

37 MM XRF FILTER HOLDER

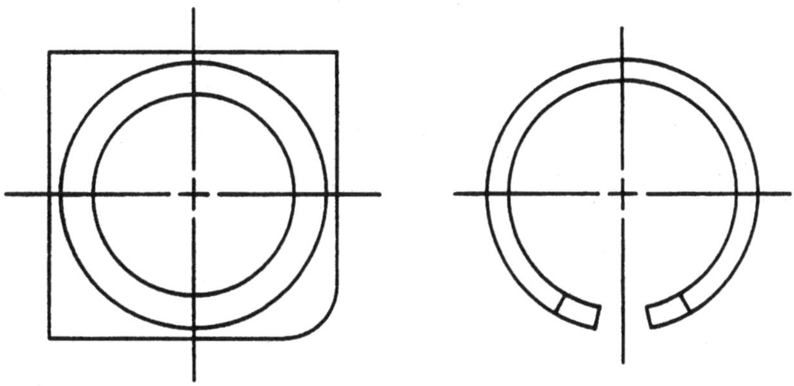

47 MM XRF FILTER HOLDER

Figure 3. Diagram of DRI 37 and 47 mm XRF sample holders for the Kevex analyzer.

control (QC) standards. NIST standards are the definitive standard reference material (SRM), but these are only available for the species Al, Ca, Co, Cu, Mn, and Si (SRM 1832) and Fe, Pb, K, Si, Ti, and Zn (SRM 1833). A separate Micromatter thin-film standard is used to calibrate the system for each element.

The Micromatter standards are vacuum deposited elements or salts on 37 mm Nuclepore filters. The standards are stored in Petri slides, deposit sides up, in a locked sample freezer to retard oxidation and aging. These samples are loaded carefully into 37 mm holders and analyzed. Two to four Nuclepore blanks are analyzed in the first run for use in blank subtraction. Calibrations should be performed every three months or when results from the QA standard that is analyzed in every run consistently falls outside the $\pm 5\%$ intervals.

Sensitivity factors, the number of X-ray counts/μg/cm^2 of the element, are determined for each excitation condition. These factors are then adjusted for absorption of the incident and emitted radiation in the thin film. They are plotted as a function of atomic number and a smooth curve is fitted to the experimental values. The calibration sensitivities are then read from these curves for the atomic numbers of each element in each excitation condition. The polymer film and NIST standards are analyzed on a periodic basis using these sensitivity factors to verify both the standards and stability of the instrument response.

Spectrum Processing

A typical energy spectrum obtained from an ambient air sample analysis is shown in Figure 4. The sensitivity factors obtained from calibration are multiplied by the net peak intensities yielded by ambient samples to obtain

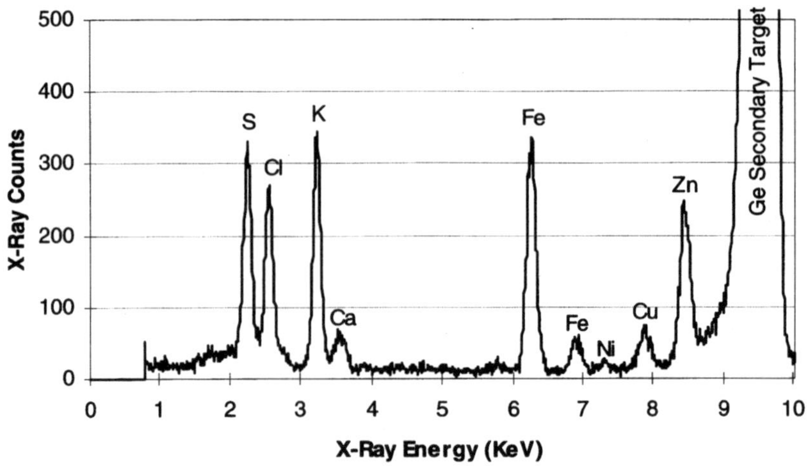

Figure 4. Example of an XRF spectrum obtained under Condition 3.

the μg/cm^2 deposit for each element. These intensities are obtained by subtracting background radiation and spectral interferences, and adjusting for X-ray absorption.

The elemental X-ray peaks reside on a background of radiation scattered from the sampling substrate. A model background is formed by averaging spectra obtained from 10 filters of the same type used in ambient sampling. This model background has the shape and features of the sample spectrum, minus the elemental peaks, when the deposit mass is small relative to the substrate mass (Russ, 1977). It is normalized to an excitation radiation scatter peak in each sample spectrum to account for the difference in scatter intensity due to different masses. Blank normalization that uses a scatter peak from a low-energy region (e.g., the Ti secondary target scatter peak from Condition 2) can lead to unreliable background estimation. Therefore, the Mo scatter peak is used for Conditions 1 and 2, and the Ge scatter peak is used for Conditions 3–5 (see Table 2).

The number and spacing of the characteristic X-ray lines relative to detector resolution are such that the peaks from one element can interfere with a peak from another element (Dzubay, 1986). Several methods have been applied to subtract these peak overlaps (Arine *et al.*, 1977; Drane *et al.*, 1983; Parkes *et al.*, 1979) including: least squares fitting to library spectra, Gaussian and other mathematical functions, and the use of peak overlap coefficients. Peak overlap coefficients are applied to aerosol deposits. The most important of these overlaps are the K-beta to K-alpha overlaps of elements from potassium (K) to zirconium (Zr), the lead (Pb) L-alpha to arsenic (As) K-alpha interference, and the lead (Pb) M line to sulfur (S) K line interference. The ratios of overlap peaks to the primary peak are determined from the thin film standards for each element for the spectral regions of the remaining elements. These ratios are multiplied by the net peak intensity of the primary peak and subtracted from the spectral regions of other elements.

The ability of an X-ray to penetrate an aerosol deposit depends on the energy of the X-ray and composition and thickness of the deposit. In general, lower energy X-rays, characteristic of light elements, are absorbed in matter to a much greater degree than higher energy X-rays. Larger particles collected during aerosol sampling have sufficient size to cause absorption of X-rays within the particles. Attenuation factors for PM-2.5 samples are generally negligible (Criss, 1976), even for the lightest elements. However these attenuations can be significant for coarse fraction particles (particles with aerodynamic diameter from 2.5 to 10 μm). Adjustment factors have been derived by Dzubay and Nelson (1975), and should be applied to PM-10 samples. PM-2.5 samples contain particles small enough so that no self-absorption corrections are needed.

Procedures

The processes followed during a typical XRF analysis run are illustrated in Figure 5. Filters are removed from their Petri slides and placed with their deposit sides downward into polycarbonate filter cassettes. A polycarbonate retainer ring keeps the filter flat against the bottom of the cassette. These cassettes are loaded into a carousel in the X-ray chamber containing 16 openings. The filter IDs are recorded on a data sheet to correspond to numbered positions in the carousel. The sample chamber is evacuated to 10^{-3} torr and a computer program controls the positioning of the samples and the excitation conditions. Complete analysis of 16 samples under five excitation conditions requires approximately six hours for protocol A. The vacuum in the X-ray chamber and the heat induced by the absorption of X-rays can cause certain materials to volatilize. For this reason, labile species such as nitrate and organic carbon are measured on a quartz-fiber filter, rather than on the Teflon® membrane filter which is subjected to XRF analysis.

A QC standard and a replicate from a previous batch are analyzed with each set of 14 samples. When a QC value differs from specifications by more than ±5% or when a replicate concentration differs from the original value (when values exceed 10 times detection limits) by more than ±10%, the samples are re-analyzed. If further tests of standards show that the system calibration has changed by more than ±2%, the instrument is recalibrated as described above. An example of the QC chart that should be updated at the end of each sample run is shown in Figure 6.

When the sample analysis lists are organized, lab blanks appropriate for the samples to be analyzed should be included. If possible analyze between two and five filters from each filter lot, particularly if the filters are glass fiber or quartz or are from a manufacturer known to have high levels of contaminants or large background variabilities. In order to allow the blank spectra to be developed and the subsequent calculations to proceed, analyze the lab blanks in the first run for any project.

Blank spectra are created as simple averages of the lab blank spectra for each condition, with corrections for escape peaks on Conditions 4 and 5. The calculated blank spectra must be verified manually for unusual levels of contaminants or other abnormalities. The spectrum for each condition is recalled and examined for unusual peaks or high levels of typically occurring peaks.

As laboratory IDs are entered into the processing and control programs, apply flags to indicate the nature of the samples being analyzed. This

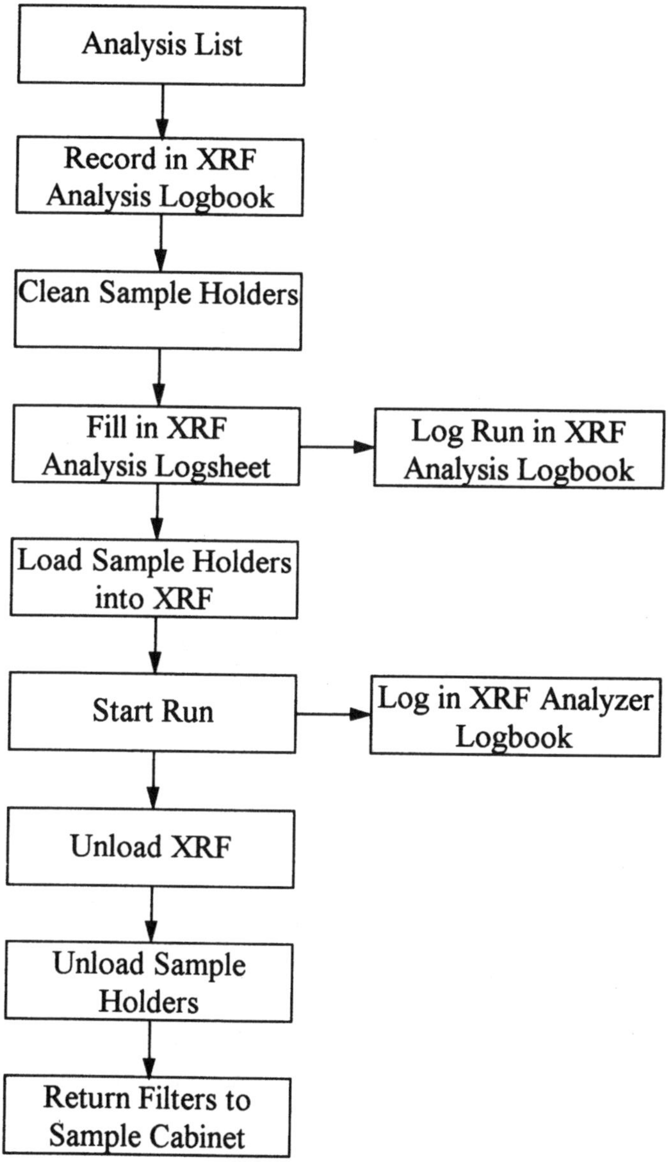

Figure 5. X-ray fluorescence analysis procedures.

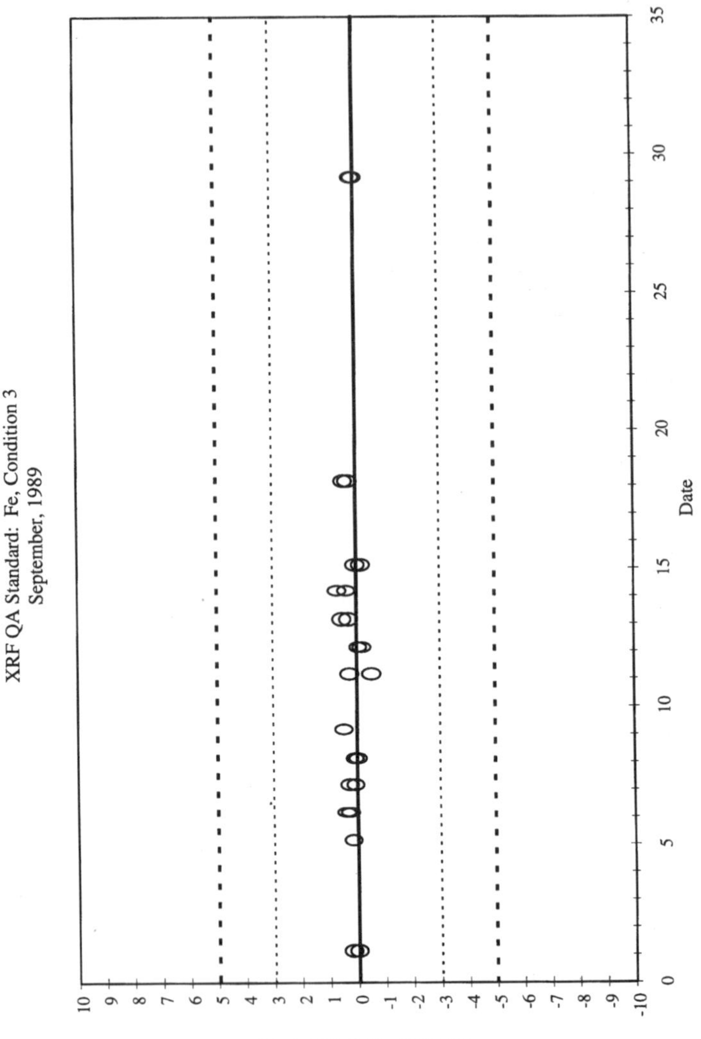

Figure 6. Example of an XRF QA plot.

facilitates their classification for further process. Typical flags might include:

b1 — field blank
b2 — laboratory blank
f1 — filter damaged, outside analysis area
f2 — filter damaged, inside analysis area
i — unspecified deposit problems
i1 — inhomogeneous filter deposit
i2 — deposit smeared or scraped after sampling
i4 — foreign objects on deposit surface
q1 — routine QC standard
r1 — first replicate analysis
r2 — second replicate analysis
r3 — third replicate analysis

Precautions to observe during filter loading and unloading are:

- Ensure that all large, loose particles on a filter deposit are removed before analysis or that filters with large amounts of potentially loose particles are placed between Mylar sheets during analysis.
- Analyzing filters with high levels of sulfur or other acidic species may cause acid to evaporate and destroy detector components. Place these samples in hermetically sealed Mylar bags prior to analysis.
- Place the QA sample in the carousel position that normally covers the detector and do not remove it when the chamber cover is open. This sample serves to cover the detector window when the cabinet cover is open and affords some degree of protection to the beryllium window. Besides the health concerns of inhaling beryllium, a broken Be window requires that the entire detector and cryostat be shipped to the manufacturer for repair. Replacement times for broken windows are estimated at three to six months. Therefore this would be catastrophic for the timely analysis of samples.

Maintenance

Regular maintenance for the XRF analyzer involves:

- Weekly fill liquid nitrogen (LN2) dewar. A plastic funnel is placed in the inlet tube at the back of the X-ray cabinet, and a piece of cardboard is placed over the cables connected to the back of the instrument to protect them from spilled LN2. Wearing gloves and safety goggles, fill the small transport dewar to 1/4" from the top from the large LN2 tank. This represents four liters of LN2. Slowly pour the LN2 into the funnel, taking care to add LN2 no faster than the funnel can accept it. NOTE: when the funnel and inlet tubing are warm, the LN2 will boil vigorously. Therefore avoid adding LN2 too fast until the funnel and inlet tubing reach LN2 temperature. Continue filling the XRF with the small transport dewar until LN2 begins to spit from the exhaust tube.

- Clean the sample chamber monthly, or as needed. The detector window is extremely fragile and susceptible to breakage, hence observe caution during maintenance.
- Check level of vacuum pump oil weekly and replace the vacuum pump oil yearly. Analysis of samples that de-gas or have high moisture content may require more frequent vacuum pump oil changes. If a significant amount of smoke is generated when the sample chamber is evacuated, the pump oil needs to be changed.
- Check level of antifreeze coolant weekly and replace the antifreeze coolant yearly.

EXAMPLES OF XRF ANALYSES IN AIR QUALITY STUDIES

The best way to evaluate the utility of XRF and other aerosol analysis methods is to examine how they have been used in previous air quality studies. Lioy *et al.* (1980), Chow and Watson (1989), and Watson and Chow (1992) describe studies of source emissions and ambient aerosol from many different parts of the world. They specify certain conditions that should be considered when an XRF data base is being constructed.

Several of these data sets represent distinct historical periods. Cities such as Los Angeles, California (Chow *et al.*, 1992; 1994); Bakersfield, California (Chow *et al.*, 1992); Fresno, California (Chow *et al.*, 1993); Portland, Oregon (Watson, 1979); Denver, Colorado (Watson *et al.*, 1988); and Steubenville, Ohio (Skidmore *et al.*, 1992) have data sets on particle sizes and chemical composition of ambient aerosols from the late 1970s as well as from the late 1980s and early 1990s. Today's concentrations for certain species might be compared with those measured more than a decade ago to evaluate the effects of emissions reductions in lead, sulfur, and other species which took place over those periods. Similar comparisons could be made for several Class I areas in which samples were taken to evaluate visibility impairment.

Many of these data sets contain particulate concentrations for toxic elements such as P, Zn, Cu, Mn, Cr, Ni, V, Cd, As, Hg, and Pb. Although these are minor components of PM-10, they are often measured with detection levels less than 0.010 $\mu g/m^3$ by XRF or other methods and are contained in these data sets. A comprehensive picture of toxic exposures in many different parts of the United States might be derived from these data sets without the need for extensive new measurements. Secondary aerosols such as sulfate and nitrate have been found to be important contributors to PM-10 in several areas. Though EDXRF cannot quantify nitrate, it provides a very precise sulfur measurement that represents sulfate concentrations when multiplied by a factor of three. Many samples have been taken of the effluents of different source types and submitted to XRF and other analyses (e.g. Chow *et al.*, 1994; Watson *et al.*, 1994a,b). Substantial differences in chemical composition exist for these emitters. For wood burning in fireplaces and stoves, organic carbon is by far the most abundant constituent, followed by elemental carbon. Chlorine and potassium also approach an abundance of 1% in these emissions, as do sulfate and nitrate.

Road dust and other geological materials contain aluminum, silicon, potassium, calcium, and iron and are also the major components of these emissions. Though total potassium is abundant in road dust, water-soluble potassium constitutes less than one tenth of the total. Strontium and lead are also present at detectable levels. Motor vehicle exhaust contains high concentrations of organic and elemental carbon, but their ratios are much different from those found in wood combustion with the abundance of elemental carbon being nearly equal to the organic carbon abundance. Bromine and lead are also much larger components in vehicle exhaust than they are in other source profiles. Bromine and lead have been phased out of most United States gasolines, but they are still found in gasolines from other countries.

Primary particles emitted by coal fired power generators have several chemical abundances that are similar to those of road dust (Watson *et al.*, 1996). The sulfate level is much higher in these emissions than in the other profiles, and the organic and elemental carbon fractions are much lower. Selenium is clearly detectable in most power station profiles, whereas it is below quantifiable limits in the other profiles. Other industrial sources also contain unique elements in abundance. Calcium is enriched in cement plant emissions, iron and manganese are found in steel mill operations, arsenic, lead, and copper characterize smelter emissions, and vanadium and nickel are often noted in emissions from residual oil combustion.

CONCLUSIONS

Energy Dispersive X-ray Fluorescence (XRF) spectroscopy is one of several X-ray methods applicable to ambient air samples. The method can not be used for all filter samples, but when sampling methods and filter media are optimized for XRF, very precise and low level elemental concentrations can be measured. Commercially available hardware, though not specifically designed for air filter analysis, can be readily adapted for quantitative measurements. XRF provides rapid analysis for many elements with minimal operator time and without destroying valuable samples. The results of these analyses can be used to verify mass concentrations, determine exposures to toxic substances, and identify contributions from different pollution sources.

REFERENCES

Arine, F., L. Wielopolski and R.P. Gardner, "The Linear Least-Squares Analysis of X-Ray Fluorescence Spectra of Aerosol Samples Using Pure Element Library Standards and Photon Excitation", in *X-Ray Fluorescence Analysis of Environmental Samples*, T.G. Dzubay (ed.), Ann Arbor Science Publishers, Ann Arbor, MI, 1977, p. 277.

Bacon, J.R., S.R.N. Chenery, A.T. Ellis, A. Fisher, A.W. McMahon, P.J. Potts and J.G. Williams, *J. Anal. Atomic Spectrometry*, **10** (1995) 253R.

Bertin, E.P., *Principles and Practice of X-Ray Spectrometric Analysis*, Plenum, New York, NY, 1975.

Buzzard, G.H. and J.P. Bell, *J. Aerosol Sci.*, **11** (1980) 435.

Cahill, T.A. *Ann. Rev. Nucl. Part. Sci.*, **30** (1980) 211.

Cahill, T.A. *Nucl.Instr. Meth. Phys. Res.*, **B75** (1993) 99.

Cahill, T.A., *Particle-Induced X-ray Emission Spectrometry*, **133** (1995) 237.

Cahill, T.A., D.W. McColm and B.H. Kusko, *Nucl. Instr. Meth.*, **B14** (1986) 38.

Charlson, R.J., S.E. Schwartz, J.M. Hales, R.D. Cess, R.A. Coakley, Jr., J.E. Hansen and D.J. Hofmann, *Science*, **5** (1992) 423.

Chow, J.C., *J. Air Waste Manag. Assoc.*, **45** (1995) 320.

Chow, J.C. and J.G. Watson, "Summary of Particulate Data Bases for Receptor Modeling in the United States", in *Transactions: Receptor Models in Air Resources Management*, Air & Waste Management Assoc., Pittsburgh, PA (1989) pp. 108–133.

Chow, J.C., C.S. Lu, J. Cassmassi, J.G. Watson, Z. Lu and L.C. Pritchett, *Atmos. Environ.*, **26A** (1992) 693.

Chow, J.C., J.G. Watson, D.H. Lowenthal, P.A. Solomon, K. Magliano, S.D. Ziman and L.W. Richards, *Aerosol Sci. Technol.*, **18** (1993) 105.

Chow, J.C., E.M. Fujita, J.G. Watson, Z. Lu, D.R. Lawson and L.L. Ashbaugh. *Environ. Mon. Assess.*, **30** (1994) 49.

Criss, J.W., *Anal. Chem.*, **48** (1976) 179.

Drane, E.A., D.G. Branton, S.H. Tysinger and W.J. Courtney, *Data Processing Procedures for Elemental Analysis of Atmospheric Aerosols by X-Ray Fluorescence*, Document # TR-83–01, Northrop Services, Inc., Research Triangle Park, NC (1983).

Dzubay, T.G., *Analysis of Aerosol Samples by X-Ray Fluorescence*, U.S. EPA, Research Triangle Park, NC (1986).

Dzubay, T.G. and R.K. Barbour, *J. Air Pollut. Control Assoc.*, **33** (1983) 692.

Dzubay, T.G. and R.O. Nelson, *Advances in X-Ray Analysis*, **18** (1975) 619.

Dzubay, T.G. and R.K. Stevens, *Environ. Sci. Technol.*, **9** (1975) 663.

Dzubay, T.G., N. Morosoff, G.L. Whitaker and H. Yasuda, "Evaluation of Polymer Films as Standards for X-ray Fluorescence Spectrometers", in *Electron Microscopy and X-Ray Applications to Environmental and Occupational Health Analysis*, Ann Arbor Science Publishers, Inc., Ann Arbor, MI (1981).

Eldred, R.A. *Personal communication*, Crocker Nuclear Laboratory, University of California, Davis, CA (1993).

Engelbrecht, D.R., T.A. Cahill and P.J. Feeney, *J. Air Pollut. Control Assoc.*, **30** (1980) 391.

Evans, J.S. and P.B. Ryan, *Aerosol Sci. Technol.*, **2** (1983) 531.

Federal Register, *Ambient Air Quality Standards for Particulate Matter: Final Rule*, **52** (1987) 24634.

Fernandez, F.J., "Atomic Absorption Spectroscopy", in *Methods of Air Sampling and Analysis*, 3rd edn., J.P. Lodge (ed.), Lewis Publishers, Chelsea, MI, (1989), pp. 143–150.

Ferris Jr., B.G., F.E. Speizer, J.D. Spengler, D.W. Dockery, Y.M. Bishop, J.M. Wolfson and C. Humble, *American Review of Respiratory Disease*, **120** (1979) 767.

Fujita, E.M. and J.F. Collins, *Quality Assurance for the Southern California Air Quality Study*, Presented at 82nd Annual Meeting of the Air & Waste Management Association, Anaheim, CA (1989).

Giauque, R.D., R.B. Garrett and L.Y. Goda, in *X-Ray Fluorescence Analysis of Environmental Samples*, T.G. Dzubay (ed.), Ann Arbor Science, Ann Arbor, MI (1973).

Harman, J.N., "Inductively Coupled Plasma Emission Spectroscopy", in *Methods of Air Sampling and Analysis*, 3rd edn., J.P. Lodge (ed.), Lewis Publishers, Chelsea, MI (1989), pp. 88–92.

Jaklevic, J.M., B.W. Loo and F.S. Goulding, "Photon-Induced X-Ray Fluorescence Analysis Using Energy-Dispersive Detection and Dichotomous Sampler, in *X-Ray Fluorescence Analysis of Environmental Samples*, 2nd edn., T.G. Dzubay (ed.), Ann Arbor Science, Ann Arbor, MI (1977).

Lancaster, N.J. and W.G. Nickling, "Aeolian Sediment Transport", in *Geomorphology of Desert Environments*, A. Abrahams and A.J. Parsons (eds.), Chapman and Hall, Publishers (1993).

Lippman, M., "Filters and Filter Holders," in *Air Sampling Instruments for Evaluation of Atmospheric Contaminants*, 8th edn., B. Cohen, S.V. Hering (eds.), American Conference of Government Industrial Hygienists, Inc., Cincinnati, OH (1995).

Lioy, P.J., J.G. Watson and J.D. Spengler, *J. Air Pollut. Control Assoc.*, **30** (1980) 1126.

Liu, B.Y.H. and K.W. Lee, *Environ. Sci. Technol.*, **10** (1976) 345.

Malm, W.C., *Visibility: A Physical Perspective,* Presented at Workshop on Visibility Values, Fort Collins, CO (1979).

Miller, F.J., D.E. Gardner, J.A. Graham, R.E. Lee Jr., W.E. Wilson and J.D. Bachmann, *J. Air Pollut.Control Assoc.*, **29** (1979) 610.

Mueller, P.K., G.M. Hidy, R.L. Baskett, K.K. Fung, R.C. Henry, T.F. Lavery, N.J. Nordi, A.C. Lloyd, J.W. Thrasher, K.K. Warren and J.G. Watson, *Sulfate Regional Experiment (SURE): Report of Findings*, Vol.1, 2, and 3, Report No. EA-1901, Electric Power Research Institute, Palo Alto, CA (1983).

Olmez, I., "Instrumental Neutron Activation Analysis of Atmospheric Particulate Matter" in *Methods of Air Sampling and Analysis*, 3rd edn., J.P. Lodge (ed.), Lewis Publishers, Chelsea, MI (1989), pp. 143–150.

Parkes, J., L.J. Rabbitt and M.J. Hamshire, *Anal. Chem.*, **46** (1979) 1830.

Potts, P.J. and Webb, P.C., *J. of Geochem. Exploration*, **44** (1992) 251.

Rhodes, J.R., *American Lab.*, **5** (1973) 57.

Russ, J.C., *X-Ray Spectrometry*, **6** (1977) 37.

Skidmore, L.W., J.C. Chow and T.T. Tucker, "PM-10 Air Quality Assessment for the Jefferson County, Ohio Air Quality Control Region", in *Transactions, PM-10 Standards and Nontraditional Particulate Source Controls*, J.C. Chow and D.M. Ono (eds.), Air & Waste Management Assoc., Pittsburgh, PA (1992), pp. 1016–1031.

Streli, C., *J. Trace and Microprobe Techniques*, **13** (1995) 109.

Streli, C., P. Wobrauschek, W. Ladisich, R. Rieder and H. Aiginger, *X-Ray Spectrometry*, **24** (1995a) 137.

Streli, C., P. Wobrauschek, G. Randolf, R. Rieder, W. Ladisich and H. Aiginger, *Anal. Sci.*, **11** (1995b) 477.

Watson, J.G., *Chemical Element Balance Receptor Model Methodology for Assessing the Sources of Fine and Total Suspended Particulate Matter in Portland, OR.*, Ph.D. Dissertation, Oregon Graduate Center, Beaverton, OR (1979).

Watson. J.G. and J.C. Chow, "Data Bases for PM-10 and PM-2.5 Chemical Compositions and Source Profiles" in *Transactions: PM-10 Standards and Nontraditional Particulate*

Source Controls, J.C. Chow, D.M. Ono (eds.), Air & Waste Management Assoc., Pittsburgh, PA (1992), pp. 61–91.

Watson, J.G. and J.C. Chow, *Annual Rev. Energy Environ.*, **19** (1994) 241.

Watson, J.G., J.C. Chow, L.W. Richards, W.D. Neff, S.R. Andersen, D.L. Dietrich, J.E. Houck and I. Olmez, *The 1987–88 Metro Denver Brown Cloud Study, Volume III: Data Interpretation. DRI Document No. 8810.1F3*, Report prepared for the 1987–88 Metro Denver Brown Cloud Study, Inc., Denver, CO, by the Desert Research Institute, Reno, NV (1988).

Watson, J.G., P.J. Lioy and P.K. Mueller, "The Measurement Process: Precision, Accuracy, and Validity" in *Air Sampling Instruments for Evaluation of Atmospheric Contaminants*, 7th edn., S.V. Hering, (ed.), American Conference of Governmental Industrial Hygienists, Cincinnati, OH (1989), pp. 51–57.

Watson, J.G., N.F. Robinson, J.C. Chow, R.C. Henry, B. Kim, E.L. Meyer and T.G. Pace, *Receptor Model Technical Series, Volume III (1989 Revision) CMB7 User's Manual*, Document No. EPA-450/4-90-004, U.S. EPA, Office of Air and Radiation Research Triangle Park, NC, January 1990.

Watson, J.G., J.C. Chow and T.G. Pace, "Chemical Mass Balance," in *Data Handling in Science and Technology, Vol. 7, Receptor Modeling for Air Quality Management*, P.K. Hopke (ed.), Elsevier Press, New York, NY (1991), pp. 83–116.

Watson, J.G., J.C. Chow, D.H. Lowenthal, L.C. Pritchett, C.A. Frazier, G.R. Neuroth and R. Robbins, *Atmos. Environ.*, **28**(15) (1994a) 2493.

Watson, J.G., J.C. Chow, Z. Lu, E.M. Fujita, D.H. Lowenthal, D.R. Lawson and L.L. Ashbaugh, *Aerosol Sci. Technol.*, **21** (1994b) 1.

Watson, J.G., D. Blumenthal, J.C. Chow, C. Cahill, L.W. Richards, D. Dietrich, R. Morris, J. Houck, R.J. Dickson and S. Andersen, *Mt. Zirkel Wilderness Area Reasonable Attribution Study of Visibility Impairment — Volume II: Results of Data Analysis and Modeling — Final Report*, Prepared for the Technical Steering Committee, Colorado Department of Public Health and Environment, Air Pollution Control Division, Denver, CO, by the Desert Research Institute, Reno, NV. July 1996.

Wobrauschek, P., P. Kregsamer, C. Streli and H. Aiginger, *Advances in X-Ray Anal.*, **34** (1991) 1.

3 ION CHROMATOGRAPHY IN ELEMENTAL ANALYSIS OF AIRBORNE PARTICLES

JUDITH C. CHOW and JOHN G. WATSON

Desert Research Institute, P.O. Box 60220,
Reno, Nevada, 89506, USA

INTRODUCTION

Ion chromatography (IC), first developed in the mid-1970s by Small *et al.* (1975), is a physio-chemical separation technique that provides quantitative analysis of inorganic and organic ions from a complex mixture. IC combines features of ion exchange, liquid chromatography, eluent suppression, and quantitative detection.

Since 1975, IC has been widely applied in environmental analysis: airborne particles, acid rain, waste water, pesticides/fertilizers, and soil/plant extracts (Appel, 1981; Butler *et al.*, 1978; Fulmer *et al.*, 1979; Kapelner *et al.*, 1979; Mulik and Sawicki, 1979; Rich *et al.*, 1978; Sawicki, 1978; Tabatabai and Dick, 1979); medical/clinical research: saliva, urine, and serum (Sawicki, 1979); and a variety of industrial applications: pharmaceuticals, semiconductors, electroplating, detergent/household products, and food/beverage (Arar *et al.*, 1992; Barkley *et al.*, 1979; Dempsey *et al.*, 1979; Fratz, 1978; Holm and Barksdale, 1978; Kim *et al.*, 1979; Lash and Hill, 1979; Mansfield, 1979; Mizisin *et al.*, 1979; Pohlandt and Steel, 1980; Rawa, 1979; Speitel *et al.*, 1979; Steiber and Statnick, 1978; Tejada *et al.*, 1978). An array of inorganic/organic anions: fluoride, chloride, nitrate, phosphate, bromide, sulfite, sulfate, weak organic acids, alcohols, aldehydes, carbohydrates, and metal complex (Appel and Wehrmeister, 1979; Durst *et al.*, 1991; Mason and Miller, 1979; Otterson, 1978; Pyen and Fishman, 1979); and cations: sodium, ammonium, potassium, magnesium, calcium, amines, amino acids, and transition/heavy metals (Basta and Tabatabai, 1985; Bogen and Nagourney, 1979; Glaser *et al.*, 1981; Han and Koch, 1988; Jantzen, 1990; Kitsa *et al.*, 1992; Mulik *et al.*, 1978a, b; Noller and Currey, 1990; Papoff *et al.*, 1994; Sinner *et al.*, 1993; Zweidiger *et al.*, 1978) can be determined by combining different separation and detection schemes (e.g., Buck *et al.*, 1992; Hansen *et al.*, 1979; Holcombe *et al.*, 1979; Legrand *et al.*, 1984; Mulik and Sawicki, 1979; Pickerell *et al.*, 1979; Schiff *et al.*, 1979; Simon and Dasgupta, 1995a, b; Slanina *et al.*, 1979; Tyree *et al.*, 1979).

Other wet-chemical analysis methods such as titration, gravimetry, turbidimetry, flame photometry, and photoelectric colorimetry are often more labor intensive than IC, and have poorer detection limits (Willard *et al.*, 1988). IC has multi-species capabilities, small sample requirements, and the detectors have high sensitivity, variable selectivity, good stability, and high linearity of response. IC methods for water and air pollutants approved or recommended by the U.S. Environmental Protection Agency (EPA), the National Institute of Occupational Safety and Health (NIOSH), and the Occupational Safety and Health Administration (OSHA) are summarized in Table 1. These techniques provide detailed procedures that can be applied to different types of environmental samples.

The objective of this chapter is to focus on the basic components and operating principles of ion chromatography, and its application in the analysis of airborne particles. Issues such as sample preparation, filter extraction, standardization and instrument calibration, peak quantification, data validation, quality control, and quality assurance are discussed. These practical considerations are often omitted from published analytical procedures.

OPERATING PRINCIPLES

During IC analysis a filtered aliquot of sample, an analyte, enters an eluent stream via a valve injector. The mixture of sample and eluent are then transported to the separator column by the pumping system. Sample injection can be carried out at or above, atmospheric pressure with a typical volume of 0.1–2 mL. The eluent ions from the separator column are then neutralized in the succeeding suppressor column and sample ions are converted to their corresponding strong acidic or basic form for quantitative detection. Each ion is identified by its retention time within the ion-exchange column (Smith and Chang, 1983). The anions and cations in airborne particles that

Table 1. Summary of ion chromatographic compliance testing methods.

Testing Methods	Matrix	Chemical species	Concentration range	References
US EPA[a] Method 300A[b]	Water	Fluoride (F^-)	0.01[d]–5.0 μg/mL	US EPA (1991)
	Water	Chloride (Cl^-)	0.02[d]–2.0 μg/mL	
	Water	Nitrite (NO_2^-)	0.004[d]–10.0 μg/mL	
	Water	Bromide (Br^-)	0.01[d]–5.0 μg/mL	
	Water	Nitrate (NO_3^-)	0.002[d]–10.0 μg/mL	
	Water	Phosphate (PO_4^{3-})	0.003[d]–5.0 μg/mL	
	Water	Sulfate (SO_4^{2-})	0.02[d]–50.0 μg/mL	
US EPA Method 300B[c]	Water	Chlorite (ClO_3^-)	0.01[d]–5 μg/ml	US EPA (1991)
	Water	Chlorate (ClO_2^-)	0.02[d]–5 μg/ml	Pfaff *et al.* (1991)
	Water	Bromate (BrO_3^-)	0.03[d]–5 μg/ml	
US EPA Method 300.7[b]	Water	Sodium (Na^{2+})	0.03[d]–1.00	US EPA (1985; 1986a, b)
	Water	Magnesium (Mg^{2+})	0.02[d]–1.00	Bachman *et al.* (1986)
	Water	Potassium (K^+)	0.01[d]–1.00	
	Water	Calcium (Ca^{2+})	0.02[d]–3.00	
	Water	Ammonium (NH_4^+)	0.03[d]–2.00	
US EPA Method 218.6 and US EPA Method 1636	Water	Hexavalent Chromine (Cr IV) determined as CrO_4^{2-}	0.001–5 μg/mL	US EPA (1994; 1995)
NIOSH[e] 6004	Air	Sulfur Dioxide (SO_2) determined as sulfate (SO_4^{2-}) sulfite (CrO_4^{2-})	1.1–20 μg/mL	*NIOSH Manual of Analytical Methods* (1994a)

Table 1. (continued).

Testing Methods	Matrix	Chemical species	Concentration range	References
NIOSH 6011	Air	Chlorine (Cl)	0.06–1.5 μg/mL	*NIOSH Manual of*
		determined as chloride (Cl^-)		*Analytical Methods*
		Bromine (Br)	0.06–1.5 μg/mL	(1994b)
		determined as bromide (Br^-)		
NIOSH Method 7903	Air	Inorganic acids		*NIOSH Manual of*
		determined as common anions:		*Analytical Methods*
		Hydrofluoric acid (HF) as F^-,	0.03–13.3 μg/mL	(1994c)
		Hydrochloric acid (HCl) as Al^-,	0.03–13.3 μg/mL	
		Phosphoric acid (H_3PO_4) as PO_4^{2-},	0.2–6.7 μg/mL	
		Hydrobromic acid (HBr) as Br^-,	0.2–64 μg/mL	
		Nitric acid (HNO_3) as NO_3^-,	0.2–33.3 μg/mL	
		Sulfuric acid ($H_2SO_4^{2-}$) as SO_4^{2-}	0.2–6.7 μg/mL	
OSHA[f] Method ID-182	Air	Nitrogen dioxide		
		determined as nitrite (NO_2^-)		
OSHA Method ID-190	Air	Nitric oxide		
		determined as nitrite (NO_2^-)		

[a]US Environmental Protection Agency.
[b]For drinking water, surface water, mixed domestic and industrial waste water, ground water, reagent water, solids, and leachates.
[c]For disinfection byproducts in drinking and reagent water.
[d]Minimum detection limit (MDL) obtained with reagent water is defined as the minimum concentration of a substance that can be measured and reported with 99% confidence that the value exceeds zero.
[e]National Institute of Occupational Safety and Health.
[f]Occupational Safety and Health Administration.

are most frequently analyzed by ion chromatography, along with minimum detection limits, lower quantifiable limits, and typical concentration ranges in urban and nonurban areas are listed in Table 2.

The minimum detection limit (MDL) is defined as the concentration at which instrument response equals three times the standard deviation of the response to a known concentration of zero. The lower quantifiable limit (LQL) equals three times the standard deviation of a dynamic field blank acquired by passive deposition from each airborne particle sampling program. The manufacturer's specifications of MDL for most species determined by ion chromatography are in the 0.001–0.002 μg/mL range. Different researchers (Bergin *et al.*, 1995; Buck *et al.*, 1992; Durst *et al.*, 1991; Hoover and Yager, 1984; Koch and Liggett, 1988; Legrand *et al.*, 1984) report MDLs from 0.0001 to 0.5 μg/mL for wet deposition samples. The MDL of 0.05 μg/mL reported in Table 2 applies to airborne particulate samples taken on filters and subsequently extracted in deionized distilled water.

The detection limit reflects the performance of an instrument at low analyte concentrations. This indication can be used as a guide to instrument optimization, gauge of the suitability of an instrument for a specific application, or criterion for the interpretation of low concentration

Table 2. Detection limits and typical range of ion chromatography measurements of airborne particles.

Ionic species	Minimum detection Limit[a] (μg/mL)	Lower quantifiable Limit[b] ($\mu g/m^3$)	Concentrations range ($\mu g/m^3$)	
			Urban	Non-urban
Chloride (Cl^-)	0.05	0.34[c]	0.1–5.0[d]	0.05–0.13[g]
Nitrate (NO_3^-)	0.05	0.34[c]	0.5–100[d,e]	0.05–0.58[g]
Sulphate (SO_4^{2-})	0.04	0.34[c]	0.5–20[d]	0.1–5.0[h]
Sodium (Na^+)	0.025	0.1[c]	0.1–4.0[f]	0.07–0.3[g]
Magnesium (Mg^{2+})	0.0004	0.1[c]	0.07–0.7[g]	0.03–0.1[g]
Potassium (K^+)	0.035	0.0140[c]	0.01–4.0[f]	0.07–0.09[g]
Calcium (Ca^{2+})	0.025	0.1[c]	0.11–1.3[g]	0.1–0.4[g]
Ammonium (NH_4^+)	0.05	0.086[c]	0.5–35[g]	0.03–0.9[g]

[a]MDL is defined as the concentration at which instrument response equals three times the standard deviation of the response to a known concentration of zero.
[b]LQL equals three times the standard deviation of dynamic field blanks.
[c]Chow *et al.* (1993) (calculation is based on a nominal 14.5 m^2 sample volume).
[d]Chow *et al.* (1994b).
[e]Chow *et al.* (1992).
[f]Chow *et al.* (1995).
[g]Watson *et al.* (1991).
[h]Watson *et al.* (1996).

measurements (Koch and Liggett, 1988). Different optimization techniques used to improve the detection limits are discussed in this chapter.

Basic Components

The historical background and perspective of chromatographic techniques since their discovery in the early 1900s (Tswett, 1906) have been well documented (Small, 1989; Weiss, 1995). The basic components of an ion chromatography system are shown schematically in Figure 1. Commercially

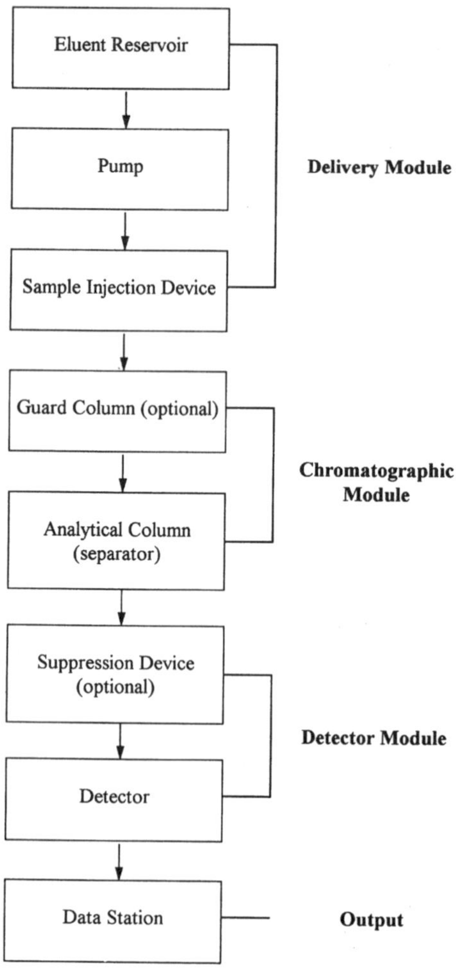

Figure 1. Basic components of the ion chromatography system.

available instruments may differ somewhat in their configurations. The analytical system includes: a sample injection valve, flow moving device (i.e. pump) which introduces the sample and eluent to a sample loop for mixing, followed by an ion-exchange separator column, background ion-suppressor column (optional), and detection system. Definitions or descriptions of these basic components of the IC are provided in Table 3.

Table 3. Commonly applied terminology in ion chromatography.

Term	Definition/Description
Analytical column/Separator column	The column used to perform the actual chromatographic separation.
Column capacity	Related to the total ionic strength of the sample, it is the maximum concentration of sample that may be loaded onto a separator column before overloading.
Conductivity	A measure of the characteristic of ions in solution to carry electrical current through a liquid between two oppositely-charged electrodes. Absolute conductivity is a function of flow rate of the regenerant and eluent strength
Eluent	The ionic liquid mobile phase used to transport the sample through the ion exchange column.
Guard column/Pre-column	A small column which prevents poisoning and contamination of the separator column by removing particles and absorbing organic compounds.
Ion exchange	A reversible process by which ions are interchanged between an insoluble material and a liquid with no substantial structural changes of materials (*Annual Book of ASTM Standards*, 1982).
Ion exchange capacity	The number of active ion-exchange sites in a given weight or volume of resin, often expressed in milliequivalents per gram (meq/g) or milliequivalents per milliliter (meq/mL).
Ion exchange resin	An insoluble carbon-based polymer matrix containing charged exchange sites (anionic or cationic). The resins used in Ion Chromatography are formed into small spherical particles (i.e., beads).

Table 3. (continued).

Term	Definition/Description
MicroMembrane suppressor	A device used to continuously minimize eluent conductivity and convert sample species to a high conductivity form, thus increasing detection sensitivity.
Regenerant	A solution that converts and maintains an active form of the suppressor. Suppressors (e.g., MicroMembrane and Fiber Suppressors) can be regenerated continuously to minimize the system down time.
Resolution	The ability of a column to separate constituents under specified test conditions. Peak resolution is a function of column efficiency, selectivity, and capacity.
Retention time	The interval measured from the point of sample injection to the point of maximum peak height or area; the basis for identification of a species in chromatography.
Theoretical plates/Column efficiency	A measure of peak dispersion as flow moves through a column proportional to column length for a given column diameter and resin.

Analytical Column

The main component of the IC system is the analytical or separator or ion exchange column. Chromatographic separation of ions by ion exchange is based on the affinities of the analyte ions relative to the function group on the ion exchange resin, that resolves components of a mixture into separate bands, peaks or signals in the chromatogram, based on their migration velocities. The difference in the equilibrium distribution of analyte ions between the mobile phase (analyte eluent flow) and stationary phase (ion-exchange column) results in this separation. The choice of ion exchange resin and the selection of chromatographic conditions determines the quality of the analysis (Colenutt and Trench, 1985; Lathouse and Heffelfinger, 1979; Lodge, 1989; Posner and Schoffman, 1979; Sawicki, 1978; Small, 1978).

Both anion and cation exchanges are classified according to the types of substrate or supporting material, pore size, and ion-exchange capacities (Rieman and Walton, 1970; Samuelson, 1963; Walton and Rocklin, 1990).

Although organic polymers are predominantly employed as packing material in ion exchange columns, silica based or latex agglomerated resins also exhibit high chromatographic efficiencies. The ion exchange resin carries functional groups with fixed changes. Organic polymers function directly at their surface, whereas latex beads provide full functionality. Alkyd or alkanol quaternary ammonium and sulfonic or carboxylic acid are frequently used as the functional group for anion and cation exchangers, respectively.

The separation mechanisms are described in Table 4. They include: ion exchange and exclusion (Rieman and Walton, 1970; Snyder and Kirkland, 1983); and ion pairing and reverse phase chromatography (Poole and Poole, 1991; Robards *et al.*, 1994; Rossner *et al.*, 1987; Weiss, 1995). Over 50 analytical columns are commercially available with various particle or latex diameters, substrate materials, degrees of cross linking, functional groups, pH ranges, solvent stabilities, hydrophobicities, maximum operating pressures, maximum flow rates, and column capacities. For specific applications, detailed information should be requested directly from the manufacturers.

The analytical or separator column contains a strong anion exchange group for anion analysis, and a strong cation exchange group for cation analysis. Depending on the separation requirements, column dimensions vary from 50 to 250 mm in length, and from four to five mm inside diameter. For determination of low level analytes, ppb or ng/mL, a preconcentration technique can be applied (Buhr *et al.*, 1995; Han and Koch, 1988; Hansen *et al.*, 1979; Hoke and Baxter, 1985; Hoover and Yager, 1984; Simon *et al.*, 1991; Simon and Dasgupta, 1993; US EPA, 1980; Wetzel *et al.*, 1979). A concentration or guard column (see Table 3) can be installed preceding the separator column to remove trace impurities in the analyte.

Detector

Both electrochemical and spectroscopic detection methods have been applied in IC. Electrochemical techniques include conductometric and amperometric detection, whereas spectroscopic methods apply ultraviolet/visible absorbance, fluorescence, or refractive index detection (Fritz *et al.*, 1982; Haddad, 1994; Rich, 1977; Rich *et al.*, 1979; Van Os, 1987; Weiss, 1995).

Detection methods are a function of separation methods, analyte ions, and eluent (Gjerde *et al.*, 1979; Kyriacon and Kempster, 1993). Detector performance is evaluated with respect to sensitivity, linearity, resolution (volume of the detector cell), and signal-to-noise ratio (detection limit).

Table 4. Separation methods in ion chromatography.

Separation mechanism	Principle	Applicability	Function group	Detection mode
Ion-exchange chromatography (high-performance ion chromatography)	Based on an ion-exchange process occurring between the mobile phase (analyte/eluent) and the stationary phase (ion-exchange group bounded to the supporting/packing material); non-ionic adsorption process contributes to the separation mechanism.	• Organic and inorganic anions • Organic and inorganic cations	• Quaternary ammonium • Sulfonic acid	• Suppressed conductivity • Amperometry • Direct ultraviolet (UV) absorbance • Post-column reaction/visible absorbance/fluorescence
Ion-exclusion chromatography (high-performance ion chromatography exclusion)	Based on Donnan exclusion, Sterie exclusion, and adsorption processes.	• Weak inorganic or organic acids • Amino acids • Alcohol • Aldehydes	• Fully-functionalized sulfonic or carboxylic acid	• Suppressed conductivity • Amperometry

Ion-pair chromatography (mobile-phase ion chromatography)	Based on adsorption. The selectivity of the separation is determined by the makeup of the mobile phase to separate the compounds. Uses supporting material with low polarity and high specific surface area.	• Surface-active anions and cations • Transitional method	(none)	• Suppressed conductivity • Direct UV absorbance
Reverse-phase liquid chromatography	Based on ion suppression.	• Higher polar and ionic species	(none)	• Amperometry • Direct UV adsorbance • Post-column reaction/ fluorescence

Photometric methods such as UV/visible spectroscopic analysis require the analyte ions to absorb light (Small and Miller, 1982). Therefore UV/visible detection is limited to those ions that can absorb light to a measurable extent. Suppressed IC with tandem UV and conductivity detection provides a combination of sensitivity and selectivity (Večeřa and Dasgupta, 1991a, b). Refractive index detection requires substantial variations between the refractive indices of analyte ions and eluent ions, and these differences are often too small to be resolved (Mulik and Sawicki, 1979). Conductimetric detection is the most widely employed detection method in IC because all ionic solutions conduct electricity, and this conductivity is linear with ion concentration at levels typical of environmental samples (Cameron and Pohlandt, 1987; Mulik and Sawicki, 1979; Pinschmidt, 1979; Pohlandt and Steele, 1980; Simon and Dasgupta, 1995b).

Suppressor Column

The effluent from the ion exchange separator column is often passed through a post column reactor prior to detection. This suppressor column chemically reduces the background conductance of the eluent, while simultaneously increasing the electrical conductance of the analyte ions and eliminating sample counter ions. For the analysis of anions, the suppressor column is packed with cation exchange resin in its hydrogen form. Background suppression occurs as the eluent, usually a mixture of sodium bicarbonate ($NaHCO_3$) and sodium carbonate (Na_2CO_3), reacts with analyte ions, so that:

$$R^-H^+ + Na^+HCO_3 \rightarrow R^-Na^+ + H_2CO_3 \quad (1)$$

$$2R^-H^+ + Na_2CO_3 \rightarrow 2(R^- - Na^+) + H_2CO_3 \quad (2)$$

$$R^-H^+ + Na^+A^- \rightarrow R^-Na^+ + H^+A^- \quad (3)$$

where:

R^- = functional group of a resin
A^- = anionic species

In Equations (1) and (2), both sodium bicarbonate and sodium carbonate are converted to carbonic acid (H_2CO_3). This is a weak acid with much lower conductivity than sodium bicarbonate or carbonate in the eluent, hence, the term of eluent suppression. The eluent used in the analysis can be either chemically or electronically suppressed for conductivity detection. The signal is enhanced as the analyte anion becomes its acid form in a weak eluent because the equivalent conductance for H^+ is approximately seven times that for Na^+ (Karmarkar, 1995).

For the analysis of cations, the suppressor column is packed with anion exchange resin in its hydroxide form. Hydrochloric acid (HCl) is commonly used as the eluent. The reaction can be illustrated as follows:

$$R^{+}OH^{-} + HCl \rightarrow R^{+}Cl + H_2O \quad (4)$$

$$R^{+}OH^{-} + C^{+}Cl^{-} \rightarrow R^{+}Cl^{-} + COH. \quad (5)$$

In Equations (4) and (5), the strong acid eluent HCl is converted to H_2O and the analyte cation C^+ is converted to its strong conductive hydroxide form COH, in a background of water rather than acid. This process greatly enhances the signals in the detector.

The suppressor column capacity allows large numbers of analyte ions to be separated and analyzed before regeneration is necessary. This regeneration process involves the pumping of chemical regenerant such as diluted acid for anion analysis or diluted base for cation analysis, through the suppressor column followed by a rinse with distilled deionized water in the opposite direction of the normal flow.

Chemical suppressor technology has evolved from packed resin beds to fiber- or flat-membrane-based beds. Commercially-available automatic suppressant devices allow the operation of suppressor systems in the following modes: external distilled deionized water, chemical suppression using a chemical regenerant, recycled, self regenerant, or post-column addition of colloidal suspension to ensure optimal column performance.

The advantages of the suppressor technique are its sensitivity and specificity. The disadvantage is that the suppressor column must be regenerated or replaced periodically, since it accumulates the unwanted ions that were removed from the eluent.

Elution Technique

The selection of eluent in IC depends mainly on the detection system. Most of the applications in IC for common anions and cations use the isocratic elution technique. In this process, the distribution of analyte and eluent ions between the mobile, analyte/eluent, and stationary, ion exchange column, phases defines the equilibrium constant for the ion exchange reaction. Isocratic elution employs a single type of eluent, operating at a fixed concentration level and preset flow rate. This technique maintains a constant baseline during analysis.

Because the analyte ions exhibit similar retention times (halide ions), extended analysis times over 30 min per run are required to quantify these species. Therefore, in order to elute ions with widely different retention times in a single run, it is necessary to apply gradient elution. The composition or concentration of the eluent is changed continuously during the analysis in gradient elution. Chemical species are displaced with different affinities that begin with an eluent of low displacement power and progresses either continuously or in a stepwise function to an eluent of greater potency. Gradient elution varies the concentration strength, flow rate, or composition of the eluent (Buck *et al.*, 1992; Rieman and Walton, 1970; Rocklin *et al.*, 1987; Tarter, 1984). In concentration gradient elution, the concentration of the eluent ions is altered during the analysis run; and in composition gradient elution, the composition of the eluent ions is changed by replacing weakly retained eluent ions with stronger ones. Analyte species with similar retention times can be separated in a single sample injection by varying the concentration or composition of the eluent.

Application of gradient elution has been used to separate halide ions: F^-, Cl^-, Br^-, I^- (Wang *et al.*, 1983), identify sulfate anions: SO_3^{2-}, SO_4^{2-}, $S_2O_3^{2-}$ (Sunden *et al.*, 1988), and quantify more complex inorganic ions: I^-, SCN^-, $Cr_2O_7^{2-}$, $S_2O_3^{2-}$ (Tarter, 1984). Rocklin *et al.* (1987) developed a gradient elution technique in which 36 ions were analyzed simultaneously in a single sample injection. Although gradient elution is a powerful technique that is widely applied in reverse-phase high-performance liquid chromatography (HPLC) with UV-photometric detection (Willard *et al.*, 1988), it is not commonly used in IC analysis. Severe baseline shifts and the presence of trace impurities or contaminants in the eluent, resulting in extraneous peaks has made gradient elution in IC difficult.

AEROSOL FILTER SUBSTRATES AND FILTER EXTRACTION

Along with the IC analysis for airborne particles, sample collection, pretreatment, and filter extraction are important considerations as. The sample analyte must be in solution, without the addition of acid or base, and filterable (Koch, 1979a).

Sampling Media

Airborne particles and their precursor gases such as nitric acid (HNO_3), sulfur dioxide (SO_2), and ammonia (NH_3), are generally collected by drawing ambient air through a filter substrate with a flow mover (Chow and Watson, 1993; 1994; Chow *et al.*, 1996). Some precursor gases such as sulfur dioxide, oxides of nitrogen, and ammonia, undergo gas to particle transformation after source emission and form ammonium sulfate and nitrate. These secondary aerosols are often major constituents of airborne particles, and their ions the most commonly measured.

A particle sampling filter consists of a tightly woven fibrous mat or plastic membrane that has been penetrated by microscopic pores. Several characteristics are important in selecting filter media for aerosol sampling, including: particle sampling efficiency, mechanical, chemical, and temperature stability, blank concentrations, flow resistance and loading capacity, and cost and availability (Chow, 1995; Chow *et al.*, 1996). The physical and chemical characteristics of gas/particle sampling filter media suitable for obtaining air samples for IC analysis are summarized in Table 5. Quartz and cellulose fiber filters report clean and consistent blank levels for ionic species (Chow and Watson, 1994) and exhibit low flow resistance. In order to sample gaseous components, these filters can be impregnated with gas absorbing chemicals prior to sampling. Active agents applied to precursor gas collection are listed in Table 5. These gaseous precursors are converted to their ionic counterparts upon interaction with the chemical impregnant.

Filter Extraction

Aerosol ions refer to chemical compounds that are water soluble. Analysis of airborne particles involves the quantification of both simple ions: fluoride, chloride, sodium, and potassium; and polyatomic ions: sulfate, nitrate, and ammonium. This requires the application of both anion and cation chromatography.

Prior to IC analysis, a fraction of the filter is extracted in solution (distilled deionized water for airborne particles), and filtered to remove insoluble residues (Stevens and Dzubay, 1978). Fibers from the filter substrate are desegregated during extraction. It is essential to provide filtration prior to sample injection to prevent system clogging. Distilled deionized water is the most commonly used extraction solution for the analysis of airborne particles. Other extraction solutions can be used with various chemical-impregnated filters as shown in Table 6.

The extraction volume needs to be as small as possible, lest the solution become too dilute to detect the desired constituents at levels typical of those found in airborne particles. Each square centimeter of filter should be extracted in no more than two mL of solvent for typical sampler flow rates of 20–30 L/min and sample durations of 24 hours. This often results in no more than 20 mL of extract which can then be submitted to different wet chemistry analytical methods. Sufficient sample deposit must be acquired to account for the solution volume required by each method.

When other analyses are to be performed on the same filter, the filter is first sectioned using a precision positioning jig attached to a paper cutter. For rectangular filters, typically $20.32\,cm \times 25.40\,cm$, a $2.54 \times 20.32\,cm$

Table 5. Commonly used filter media for particulate sampling and analysis.

Filter medium (major manufacturer and catalog No. or trade name)	Chemical characteristics	Physical characteristics	Filter size	Suitable for precursor gas collection		
				SO_2	HNO_3	NH_3
• Ringed Teflon-membrane (Gelman Scientific; Ann Arbor, MI; Teflo®, R2PJ047, R@PJ037)	• Usually low blank levels, but several contaminated batches have been found. Made of carbon-based material, so inappropriate for Carbon analysis. • Inert to adsorption of gases. • Low hygroscopicity. • Low blank weight.	• Thin membrance stretched between polymethylpentane ring. • White surface, nearly transparent. • Minimal diffusion of transmitted light. • High particle collection efficiencies. • Cannot be accurately sectioned. • 1.2, 2.0, 3.0, 5.0 and 10 μm pore sizes (determined from liquid filtration). • Melts at ~60°C. • High flow resistance.	25 mm 37 mm 47 mm	NA[e]	NA	NA
• Backed Teflon membrane, (Gelman Scientific, Ann Arbor, MI; "Zefluor",	• Usually low blank levels. Made of carbon-based material, so inappropriate for carbon analysis. • Inert to adsorption of gases. • Higher background levels for XRF	• Thin membrane mounted on thick polypropylene backing. • White opaque surface, diffuses transmitted light. • High particle collection efficiencies. • Melts at ~60°C.	47 mm	NA	NA	NA

F2996-25)	and PIXE than Teflo® owing to greater filter thickness. • Low hygroscopicity. • High blank weight.	• High flow resistance.				
• Nylon membrane, (Gelman Scientific, Ann Arbor, MI; "Nylosorb", #66509)	• High HNO_3 collection efficiency. • Passively adsorbs low levels of NO, NO_2, PAN, and SO_2. • Low hygroscopicity. • Low blank weight.	• Thin membrane of pure nylon. • White opaque surface, diffuses transmitted light. • 1 µm pore size. • Melts at ~60°C. • High flow resistance.	25 mm 37 mm 47 mm	NA	NA	NA
• Pure quartz-fiber (Pallflex Corp., Putnam, CT; 2500 QAT-UP)	• Pre-washed during manufacture-low blank levels for ions. • Contains large and variable quantities of Al and Si. Some batches contain other metals. • Passively adsorbs organic vapors. Adsorbs little HNO_3, NO_2, and SO_2. • Low hygroscopicity.	• Mat of pure quartz fibers. • White opaque surface, diffuses transmitted light. • High particle collection efficiencies. • Soft and friable edges flake in most filter holders. • Melts at >900°C. • Moderate flow resistance.	25 mm 37 mm 47 mm 20.3 × 25.4 cm	x	x	x
• Mixed quartz-fiber (Whatman Corp., Hillsboro, OR; QM/A #1861865)	• High blank weight. • Contains large and variable quantities of Na, Al, and Si in all batches. Variables levels of other metals are found in many batches.	• Quartz (SiO_2) fibers with ~5% borosilicate content. • White opaque surface, diffuses transmitted light. • High particle collection efficiencies. • Some batches can melt at ~500°C.	20.3 × 25.4 cm	NA	NA	NA

Table 5. (Continued).

Filter medium (major manufacturer and catalog No. or trade name)	Chemical characteristics	Physical characteristics	Filter size	Suitable for precursor gas collection		
				SO_2	HNO_3	NH_3
	• Passively adsorbs organic vapors. Adsorbs little HNO_3, NO_2, and SO_2. • Low hygroscopicity. • High blank weight.	Effects on thermal carbon analysis are unknown. • Becomes brittle when heated. • Low flow resistance.				
• Cellulose-fiber, (Whatman Corp., Hillsboro, OR; #1441047)	• High purity, low blank levels. Made of carbon-based mateiral, so inappropriate for carbon analysis. • Adsorbs gases, especially water vapor. • Most appropriate for adsorbing gases such as HNO_3, SO_2, NH_3, and NO_2 when impregnated with reactive chemicals. • High hygroscopicity. • High blank weight.	• Thick mat of cellulose fibers, often called a "paper" filter. • White opaque surface, diffuses transmitted light. • Low particle collection efficiencies, $<70\%$ for some variations of this filter. • High mechanical strength. • Burns at elevated temperatures (~150°C, exact temperature depends on nature of particle deposit). • Variable flow resistance.	25 mm 37 mm 47 mm	x	x	x
• Teflon-coated	• Low blank levels for ions (glass	• Thick mat of borosilicate glass fiber	37 mm	NA	NA	NA

glass-fiber (Pallflex, Putnam, CT; TX40HI20	backing and carbon content make it less suitable for elemental and carbon analyses). • Inert to adsorption of HNO_3, NO_2, and SO_2. • Low hygroscopicity. • High blank weight.	with a layer of Teflon on the surface. • Glass fiber supporting Teflon is shiny. • High particle collection efficiencies. • Glass melts at ~500°C. Teflon melts at ~60°C. • Low flow resistance.	47 mm			
• Glass fiber (Gelman Scientific, Ann Arbor, MI; Type A/E)	• High blank levels. • Adsorbs HNO_3, NO_2, SO_2, and organic vapors. • Low hygroscopicity. • High blank weight.	• Borosilicate glass fiber. • White opaque surface, diffuses transmitted light. • High particle collection efficiencies. • Melts at ~500°C. • Low flow resistance.	20.3 × 25.4 cm	NA	NA	NA

[a] AAS: Atomic Absorption Spectrophotometry; AC: Automated Colorimetry; IC: Ion Chromatography; ICP/AES: Inductively-Coupled Plasma with Atomic Emission Spectrophotometry; ICP/MS: Inductively-Coupled Plasma with Mass Spectrophotometry; INAA: Instrumental Neutron Activation Analysis; OA: Optical Absorption or Light Transmission (b_{abs}); OM: Optical Microscopy; PIXE: Proton-Induced X-Ray Emissions; SEM: Scanning Electron Microscopy; T: Thermal Carbon Analysis; TEM: Transmission Electron Microscopy; TMO: Thermal Manganese Oxidation Carbon Analysis; TOR: Thermal/Optical Reflectance Carbon Analysis; TOT: Thermal Optical Transmission Carbon Analysis; XRD: X-Ray Diffraction; XRF: X-Ray Fluorescence.
[b] Available in 0.45 μm pore size.
[c] Available in 0.80 μm pore size.
[d] Filter disc is available in-size between 13 to 293 mm depending on the pore size.
[e] Not applicable.
[f] Impregnated with potassium carbonate or sodium carbonate.
[g] Impregnated with triethanolamine (TEA).
[h] Impregnated with sodium chloride or sodium carbonate.
[i] Impregnated with citric acid, sulfuric acid, oxalic acid, phosphoric acid, or sodium carbonate.

Table 6. Example of filter impregnation and extraction solutions applied in ion chromatography analysis.

Filter medium	Impregnant solution	Extraction solution	Commonly analyzed ionic species	Comments
Teflon-membrane filter	None	Distilled deionized water (DDW)	F^-, Cl^-, Br^-, NO_2, NO_3^-, PO_4^{2-}, SO_4^{2-}, Na^+, Mg^{2+}, K^+, Ca^{2+}, NH_4^+	Add 200 μL of ethanol as wetting solution to filter surface prior to extraction
Nylon-membrane filter	None	Sodium carbonate/ Sodium bicarbonate	NO_3^-	
Quartz-fiber filter	None	Distilled deionized water	F^-, Cl^-, Br^-, NO_2^-, NO_3^-, PO_4^{2-}, SO_4^{2-}, Na^+, Mg^{2+}, K^+, Ca^{++}, NH_4^+	
Impregnated quartz-fiber or cellulose-fiber filter	Potassium carbonate with 5% glycerol	First step: Add 0.1% Hydrogen peroxide (H_2O_2) in 10 mL distilled deionized water Second step: Perform 1 to 11 dilution on the 0.1% H_2O_2 extract	SO_4^{2-} as SO_2	Store 0.1% H_2O_2 solution for two days in the refrigerator to ensure complete SO_2 and SO_3^- oxidation
Impregnated quartz-fiber or cellulose-fiber filter	Citric acid with 5% glycerol	Distilled deionized water	NH_3 as NH_4^+	
Impregnated quartz-fiber or cellulose-fiber filter	Sodium chloride with 5% glycerol	Sodium carbonate/ Sodium bicarbonate	HNO_3 as NO_3^-	
Impregnated quartz-fiber or cellulose-fiber filter	Triethanolamine (TEA) with 5% glycerol	Sodium carbonate/ Sodium bicarbonate	NO_2 as NO_2^-	

wide strip is cut from the center two thirds of the filter. High volume PM-10 samplers have a 1.27 cm border around them, making the exposed area ~406 cm^2 and the area of the deposit on the filter strip 33.9 cm^2. It is necessary to multiply the analysis results by approximately 12.0 to estimate the ion deposit on the entire filter, and verify these values, since various filter frames may have different measurements. Circular filters are often cut in half for these analyses, thus the results need to be multiplied by two to obtain the deposit on the entire filter. Retention of half of the filter provides the opportunity to reanalyze the sample if discrepancies need to be resolved. The choice of filter materials for these analyses is done so that it can be easily sectioned without causing damage to the filter or the deposit. The deposit must also be uniform across the exposed area, so that the measurement on a portion of the filter can be scaled to the entire exposed area. The cutting blade must be cleaned between each filter sectioning.

The filter section is placed in an extraction vial which is capable of allowing it to be fully immersed in approximately 15 mL of solvent. Properly labelling each vial with the sample ID is crucial. Since much of the deposit is inside a fiber filter, agitation is needed to extract the water-soluble particles into solution. Experiments show that sonication for one hour and shaking for one hour at room temperature, followed by aging for a minimum of 12 hours at low temperature (e.g., refrigeration), assures complete extraction of the deposited material in the solvent. A flow diagram for aerosol filter extraction is provided in Figure 2. The sonicator bath water should be periodically replaced or recirculated to prevent temperature increases from the dissipation of ultrasonic energy in the water. Never place frozen or very cold samples directly into the ultrasonic bath, as heating will cause the vials to crack and the analyte to be contaminated. Return unused filter sections to their labeled containers and store under refrigeration. These can be used for other analyses or they can serve as a backup if the original sample becomes contaminated or is insufficient for the planned ionic analyses.

Teflon® membrane filters exhibit low hygroscopicity. Applying a wetting solution of 200 μL ethanol or methanol onto the Teflon® filter surface prior to the addition of extraction solution will ensure complete dissociation of the analyte ions. Teflon® filters also tend to float in the extraction vial unless secured below the surface of the extracting solution (Appel, 1994).

Zefluor filters (see Table 5) have a porous Teflon® backing that requires careful examination to distinguish it from the membrane surface. Often aerosol constituents are incorrectly sampled onto the back filter surface where particles become trapped between the Teflon® backing and the membrane surface. Consequently this hinders extraction efficiency. In this

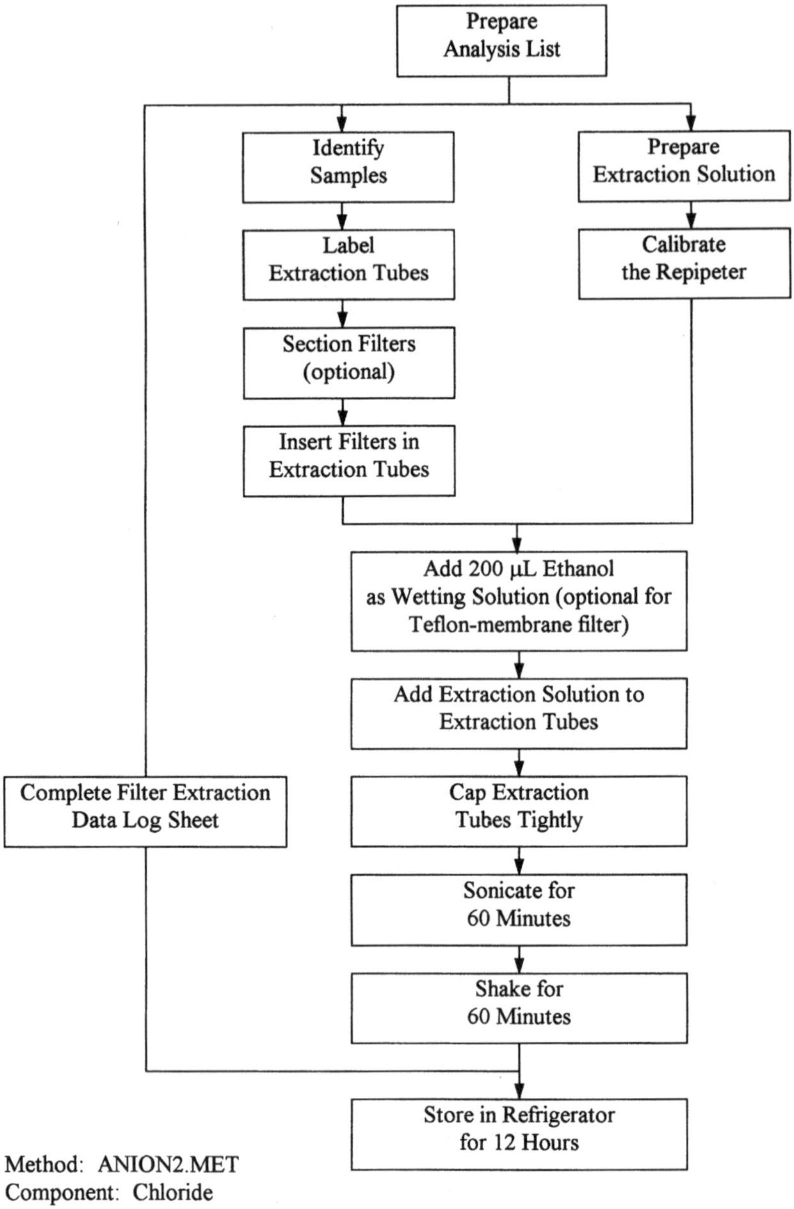

Figure 2. Flow diagram of filter extraction procedure.

event the membrane surface of each filter needs to be manually separated from the backing with a forceps. Then both portions must be extracted together. Nitric acid and nitrate tend to bind to nylon membrane filters and are not efficiently recovered by extraction in distilled deionized water solution (Sickles *et al.*, 1986). Use an alkaline solution such as potassium carbonate (Table 6) to provide an efficient extraction (Appel *et al.*, 1988; Herring *et al.*, 1988).

INSTRUMENT CALIBRATION AND STANDARDIZATION

Depending on the dissociation of the species, the linear response range of the ion chromatography system can be theoretically extended from 0.01 to 100 µg/mL (Mulik and Sawicki, 1979). Strong acids or bases that are highly dissociated or ionized can be easily assayed by IC, whereas weak acids and bases that lack sufficient ionic character are more difficult to quantify. In practice, the linear response can only be assured within one to two orders of magnitude from the lowest calibration range. Overextending the concentration range will result in an overestimation of the low concentrations or underestimation of the high concentrations. Calibration standards should span the entire range of sample concentrations, and separate calibration curves are required for samples with low and high concentrations.

Calibration Range

Determine optimal calibration levels based on the typical concentration ranges in each area from which samples are acquired. For airborne particles, a six-point calibration curve is needed with concentrations of 0 (reagent blank), 0.05, 0.10, 0.20, 0.50, and 1.00 µg of analyte per mL of standard solution. Adequate linear response can be established within this concentration range. Over 80% of the airborne particle samples collected in urban and nonurban areas can be assayed within this concentration range without further dilution, or with higher or lower standard concentrations.

A high concentration calibration curve (e.g., 0, 1.0, 2.0, 5.0, 7.5, and 10.0 µg/mL) can be established for heavily-loaded ambient or source samples. Since the lowest point is 1.0 µg/mL instead of 0.05 µg/mL, this approach compromises the accuracy of low concentration samples. It is advisable to reanalyze samples with concentrations lower than 1.0 µg/mL using low concentration calibration curves, 0–1.0 µg/mL. For pristine environments with ultra low ionic concentrations, standards of 0, 0.01, 0.02, 0.05, 0.10, and 0.50 µg/mL usually correspond to these airborne particle levels.

Calibration Standards

Standard stock solutions, typically 1000 μg/mL, can be purchased or prepared with American Chemical Society (ACS) reagent grade material. Solid reagent chemicals should be dried at 105°C for one hour and cooled in a dessicator. Care must be taken when performing gravimetric analysis of these chemicals to the nearest 0.1 mg, in order to minimize static charges and to prevent loss of solid chemicals after weighing. The standard stock solution is stored in a refrigerator and replaced at least annually for stable compounds such as sulfate, and more frequently for volatile compounds such as ammonium. Working standards of 10 and 100 μg/mL are used to prepare calibration standards on a monthly basis.

Calibration standards are prepared weekly or biweekly in routine operations. Verify new calibration standards with those from the previous week to ensure consistency among the measurements. During instrument calibration, if any calibration point varies by more than $\pm 5\%$ of the specified value, recheck the system or prepare a new calibration standard to assure the accuracy of the analysis. Retain calibration data for reference and generate calibration curves for visual inspection. An example of a six-point calibration curve is displayed in Figure 3.

DATA PROCESSING AND DATA VALIDATION

In most commercially available instruments the peaks in the chromatograms are identified, and area counts are converted to concentrations in μg/mL as shown in Figure 4. Dilute and reanalyze samples with area counts above the highest standard, or group them with a higher concentration instrument calibration. Methods for chromatogram peak quantification, precision calculation, and data validation are documented in the following subsections.

Peak Quantification

Concentrations of the ionic species on the filter, as well as their precision estimates, can be calculated using a computer spreadsheet program according to the following steps:

- Determine the linear regression slope and intercept of the relation between standard concentration and detector response. The calibration equation is given by the following relationship:

$$\mu g/mL = \text{slope (integrator reading)} + \text{intercept} \qquad (6)$$

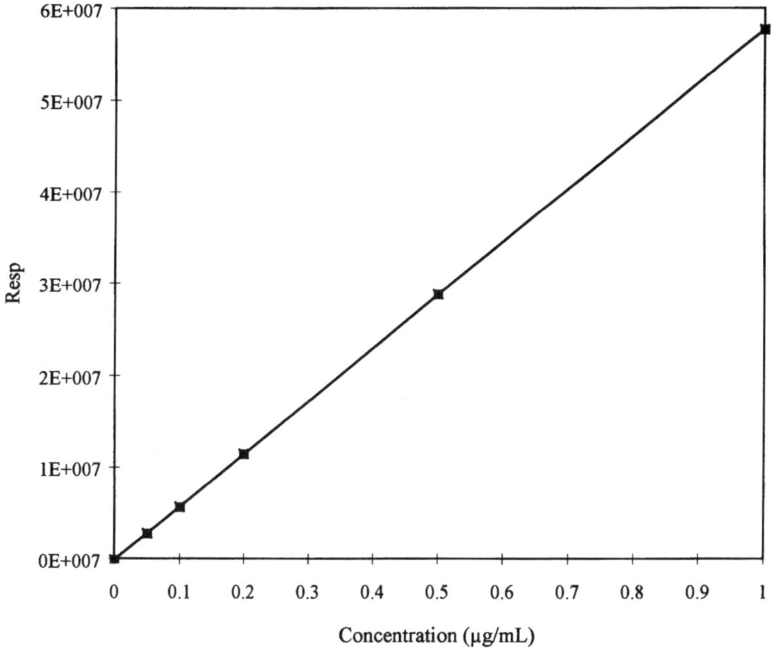

Figure 3. Example of ion chromatography calibration curve.
where:

r^2: 0.999590
Conc = Resp * 1.73E − 008 + 0.002335
Resp = Conc * 5.779E + 007 + − 1.349E + 005
Standardization: External
Calibration: Area

- Calculate concentrations of the sample extract corresponding to the area counts according to the calibration curve.
- Calculate the total amount of ionic species in each sample by applying the dilution and scaling factors:

$$\mu\text{g/filter} = \frac{(\mu\text{g/mL}) \times V_e \times D}{F} \qquad (7)$$

where:

V_e = volume of extract
D = sample extract dilution factor
F = fraction of filter area being analyzed

```
=============================================================================
| Sample Name: STD-F                        Date: Mon Oct 16 12:14:09 1989|
| Data File   : A:\1018T1.D04                                              |
| Method      : C:\DX\METHOD\ANION1.MET                                    |
| ACI Address: 1          System : 1        Inject#:   4     Detector: CDM-2 |
=============================================================================
**********************  EXTERNAL STANDARD REPORT  *************************

Stop time =  5.02 Minutes            Number of Data Points =  3010
Area reject = 100             One Data Point per 0.1 seconds
Amount Injected = 1          Dilution factor = 1.

PEAK  RET   PEAK              CONC. in                                REF   % DELT
NUM   TIME  NAME              ug/ml          AREA       HEIGHT  BL  PEAK  RET TI
-------------------------------------------------------------------------------
 1   1.55 CHLORIDE          4.775e-001  3.235e+007 6602913  1    1    0.00%
 2   1.85 NITRITE           4.885e-001  1.944e+007 3811841  1    2    0.00%
 3   3.17 NITRATE           4.914e-001  1.771e+007 2137115  1    3    0.00%
 4   4.20 SULFATE           4.908e-001  2.338e+007 2278689  1    4    0.00%
```

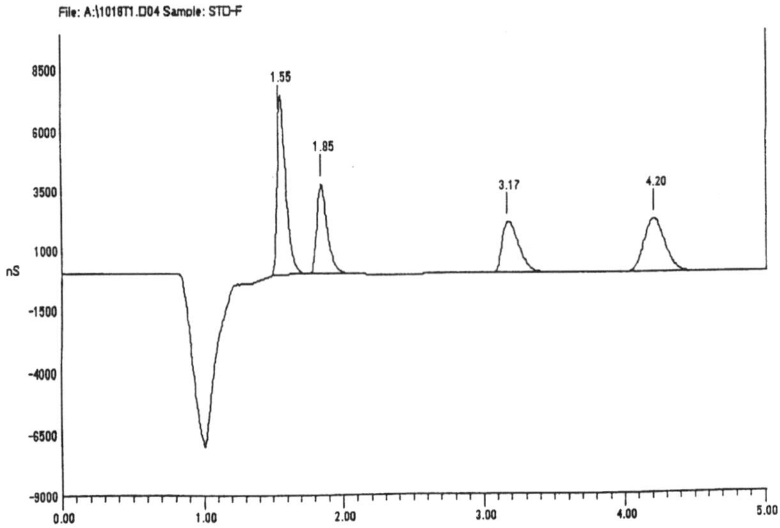

Figure 4. Example of ion chromatogram for calibration standard at 0.5 μg/mL.

and:

$$\text{Aerosol concentration } (\mu\text{g/m}^3) = \frac{\mu\text{g/filter}}{T(\text{min}) \times V_a(\text{L/min}) \times 10^3(\text{L/m}^3)} \quad (8)$$

where:

T = sampling period (minutes)
V_a = sampling flow rate (L/min)

Precision Estimates

Propagate aerosol precisions reported in the data bases from the precision of the replicated analysis and the field blank variability using the methods of Bevington (1969). The following equations describe the calculation of the precision propagation:

$$D_{M_{ij}} = M_{ijf} - M_{ijr} \tag{9}$$

$$\overline{D}_{M_i} = \frac{1}{n} \sum_{j=1}^{n} D_{M_{ij}} \tag{10}$$

$$\sigma_{M_j} = \left[\frac{1}{n} \sum_{j=1}^{n} (D_{m_{ij}} - \overline{D}_{M_i})^2 \right]^{1/2} \tag{11}$$

where:

M_{ij} = amount of species i on the sample j
M_{ijf} = amount of species i on the sample j from primary analysis
M_{ijr} = amount of species i on the sample j from replicate analysis
$D_{M_{ij}}$ = absolute difference between primary and replicate analyses
$\overline{D}_{M_j}$ = average difference between primary and replicate analyses on a batch of n samples
σ_{M_j} = precision of amount of species i on a batch of n samples

Always report a precision estimate with each ionic measurement to indicate the uncertainty of the measurement. While precision should be calculated for replicate pairs in several concentration bins, the absolute precisions often vary with concentration.

Data Validation

Chow *et al.* (1994a) define different levels of data validation to maintain the internal and external consistency of the field and laboratory measurements. For each batch of 50 to 100 ion chromatography analyses, a standard chain of custody as specified in Figure 5 is followed. Apply both Level I and II data validation to ensure the validity of the measurement (Chow *et al.*, 1994a, b).

For Level I data validation, examine each entry in the data base to verify the sample identification codes, dilution volume, filter exposure area, and filter size being analyzed. Mistakes are often made when entering dilution volume or filter area in the data base. Individually check chromatograms generated for each sample, along with their calibrated standard peaks, to make certain no anomalies exist. Difficulties encountered in the IC analysis include hardware (eluent delivery pump or electronic injection valve), system fluidics, or software (Koch, 1979b; Sunden *et al.*, 1988; Wang *et al.*, 1983). Instrument malfunctions are specific to each system. It is essential to refer to the operating manual and/or contact the manufacturer for troubleshooting advice. Measurement interferences often found in

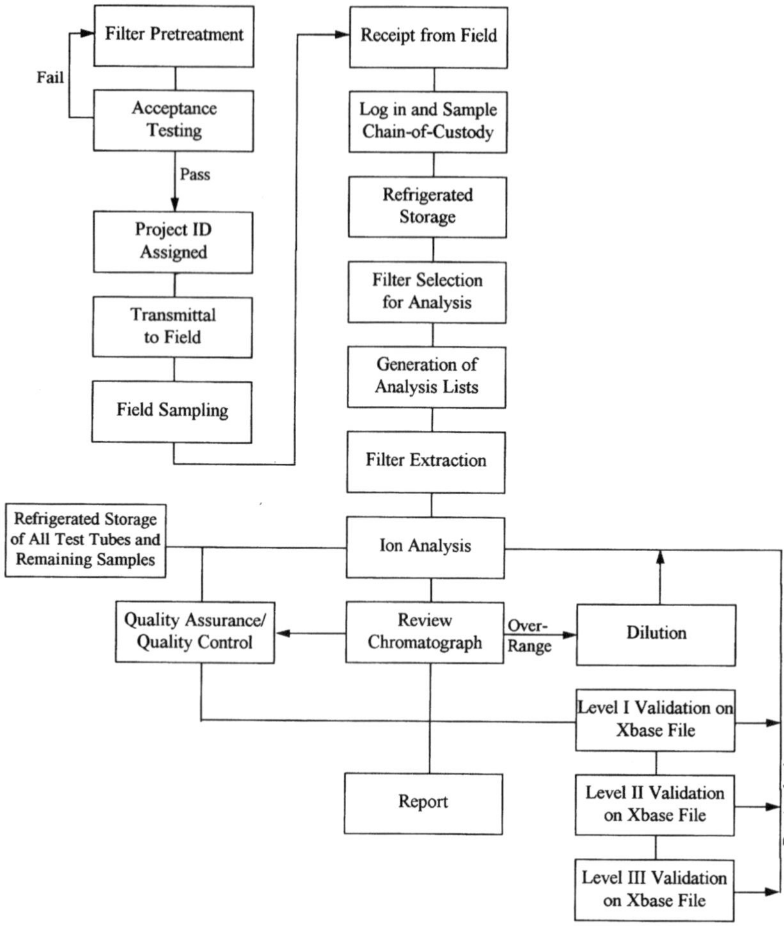

Figure 5. General sample chain of custody flow diagram.

IC analysis of airborne particulate samples are listed in Table 7. In each case, the symptom, cause, and solution are discussed.

The chloride ion elutes immediately after the deionized distilled water dip. This often results in a shift in the baseline or a distortion of the analyte peaks in the chromatogram, and an overestimation of the chloride in anion analysis. Manual adjustment of the baseline for each sample is often necessary. A drifting baseline is a sign of inadequate equilibration. Flushing eluent for 30 min at the beginning of each day will equilibrate the concentrator, guard, and separator column (Han and Koch, 1988). Unresolved peaks or co-eluting peaks may require adjustment of the

Table 7. Measurement interferences in ion chromatographic analysis.

Interference	Symptom	Cause	Solution
Water dip	Chromatogram	• Since the conductance of water is less than that of the suppressed eluent, peaks of ionseluting near the water dip (e.g., f^-, 0) are suppressed	Change eluent concentration, decrease eluent flow rate, increase the retention time, or manually adjust baseline and recalculate peak areas
Drifting baseline	Chromatogram contains drifting baseline over time	• Temperature, which affects the conductivity, is poorly controlled	Increase equilibration time, or use a column heater
Air spikes	Chromatogram baseline contains spikes	• Bubbles in columns, flow cell, or detector cell	Prevent air introduction during sample/ standard injection, equip the system with an "eluent Degas Module", or purge eluent with helium gas
Split peaks	Chromatogram contains split peaks	• Blockage or void formed in the column	Replace the front of the guard column (be careful not to disturb column packing) or the analytical column
Unresolved peaks	Chromatogram contains unexpected peaks	• The analyses were stopped or interrupted before all expected ions eluted	Ensure a stable baseline prior to sample injection
Co-eluting	Chromatogram contains unresolved adjacent peaks	• Analyte contains a higher concentration (e.g., 10–20 times higher) on one of the two adjacent peaks in the chromatogram	Reduce the sample loop, decrease the eluent concentration, or slow the eluent flow rate (this process will

Table 7. (continued).

Interference	Symptom	Cause	Solution
			lengthen the retention times)
Column overloding	Chromatogram contains non-quantifiable response	• Analyte contains ionic concentration exceeding 50 ppm • Sample loop volume greater than 250 mL	Dilute the sample aliquot prior to analysis
Retention time shift	The retention time in peak resolution of analyte ion is outside of the acceptable range	• Eluent is not prepared correctly • Accumulation of contaminants on the exchange resin • Deterioration in column performance	Prepare new eluent, clean the column resin, or replace the old column
High back pressure	Observed back Pressure is outside of the tolerant range (typical back pressure depends on flow rate, column type, and column age	• One of the system's fluid components is malfunctioning	Bypass each component in sequence until the problem is isolated

operating flow rate to increase retention times. Split peaks or shifts in retention times suggest degradation in column performance. In each case, resubmit samples or sets of samples for analysis after the system malfunction is resolved.

Level II data validation tests are applied when multiple ions are measured and independent analyses are performed on the air samples (Basta and Tabatabai, 1985; Butler *et al.*, 1979; Chow *et al.*, 1994a, b; 1996; Fung *et al.*, 1979; Kyriacou and Kempster, 1993; Lathous and Coutant, 1978; Mueller *et al.*, 1978; Peden *et al.*, 1986; Steiber and Merrill, 1979). For anion and cation analyses, ion balances can be made by summing all cation charges and comparing them with anion charges. Ammonium ion concentrations can be calculated based on the assumption that all the particulate nitrate is in the form of ammonium nitrate, and sulfate is in the form of sulfuric acid, ammonium sulfate, or ammonium bisulfate. Scatter plots are generated with their associated measurement uncertainties, and linear regression statistics are computed to establish measurement equivalence and identify deviations.

QUALITY CONTROL (QC) AND QUALITY ASSURANCE (QA)

Every measurement consists of a value, precision, accuracy, and validity. QC and quality auditing establishes the precision, accuracy, and validity of the measured values. QA integrates quality control and auditing, and the measurement process. The results of QA are data values with specified precisions, accuracies, and valdities.

Quality Control

QC is intended to prevent, identify, correct, and define the consequences of difficulties during IC analysis that might affect the precision, accuracy, and/or validity of the measurements. QC activities include:

(1) ensuring standard operating procedures (SOPs) are properly followed during chemical analysis and data processing
(2) performing regular standard calibrations
(3) analyzing a blank, standard, and replicate from a prior sample run with each set of 10 samples
(4) participating in quality audits and interlaboratory comparisons

Prior to beginning IC analysis, protocols for method development are formulated. Empirical experiments are essential to quantitative analysis. To incorporate many ionic species in a single sample injection, the analysis time may have to be increased for adequate peak separation.

External QA standards need to be purchased for independent checks of the internal calibration standards used in daily operations. Examples of concentration ranges for commonly measured ionic species are provided in Table 8. The certified simulated rain water standard, Standard Reference Method 2694-I, 2694-V from the National Institute for Standards and Technology (NIST) has consistent quality and serves as an external QA standard. The WasteWatR™ standard from Environmental Resource Associates (Arvada, Colorado) requires dilution by a factor of 200–500 for comparison with air samples, therefore introducing additional measurement uncertainties.

Replicate analysis is performed by random selection of one out of each set of 10 samples. Conduct this analysis on a different day or on a separate

Table 8. Examples of commercial quality control standards.

Constituent elements and parameters	Concentration unit	National institute of standards and technology[a] simulated rain water standards		Environmental research associates waste water quality control standards	
		SRM 2694-I	SRM 2694-II	ERT Certified	Advisory Range
Flouride	μg/mL	0.050 ± 0.002	0.098 ± 0.007	13.5[c]	11.6–15.4[c]
Chloride	μg/mL	(0.24)	1.0	243[c]	224–262[c]
Nitrate	μg/mL	NA	7.06 ± 0.15[b]	NA	NA
Nitrate plus Nitrite	μg/mL	NA	NA	10.4[d]	9.4–11.4[d]
Phosphate	μg/mL	NA	NA	4.0[d]	3.6–4.4[d]
Sulfate	μg/mL	2.73 ± 0.05	10.9 ± 0.2	249[c]	224–274[c]
Sodium	μg/mL	0.205 ± 0.009	0.419 ± 0.015	230[c]	207–253[c]
Magnesium	μg/mL	0.024 ± 0.002	0.051 ± 0.00	NA	NA
Potassium	μg/mL	0.052 ± 0.07	0.106 ± 0.008	230[c]	207–253[d]
Calcium	μg/mL	0.014 ± 0.003	0.049 ± 0.011	NA	NA
Ammonium	μg/mL	NA	1.0	NA	NA
Ammonia	μg/mL	NA	NA	7.9[d]	6.9–8.9[c]
Acidity	μg/mL	0.030 ± 0.002	0.284 ± 0.003	NA	NA
Conductivity at 29.0°C		26 ± 2	130 ± 2	NA	NA
pH at 25°C		4.27 ± 0.03	3.59 ± 0.02	9.1[c]	8.9–9.3[c]

[a]The certified values are based on proven reliable methods of analysis. The estimated uncertainties are 2 standard deviations of the certified values except for uncertainties associated with sulfate, acidity, pH, and specific conductance which are based on scientific judgment and are roughly equivalent to 2 standard deviations of the certified value.
[b]The nitrate value is not certified because of instability. It is believed that bacterial or fungal activity contributes to that instability.
[c]Minerals WasteWatR™.
[d]Nutrients WasteWatR™.

run to ensure the reproducibility of ionic concentrations from run to run and day to day. Replicate analysis performed immediately after primary analysis only confirms the stability of the instrument during the particular operation defeating the purpose of the technique. It usually results in reported precisions that are better than actual ones.

Standard Operating Procedures

SOPs codify the actions taken to implement a measurement process over a specified time period. Incorporate state of the art scientific information into the SOP with each revision. SOPs include the following:

- brief summary of the measurement method, principles of operation, expected accuracy and precision, and assumptions that must be met for it to be valid
- list of materials, equipment, regents, and suppliers. Specifications are given for each expendable item
- designation of individuals responsible for each part of the procedure
- general traceability path, designation of primary standards or reference materials, and their levels and tolerances
- start up, routine, and shut-down operating procedures, and abbreviated checklists
- copies of data forms with examples of filled out forms
- routine maintenance schedules and procedures, and troubleshooting tips
- internal calibration, performance testing procedures, and schedules
- external performance auditing schedules
- references to relevant literature and related SOPs

Quality Audits

The quality audit function consists of two components: systems and performance audits. Systems audits start with a review of the SOPs to assess whether or not they are adequate to assure valid data that meets specified levels of accuracy and precision. After reviewing the procedures, the auditor examines all phases of the measurement or data processing activity to determine if procedures are being followed and operational people are properly trained. The systems audit is intended to be a cooperative assessment resulting in improved data rather than a judgmental activity, and should be completed at least once per year.

Performance audits establish whether the predetermined specifications are being achieved in practice. It challenges the measurement/analysis system with a known standard sample that is traceable to a primary standard.

For data processing, the audit consists of independently processing selections of the data and comparing the results with those reported by routine analysis and data processing. The performance audit for IC analysis includes laboratory intercomparisons, and spiked sample analyses. This is accomplished by depositing known ionic concentrations on unexposed filters and comparing analysis results with spike values. It is recommended that these activities be conducted on a biannual basis.

SUMMARY AND CONCLUSIONS

IC has demonstrated its potential and capabilities in the analysis of airborne particles, and has been applied to environmental samples for more than 20 years. The advantages of this method are:

- rapid and simultaneous quantification of multiple ionic species in a single analysis run
- small sample volumes are required, 1 or 2 mL
- it provides high sensitivity with low minimum detection limits, variable selectivity, good stability, and quantitative detection in a wide range of concentrations — 0.01–100 μg/mL

Prior to beginning IC, protocols for method development are formulated. When incorporating many ionic species in a single sample injection, analysis times may have to be increased to ensure adequate peak resolution. Airborne particulate measurements need to be acquired on filter substrates with low background concentrations and high sampling efficiencies. Both sonication and mechanical agitation are recommended to ensure complete dissociation of the analyte ions from each other and from the filter. After sample extraction, the analyte requires additional gravitational settling of insoluble residues in a low temperature environment (i.e., refrigeration). Filtration is advised to minimize system clogging or contamination of the IC columns.

Standardization and instrument calibration at adequate concentration levels are essential for the accuracy of the ion analysis. Calibration intervals should not exceed two orders of magnitude between the lowest and highest calibration values. Five to six calibration points at approximately equivalent incremental intervals are used to establish a straight line calibration curve and external QA standards are analyzed to verify the accuracy of the internal standards. While the analysis might be rapid and automated, extensive chromatogram validation is required. Level I and Level II data validation, along with QC and QA procedures, provide accuracy, precision, and validity of each measurement. These procedures are an integral part of IC analysis.

The major limitation of IC analysis is its complexity. The technique while versatile has specific requirements:

- stringent operating conditions: eluent composition, flow rate, temperature, and pressure
- several separation mechanisms: ion exchange, ion exclusion, and ion pair
- array of column types: analytical, guard, concentrator, and suppressor columns
- column properties: capacity, particle or latex diameter, substrate material, functional group, solvent stability, and hydrophobicity
- potentials for interference or contamination
- detection methods: electrochemical versus spectroscopic

These variables need to be individually or collectively tested to achieve optimal resolution.

ACKNOWLEDGEMENTS

The authors are grateful to Dr. Kochy Fung (AtmAA, Calabasas, California) for reviewing the manuscript and supplying valuable technical advice, and thank Mr. Norman Mankim (Desert Research Institute, Reno, Nevada) for assembling and editing the manuscript.

REFERENCES

Appel, B.R. and W.J. Wehrmeister, "An Evaluation of Sulfate Analyses of Atmospheric Samples by Ion Chromatography", in *Ion Chromatographic Analysis of Environmental Pollutants*, E. Sawicki, J.D. Mulik and E. Wittgenstein, eds., Ann Arbor Science, Ann Arbor, MI (1979).

Appel, B.R., "Studies in Atmospheric Particulate Characterization Techniques", in *Air/Particulate Instrumentation and Analysis*, P.N. Cheremisinoff, ed., Ann Arbor Science, Ann Arbor, MI (1981), pp. 25–87.

Appel, B.R., Y. Tokiwa, E.L. Kothny, R. Wu and V. Povard, *Atmos. Environ.*, **22** (1988) 1565.

Appel, B.R., "Atmospheric Sample Analysis and Sampling Artifacts", in *Aerosol Measurement — Principles, Techniques, and Applications*, K. Willeke and P.A. Baron, eds. (1994) pp. 233.

Arar, E.J., S.E. Long, T.D. Martin and S. Gold, *Environ. Sci. Technol.*, **26** (1992) 1944.

ASTM, "Part 31 — Definitions of Terms Related to Water — Standard D 1129-82b", in *Annual Book of ASTM Standards* (1982), p. 4.

Bachman, S.R., J.E. Rothert, B. Kaiser, C.J. Brennan and M.A. Peden, *Method 300.7 — Dissolved Sodium, Ammonium, Potassium, Magnesium, and Calcium in Wet Deposition by Chemically Suppressed Ion Chromatography,* Report prepared for the US EPA, Cincinnati, OH, by the Illinois State Water Survey, Analytical Chemistry Unit, Champaign, IL. March 1986.

Barkley, N.C., G.L. Contner and M. Malanchuk, "Simultaneous Analysis of Anions and Cations in Diesel Exhaust using Ion Chromatography", in *Ion Chromatographic Analysis of Environmental Pollutants, Vol. 2*, J.D. Mulik and E. Sawicki, eds., Ann Arbor Science, Ann Arbor, MI (1979).

Basta, N.T. and M.A. Tabatabai, *J. Environ. Qual.*, **14** (1985) 450.

Bergin, M.H., J.L. Jaffrezo, C.I. Davidson, J.E. Dibb, S.N. Pandis, R. Hillamo, W. Maenhaut, H.D. Kuhns and T. Makela, *J. Geophys. Res.*, **100** (1995) 16275.

Bevington, P.R., *Data Reduction and Error Analysis for the Physical Sciences*, McGraw-Hill, New York, NY (1969).

Bogen, D.C. and S.J. Nagourney, "Ion Chromatographic Analysis of Cations at Baseline Precipitation Stations", in *Ion Chromatographic Analysis of Environmental Pollutants, Vol. 2*, J.D. Mulik and E. Sawicki, eds., Ann Arbor Science, Ann Arbor, MI (1979).

Buck, C.F., P.A. Mayewski, M.J. Spencer, S. Whitlow, M.S. Twickler and D. Barrett, *J. Chromatography*, **594** (1992) 225.

Buhr, S.M., M.P. Buhr, F.C. Fehsenfeld, J.S. Holloway, U. Karst, R.B. Norton, D.D. Parrish and R.E. Sievers, *Atmos. Environ.*, **19** (1995) 2609.

Butler, F.E., R.H. Jungers, L.F. Porter, A.E. Riley and F.J. Toth, "Analysis of Air Particulates by Ion Chromatography: Comparison with Accepted Methods", in *Ion Chromatographic Analysis of Environmental Pollutants*, E. Sawicki, J.D. Mulik, and E. Wittgenstein, eds., Ann Arbor Science, Ann Arbor, MI (1978).

Butler, F.E., F.J. Toth, D.J. Driscoll, J.N. Hein and R.H. Jungers, "Analysis of Fuels by Ion Chromatography: Comparison with ASTM Standards", in *Ion Chromatographic Analysis of Environmental Pollutants, Vol. 2*, J.D. Mulik and E. Sawicki, eds., Ann Arbor Science, Ann Arbor, MI (1979).

Cameron, A. and C. Pohlandt, *The Simultaneous Determination of Common Anions by Ion Chromatography: Laboratory Method No. 9/22*, Report No. M222D, Prepared by Mintek for the Council for Mineral Technology, Randburg, South Africa (1987).

Chow, J.C., *J. Air & Waste Manage. Assoc.*, **45** (1995) 320.

Chow, J.C. and J.G. Watson, "Ambient Air Sampling", in *Aerosol Measurement: Principles, Techniques and Applications*, K. Willeke, ed., Van Nostrand Reinhold, New York, NY (1993), pp. 622–639.

Chow, J.C. and J.G. Watson, "Particle and Gas Measurements on Filters", in *Sampling of Environmental Materials for Trace Analysis*, B. Markert, ed., VCH, Weinheim, Federal Republic of Germany (1994).

Chow, J.C., C.S. Liu, J. Cassmassi, J.G. Watson, Z. Lu and L.C. Pritchett, *Atmos. Environ.*, **26A** (1992) 693.

Chow, J.C., J.G. Watson, J.L. Bowen, A.W. Gertler, C.A. Frazier, K.K. Fung and L. Ashbaugh, "A Sampling System for Reactive Species in the Western U.S." in *Sampling and Analysis of Airborne Pollutants*, E. Winegar and L.H. Keith, eds., American Chemical Society, Washington, DC (1993), pp. 209-228.

Chow, J.C., E.M. Fujita, J.G. Watson, Z. Lu and D.R. Lawson, *Environ. Monitoring and Assessment*, **30** (1994a) 49.

Chow, J.C., J.G. Watson, E.M. Fujita, Z. Lu, D.R. Lawson and L.L. Ashbaugh, *Atmos. Environ.*, **28** (1994b) 2061.

Chow, J.C., D. Fairley, J.G. Watson, R. DeMandel, E.M. Fujita, D.H. Lowenthal, Z. Lu, C.A. Frazier, G. Long and J. Cordova, *J. Environ. Engineering*, **21** (1995) 378.

Chow, J.C., J.G. Watson, Z. Lu, D.H. Lowenthal, C.A. Frazier, P.A. Solomon, R.H. Thuillier and K. Magliano, *Atmos. Environ.*, **30** (1996) 2079.

Colenutt, B.A. and P. Trenchard, *Environ. Pollut.* (*Series B*), **10** (1985) 77.

Dempsey, J.H., P. Cruse and K. Yates, "The Analysis of Ions in Flue Gas Desulfurization Systems by Ion Chromatography", in *Ion Chromatographic Analysis of Environmental Pollutants, Vol. 2*, J.D. Mulik and E. Sawicki, eds., Ann Arbor Science, Ann Arbor, MI (1979).

Durst, R.A., W. Davison, K. Toth, J.E. Rothert, M.E. Peden and B. Griepink, *Pure & Appl. Chem.*, **63** (1991) 907.

Fratz, D.D., "Quantitative Determination of Inorganic Salts in Certifiable Color Additives", in *Ion Chromatographic Analysis of Environmental Pollutants*, E. Sawicki, J.D. Mulik and E. Wittgenstein, eds., Ann Arbor Science, Ann Arbor, MI (1978).

Fritz, J.S., D.T. Gjerde and C. Pohlandt, *Ion Chromatography*, Dr. Alfred Hüthig Verlag, Heidelberg, Germany (1982).

Fulmer, M.A., J. Penkrot and R.J. Nadalin, "Sodium, Potassium, Chloride, and Sulfate Analysis at PPB Levels in Water", in *Ion Chromatographic Analysis of Environmental Pollutants, Vol. 2*, J.D. Mulik and E. Sawicki, eds., Ann Arbor Science, Ann Arbor, MI (1979).

Fung, K.K., S.L. Heisler, A. Price, B.V. Nuesca and Pp.K. Mueller, "Comparison of Ion Chromatography and Automated Wet Chemical Methods for Analysis of Sulfate and Nitrate in Ambient Particulate Filter Samples", in *Ion Chromatographic Analysis of Environmental Pollutants, Vol. 2*, J.D. Mulik and E. Sawicki, eds., Ann Arbor Science, Ann Arbor, MI (1979).

Gjerde, D.T., J.S. Fritz and G. Schmuckler, *J. Chromatography*, **186** (1979) 509.

Glaser, J., D. Foerst, G. McKee, S. Quave and W. Budde, *Environ. Sci. Technol.*, **15** (1981) 1426.

Haddad, P.R., *Ion Chromatography — Principles and Applications*, Elsevier, Amsterdam, Netherlands (1994).

Han, K. and W.F. Koch, *LC-GC*, **6** (1988) 56.

Hansen, L.D., B.E. Richter, D.K. Rollins, J.D. Lamb and D.J. Eatough, *Anal. Chem.*, **51** (1979) 633.

Hering, S.V., D.R. Lawson, I. Allegrini, A. Febo, C. Perrino, M. Possanzini, J.E. Sickles II, K.G. Anlauf, A. Wiebe, B.R. Appel, W. John, J. Ondo, S. Wall, R.S. Braman, R. Sutton, G.R. Cass, P.A. Solomon, D.J. Eatough, E.C. Ellis, D. Grosjean, B.B. Hicks, J.D. Womack, J. Horrocks, K.T. Knapp, T.G. Ellestad, R.J. Paur, W.J. Mitchell, M. Pleasant, E. Peake, A. MacLean, W.R. Pierson, W. Brachaczek, H.I. Schiff, G.I. Mackay, C.W. Spicer, D.H. Stedman, A.M. Winer, H.W. Biermann and E.C. Tuazon, *Atmos. Environ.*, **22** (1988) 1519.

Hoke, S.H. and L.J. Baxter, *Automated Ion Chromatography*, Technical Report 8501, Document No. AD-A159-764, U.S. Army Medical Bioengineering Research & Development Laboratory, Fort Detrick, Frederick, MD (1985).

Holcombe, L.J., B.F. Jones, E.E. Ellsworth and F.B. Meserole, "The Quantitative Determination of Aqueous Solutions of Sulfite, Sulfate, and Thiosulfate using the Ion Chromatograph", in *Ion Chromatographic Analysis of Environmental Pollutants, Vol. 2*, J.D. Mulik and E. Sawicki, eds., Ann Arbor Science, Ann Arbor, MI (1979).

Holm, R.D. and S.A. Barksdale, "Analysis of Anions in Combustion Products", in *Ion Chromatographic Analysis of Environmental Pollutants*, E. Sawicki, J.D. Mulik and E. Wittgenstein, eds., Ann Arbor Science, Ann Arbor, MI (1978)

Hoover, T.B. and G.D. Yager, *Anal. Chem.*, **56** (1984) 221.

Jantzen, C.M., *The Determination of* Fe^{2+}/Fe^{3+} *Ratio in Simulated Nuclear Waste Glass by Ion Chromatography (U), Document No. WSRC-TR-90-528 Rev. 1, DE91-009296,* Report prepared for the U.S. Dept. of Energy, Washington, DC, by Westinghouse Savannah River Co., Aiken, SC, under Contract No. DE-AC09-89SR18035 (1990).

Kapelner, S.M., J.C. Trocciola and M.S. Freed, "Trace Level Analysis of Anions in High-Purity Water", in *Ion Chromatographic Analysis of Environmental Pollutants, Vol. 2*, J.D. Mulik and E. Sawicki, eds., Ann Arbor Science, Ann Arbor, MI (1979).

Karmarkar, S.V., "Environmental IC Analysis — Operation and Problem Solving", in *Water Environment Laboratory Solutions, Vol. 2,* Lachat Instruments, Milwaukee, WI (1995), p. 6.

Kim, W.S., C.L. Geraci Jr. and R.E. Kupel, "Sampling and Analysis of Formaldehyde in the Industrial Atmosphere", in *Ion Chromatographic Analysis of Environmental Pollutants, Vol. 2*, J.D. Mulik and E. Sawicki, eds., Ann Arbor Science, Ann Arbor, MI (1979).

Kitsa, V., P.J. Lioy, J.C. Chow, J.G. Watson, S. Shupack, T. Howell and P. Sanders, *Aerosol Sci. Technol.*, **17** (1992) 213.

Koch, W.F., *Anal. Chem.*, **51** (1979a) 1573.

Koch, W.F., *J. Research of the National Bureau of Standards,* **84** (1979b) 241.
Koch, W.F. and W.S. Liggett, "Critical Assessment of Detection Limits for Ion Chromatography", in *Detection in Analytical Chemistry — Importance, Theory, and Pracitece*, ACS Symposium Series 361, Chapt. 11., American Chemical Society, Washington, DC (1988).
Kyriacou, P. and P.L. Kempster, *Water SA*, **19** (1993) 339.
Lash, R.P. and C.J. Hill, "Ion Chromatographic Determination of Dibutylphosphoric Acid in Nuclear Fuel Reprocessing Streams", in *Ion Chromatographic Analysis of Environmental Pollutants, Vol. 2*, J.D. Mulik and E. Sawicki, eds., Ann Arbor Science, Ann Arbor, MI (1979).
Lathouse, J. and R.W. Coutant, "Practical Experience on the Use of Ion Chromatography for Determination of Anions in Filter Catch Samples", in *Ion Chromatographic Analysis of Environmental Pollutants*, E. Sawicki, J.D. Mulik and E. Wittgenstein, eds., Ann Arbor Science, Ann Arbor, MI (1978).
Lathouse, J. and R.E. Heffelfinger, "Analytical Considerations for the Ion Chromatograph", in *Ion Chromatographic Analysis of Environmental Pollutants, Vol. 2*, J.D. Mulik and E. Sawicki, eds., Ann Arbor Science, Ann Arbor, MI (1979).
Legrand, M., M. de Angelis and R.J. Delmas, *Analytica Chimica Acta*, **156** (1984) 181.
Lodge, J.P., Jr., *Methods of Air Sampling and Analysis, 3rd Edition*, Lewis Publishers, Chelsea, MI (1989).
Mansfield, G.H., "Ion Analysis of Chemical Plant Control and the Effect on Effluents" in *Ion Chromatographic Analysis of Environmental Pollutants, Vol. 2*, J.D. Mulik and E. Sawicki, eds., Ann Arbor Science, Ann Arbor, MI (1979).
Mason, D.W. and H.C. Miller, "Measurement of Ambient Sulfuric Acid Aerosol with Analysis by Ion Chromatography" in *Ion Chromatographic Analysis of Environmental Pollutants, Vol. 2*, J.D. Mulik and E. Sawicki, eds., Ann Arbor Science, Ann Arbor, MI (1979).
Mizisin, C.S., D.E. Kuivinen and D.A. Otterson, "Ion Chromatographic Determination of Sulfur in Fuels", in *Ion Chromatographic Analysis of Environmental Pollutants, Vol. 2*, J.D. Mulik and E. Sawicki, eds., Ann Arbor Science, Ann Arbor, MI (1979).
Mueller, P.K., B.V. Mendoza, J.C. Collins and E.S. Wilgus, "Application of Ion Chromatography to the Analysis of Anions Extracted from Airborne Particulate Matter", in *Ion Chromatographic Analysis of Environmental Pollutants*, E. Sawicki, J.D. Mulik and E. Wittgenstein, eds., Ann Arbor Science, Ann Arbor, MI (1978).
Mulik, J.D. and E. Sawicki, *Environ. Sci. Technol.*, **13** (1979) 804.
Mulik, J.D., E. Estes and E. Sawicki, "Ion Chromatographic Analysis of Ammonium Ion in Ambient Aerosols", in *Ion Chromatographic Analysis of Environmental Pollutants*, E. Sawicki, J.D. Mulik and E. Wittgenstein, eds., Ann Arbor Science, Ann Arbor, MI (1978a).
Mulik, J.D., G. Todd, E. Estes, R. Puckett, E. Sawicki and D. Williams, "Ion Chromatographic Determination of Atmospheric Sulfur Dioxide", in *Ion Chromatographic Analysis of Environmental Pollutants*, E. Sawicki, J.D. Mulik and E. Wittgenstein, eds., Ann Arbor Science, Ann Arbor, MI (1978b).
NIOSH, "Material safety data sheet for Sulfur Dioxide (6004)", in *NIOSH Manual of Analytical Methods* (1994a).
NIOSH, "Material safety data sheet for Chlorine (6011)", in *NIOSH Manual of Analytical Methods* (1994b).
NIOSH, "Material safety data sheet for Inorganic Acids (7903)", in *NIOSH Manual of Analytical Methods* (1994c).
Noller, B.N. and N.A. Currey, *Wat. Res.*, **24** (1990) 471.
Otterson, D.A., "Ion Chromatographic Determination of Anions Collected on Filters at Altitudes between 9.6 and 13.7 Kilometers (1978)", in *Ion Chromatographic Analysis of Environmental Pollutants*, E. Sawicki, J.D. Mulik and E. Wittgenstein, eds., Ann Arbor Science, Ann Arbor, MI (1978).
Papoff, P., M. Onor and M. Betti, *Intern. J. Environ. Anal. Chem.*, **55** (1994) 149.
Peden, M.E., S.R. Bachman, C.J. Brennan, B. Demir, K.O. James, B.W. Kaiser, J.M. Lockard, J.E. Rothert, J. Sauer, L.M. Skowron and M.J. Slater, *Development of*

Standard Methods for the Collection and Analysis of Precipitation, Document No. EPA/600/4-86/024, U.S. EPA, Inorganic Analysis Section, Physical and Chemical Methods Branch, Environmental Monitoring and Support Laboratory, Office of Research and Development, Cincinnati, OH (1986).

Pfaff, J.D., C.A. Brockhoff and J.W. O'Dell, *The Determination of Inorganic Anions in Water by Ion Chromatography — Method 300.0.,* Report prepared for the U.S. EPA, Environmental Monitoring and Systems Laboratory, Cincinnati, OH (1991).

Pickerell, D., T. Hook, T. Dolzine and J.K. Robertson, "Intensity-Weighted Sequential Sampling of Precipitation: A Technique for Monitoring Changes in Storm Chemistry during a Storm", in *Ion Chromatographic Analysis of Environmental Pollutants, Vol. 2*, J.D. Mulik and E. Sawicki, eds., Ann Arbor Science, Ann Arbor, MI (1979).

Pinschmidt, R.K., "Ion Chromatographic Analysis of Weak Acid Ions Using Resistivity Detection", in *Ion Chromatographic Analysis of Environmental Pollutants, Vol. 2*, J.D. Mulik and E. Sawicki, eds., Ann Arbor Science, Ann Arbor, MI (1979).

Pohlandt, C. and T.W. Steele, *The Separation and Determination of Anions by Ion Chromatography,* Report No. 2044 prepared by the National Institute for Metallurgy, Analytical Chemistry Division, Randburg, South Africa (1980).

Poole, C.F. and S.K. Poole, *Ion Chromatography*, Elsevier, Amsterdam, Netherlands (1991).

Posner, R.D. and A. Schoffman, "Altering the Dynamic Range of IC Columns by Prior Chemical Treatment of the Samples", in *Ion Chromatographic Analysis of Environmental Pollutants, Vol. 2*, J.D. Mulik and E. Sawicki, eds., Ann Arbor Science, Ann Arbor, MI (1979).

Pyen, G.S. and M.J. Fishman, "Determination of Anions in Pore Waters from Cores by Ion Chromatography", in *Ion Chromatographic Analysis of Environmental Pollutants, Vol. 2*, J.D. Mulik and E. Sawicki, eds., Ann Arbor Science, Ann Arbor, MI (1979).

Rawa, J.A., "Application of Ion Chromatography to Analysis of Industrial Process Waters", in *Ion Chromatographic Analysis of Environmental Pollutants, Vol. 2*, J.D. Mulik and E. Sawicki, eds., Ann Arbor Science, Ann Arbor, MI (1979).

Rich, W.E., *Analysis Instrumentation*, **15** (1977) 113.

Rich, W.E., J.A. Tillotson and R.C. Chang, "Ion Chromatography: An Analytical Perspective", in *Ion Chromatographic Analysis of Environmental Pollutants*, E. Sawicki, J.D. Mulik and E. Wittgenstein, eds., Ann Arbor Science, Ann Arbor, MI (1978).

Rich, W.E., F. Smith Jr., L. McNeil and T. Sidebottom, "Ion Exclusion Coupled to Ion Chromatography: Instrumentation and Application", in *Ion Chromatographic Analysis of Environmental Pollutants, Vol, 2*, J.D. Mulik and E. Sawicki, eds., Ann Arbor Science, Ann Arbor, MI (1979).

Rieman, W., III and H.F. Walton, Ion Exchange in Analytical Chemistry, Pergamon Press, Oxford, United Kingdom (1970).

Robards, K., P.R. Haddad and P.E. Jackson, *Principles and Practices of Modern Chromatography Methods*, Academic Press, Harcourt Brace & Co., Sydney, Australia (1994).

Rocklin, R.D., C.A. Pohl and J.A. Schibler, *J. of Chromatography,* **41** (1987) 107.

Rossner, B., J. Behnert and A. Kipplinger, *Fresenius Z Anal. Chem.,* **327** (1987) 698.

Samuelson, O., *Ion Exchange Separations in Analytical Chemistry*, John Wiley & Sons, New York, NY (1963).

Sawicki, E., "Potential of Ion Chromatography", in *Ion Chromatographic Analysis of Environmental Pollutants*, E. Sawicki, J.D. Mulik and E. Wittgenstein, eds., Ann Arbor Science, Ann Arbor, MI (1978).

Sawicki, E., "Application of Ion Chromatography to the Analysis of Environmental Genotoxicants", in *Ion Chromatographic Analysis of Environmental Pollutants, Vol, 2*, J.D. Mulik and E. Sawicki, eds., Ann Arbor Science, Ann Arbor, MI (1979).

Schiff, L.J., S.G. Pleva and E.W. Sarver, "Analysis of Phosphonic Acids by Ion Chromatography", in *Ion Chromatographic Analysis of Environmental Pollutants, Vol. 2*, J.D. Mulik and E. Sawicki, eds., Ann Arbor Science, Ann Arbor, MI (1979).

Sickles, J.E., II, C. Perrino, I. Allegrini, A. Febo, M. Possanzini and R.J. Paur, *Measurement of Selected Inorganic Air Pollutants Near Los Angeles Using an Annular Denuder System*,

Paper read at the EPA/APCA Symposium on Measurement of Toxic Air Pollutants, Raleigh, NC (1986).
Simon, P.K. and P.K. Dasgupta, *Anal. Chem*, **65** (1993) 1134.
Simon, P.K. and P.K. Dasgupta, *Environ. Sci. Technol.*, **29** (1995a) 1534.
Simon, P.K. and P.K. Dasgupta, *Anal. Chem.*, **67** (1995b) 71.
Simon, P.K., P.K. Dasgupta and Z. Večeřa, *Anal. Chem.*, **63** (1991) 1237.
Sinner, T., P. Hoffmann and H.M. Ortner, *Spectrochimica Acta,* **48B** (1993) 255.
Slanina, J., W.A. Ingerak, J.E. Ordelman, Pp. Borst and F.Pp. Bakker, "Automation of the Ion Chromatograph and the Adaptation for Rainwater Analysis", in *Ion Chromatographic Analysis of Environmental Pollutants, Vol. 2*, J.D. Mulik and E. Sawicki, eds., Ann Arbor Science, Ann Arbor, MI (1979).
Small, H., "An Introduction to Ion Chromatography", in *Ion Chromatographic Analysis of Environmental Pollutants*, E. Sawicki, J.D. Mulik and E. Wittgenstein, eds., Ann Arbor Science, Ann Arbor, MI (1978).
Small, H., *Ion Chromatography*, Plenum Press, New York, NY (1989).
Small, H. and T.E. Miller, Jr., *Anal. Chem.,* **54** (1982) 462.
Small, H., T.S. Stevens and W.C. Bauman, *Anal. Chem.*, **47** (1975) 1801.
Smith, F.C., Jr. and R.C. Chang, *The Practice of Ion Chromatography*, John Wiley & Sons, New York, NY (1983).
Snyder, L.R. and J.J Kirkland, *Introduction to Modern Liquid Chromatography*, John Wiley & Sons, New York, NY (1983).
Speitel, L.C., J.C. Spurgeon and R.A. Filipczak, "Ion Chromatographic Analysis of Thermal Decomposition Products of Aircraft Interior Materials", in *Ion Chromatographic Analysis of Environmental Pollutants, Vol. 2*, J.D. Mulik and E. Sawicki, eds., Ann Arbor Science, Ann Arbor, MI (1979).
Steiber, R. and R. Merrill, "Application of Ion Chromatography to the Analysis of Source Assessment Samples", in *Ion Chromatographic Analysis of Environmental Pollutants, Vol. 2*, J.D. Mulik and E. Sawicki, eds., Ann Arbor Science, Ann Arbor, MI (1979).
Steiber, R. and R.M. Statnick, "Application of Ion Chromatography to Stationary Source and Control Device Evaluation Studies", in *Ion Chromatographic Analysis of Environmental Pollutants*, E. Sawicki, J.D. Mulik and E. Wittgenstein, eds., Ann Arbor Science, Ann Arbor, MI (1978).
Stevens, R.K. and T.G. Dzubay, *Atmos. Environ.*, **12** (1978) 55.
Sunden, T., M. Lindgren, A. Cedergren and D.D. Siemer, *Anal. Chem.*, **55** (1988) 2.
Tabatabai, M.A. and W.A. Dick, "Ion Chromatographic Analysis of Sulfate and Nitrate in Soils", in *Ion Chromatographic Analysis of Environmental Pollutants, Vol. 2*, J.D. Mulik and E. Sawicki, eds., Ann Arbor Science, Ann Arbor, MI (1979).
Tarter, J.G., *Anal. Chem.*, **56** (1984) 1264.
Tejada, S.B., R.B. Zweidinger, J.E. Sigsby Jr. and R.L. Bradow, "Modification of an Ion Chromatograph for Automated Routine Analysis: Applications to Mobile Source Emissions", in *Ion Chromatographic Analysis of Environmental Pollutants*, E. Sawicki, J.D. Mulik and E. Wittgenstein, eds., MI, Ann Arbor Science, Ann Arbor (1978).
Tswett, M., *Ber. Deut. Botan. Ges.*, **24** (1906) 385.
Tyree, S.Y., Jr., J.M. Stouffer and M. Bollinger, "Ion Chromatographic Analysis of Simulated Rainwater", in *Ion Chromatographic Analysis of Environmental Pollutants, Vol. 2*, J.D. Mulik and E. Sawicki, eds., Ann Arbor Science, Ann Arbor, MI (1979).
U.S. Environmental Protection Agency, *Ion Chromatography of Anions,* Document No. EPA-600/4-80-020. U.S. EPA, Environmental Research Laboratory, Athens, GA (1980).
U.S. Environmental Protection Agency, *Method 300.7 — Sodium, Ammonium, Potassium, Magnesium, and Calcium in Wet Deposition by Chemically Suppressed Ion Chromatography*, Environmental Monitoring and Support Laboratory, Cincinnati, OH (1985).
U.S. Environmental Protection Agency, *Method 300.6 — Chloride, Orthophosphate, Nitrate, and Sulfate in Wet Deposition by Chemically Suppressed Ion Chromatography*, Environmental Monitoring and Support Laboratory, Cincinnati, OH (1986a).
U.S. Environmental Protection Agency, *Method 300.7 — Dissolved Sodium, Ammonium, Potassium, Magnesium, and Calcium in Wet Deposition by Chemically Suppressed Ion*

Chromatography, U.S. EPA, Inorganic Analysis Section, Physical and Chemical Methods Branch, Office of Research and Development, Environmental Monitoring and Support Laboratory, Cincinnati, OH (1986b).

U.S. Environmental Protection Agency, *Method 300.0 — Test Method: The Determination of Inorganic Anions in Water by Ion Chromatography,* U.S. EPA, Environmental Monitoring and Systems Laboratory, Cincinnati, OH (1991).

U.S. Environmental Protection Agency, Methods for Determination of Metals in Environmental Samples, Supplement I. Document No. EPA/600/R-94/111. U.S. EPA, Office of Research and Development, Washington, DC (1994).

U.S. Environmental Protection Agency, *Method 1636: Determination of Hexavalent Chromium by Ion Chromatography*, Document No. EPA/821-R-95-029, U.S. EPA, Office of Water, Research Triangle Park, NC (1995).

Van Os, M.J., *Ion Chromatography of Inorganic Trace Constituents in Water*, Document No. ECN-204, PB89-227706, Netherlands Energy Research Foundation ECN, Petten (1987).

Vecera, Z. and P.K. Dasgupta, *Environ. Sci. Technol.*, **25** (1991a) 255.

Vecera, Z. and P.K. Dasgupta, *Anal. Chem.*, **63** (1991b) 2210.

Walton, H.F. and R.D. Rocklin, *Ion Exchange in Analytical Chemistry*, CRC Press, Boca Raton, FL (1990).

Wang, C.-Y., S.D. Bunday and J.G. Tartar, *Anal. Chem.*, **55** (1983) 1617.

Watson, J.G., J.C. Chow, R.T. Egami, J.L. Bowen, C.A. Frazier, A.W. Gertler, D.H. Lowenthal and K.K. Fung, *Measurement of Dry Deposition Parameters for the California Acid Deposition Monitoring Program, Final Report*, Document No. 8068.2F1, prepared for the California Air Resources Board, Sacramento, CA, by the Desert Research Institute, Reno, NV, under Contract No. A6-076-32 (1991).

Watson, J.G., D. Blumenthal, J. Chow, C. Cahill, L.W. Richards, D. Dietrich, J. Houck, R.J. Dickson, S. Andersen and R. Morris, *Mt. Zirkel Wilderness Area Reasonable Attribution Study of Visibility Impairment —Vol. II: Results of Data Analysis and Modeling — Final Report*, Prepared for the Technical Steering Committee, Colorado Department of Public Health and Environment, Air Pollution Control Division, Denver, CO, by the Desert Research Institute, Reno, NV (1996).

Weiss, J., *Ion Chromatography*, 2nd Edition, VCH Verlagsgesellschaft mbH, Weinheim, Federal Republic of Germany (1995).

Wetzel, R.A., C.L. Anderson, H. Schleicher and G.D. Crook, *Anal. Chem.*, **51** (1979) 532.

Willard, H.H., L.L. Merritt, Jr., J.A. Dean and F.A. Settle, Jr., *Instrumental Methods of Analysis,* 7th Edition, Wadsworth Pub. Co., Belmont, CA (1988).

Zweidinger, R.B., S.B. Tejada, J.E. Sigsby Jr. and R.L. Bradow, "Application of Ion Chromatography to the Analysis of Ammonia and Alkylamines in Automobile Exhaust", in *Ion Chromatographic Analysis of Environmental Pollutants*, E. Sawicki, J.D. Mulik and E. Wittgenstein, eds., Ann Arbor Science, Ann Arbor, MI (1978).

4 ACCELERATOR BASED ION BEAM TECHNIQUES FOR TRACE ELEMENT AEROSOL ANALYSIS

D.D. COHEN

Australian Nuclear Science and Technology Organisation,
Australia Menai, New South Wales 2234, Australia

INTRODUCTION

During the past twenty years there has been a steady increase in the number and the sophistication of multi-elemental analysis techniques being applied to atmospheric aerosol studies. These techniques include neutron activation analysis (NAA), X-ray fluorescence (XRF), atomic absorption spectroscopy (AAS), inductively coupled plasma mass spectrometry (ICP-MS) and accelerator based ion beam analysis (IBA). The principles, instrumentation, and methodological aspects of several of these techniques as applied to atmospheric aerosol studies have been reviewed by Johansson and Campbell (1988), Maenhaut (1990) and in other chapters of this book.

Typically the accelerator based ion beam analysis work is performed on particle accelerators at bombarding proton or alpha energies of two to five million electronvolt (MeV), with three MeV protons being the most popular. Light ions of these energies produce sufficient X-ray, gamma-ray, or scattered particle yields for analysis with only a few nanoamps of beam current being required. Currents as small as this correspond to power densities for MeV ions of only a few mW/mm^2. This leaves the sample fully intact for further analyses by other techniques and is the main reason for the non-destructive nature of ion beam analysis techniques. Generally, these ion beam techniques are fast, non destructive, require minimal sample preparation, and can produce a multi-elemental analyses in just a few minutes of machine running time. These techniques are capable of analysing each filter for the following commonly occurring elements, H, C, N, O, F, Na, Mg, Al, Si, and most elements up to uranium (Cahill, 1992; Cohen, 1992). Hence a reasonably comprehensive picture can be put together of the urban and non-urban, and natural and anthropogenic contributions to the composition of these aerosols in our atmosphere.

The use of the accelerator based proton induced X-ray emission (PIXE) technique in aerosol studies was first proposed by the Swedish group at the University of Lund (Johansson and Johansson, 1976). Proton beams, from a several million volt particle accelerator, were fired at filter papers containing atmospheric particles. A multi-elemental analysis was performed by studying the X-rays induced by the atomic interactions of the proton beam with the particulate matter through which it passed. Cahill (1990) commented that since 1975 in North America alone, 90% of all size segregated ambient aerosol samples analysed have used the accelerator based ion beam technique of PIXE. Following the work of the late 1970s, more than 100 experimental physics groups in over 30 countries around the world began applying the PIXE technique, along with other accelerator based ion beam methods to characterise atmospheric particles on filters.

In general, analytical techniques with capabilities of detecting elemental concentrations of parts per million by weight on milligram samples will have the ability to quantify nanograms of material in the sample. Aerosol filters are such samples, with masses typically around 0.3 mg/cm^2 or less, and elemental concentrations several orders of magnitude below this. The ion beam techniques discussed here are ideally suited to perform such elemental analyses on these filters. Cahill (1990) demonstrated that IBA techniques were used to detect elemental quantities at concentrations as small as 50 pg/m^3 of air obtained from only 50 μg of sample on a filter. MeV ion accelerators are not small devices. Accelerator tanks can vary in size from 1 m in diameter and 2 m long for a 1 MV machine to 4 m in diameter and 14 m long for a 10 MV Tandem accelerator. Accelerators are not portable devices, therefore samples must be sent to the machines for analysis.

In this chapter commonly used ion beam methods, analysis and interpretation of data obtained from these techniques, as well as typical results from a large area fine particle sampling network currently operating in New South Wales (NSW), Australia, will be discussed. These results demonstrate the power of multi-elemental ion beam analysis methods to fully characterise fine particulate atmospheric aerosols.

ION BEAM TECHNIQUES

When a light ion such as a proton or helium from an accelerator interacts with an atom in the target material several reaction processes are possible. Some of these processes are shown schematically in Figure 1. The incoming swift ion from an accelerator interacts with the target atom electron cloud producing ionisation (electron ejection) and subsequent photon emission, usually an X-ray photon. This photon is characteristic of the target atom struck and the process is called particle or proton induced X-ray emission (PIXE). Inelastic interactions with the target atom nucleus may scatter the incoming ion, producing gamma-rays. This is called particle induced gamma ray emission (PIGME). Instead it can interact with the nucleus producing other light particles. This is known as nuclear reaction analysis (NRA).

Elastic interactions with the target atom nucleus may scatter the incoming ion in a "billiard ball" type interaction. This process is called Rutherford backscattering (RBS) if the ion is scattered backwards or particle elastic scattering (PESA) if the incoming ion glances off the nucleus

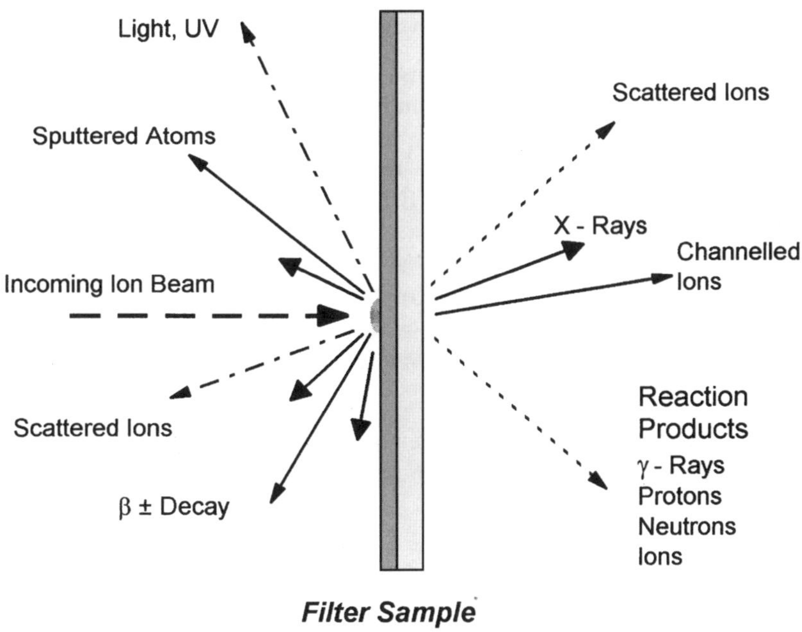

Figure 1. Schematic diagram of possible ion beam interactions with an aerosol filter.

and is forward scattered. Ion interactions with target atoms may break chemical bonds, produce light or UV, or sputter atoms from the surface itself. In addition, if the target has its own crystalline structure, it may even channel the incoming ion along the ordered rows of atoms in the target. The result of all these processes is that the ion loses energy in the target material and products are produced, and when detected can provide valuable information about the target's elemental composition with very high sensitivities. These processes are the essence of accelerator based ion beam analysis techniques and are described in detail by Cahill (1992) and Cohen (1993, 1994).

Accelerator based IBA techniques are ideal for elemental analysis of aerosols. They are fast, non-destructive, require no sample preparation, and can analyse submicrogram samples on a filter using only few minutes of accelerator machine time for all elements simultaneously. The range of possible elements covers most of the periodic table including hydrogen with more than sufficient sensitivity. Since the techniques are so fast, hundreds of samples a day can be analysed for a wide range of elements.

There have been four ion beam analyses (IBA) techniques commonly used for aerosol characterisation, and are often used simultaneously to cover the widest possible range for elemental analysis (Cohen, 1993). These four most useful IBA techniques are:

1. PIXE: Particle induced X-ray emission — useful for the analysis of elements from aluminium to uranium in concentrations from a few ng/m^3 upwards.
2. PIGME: Particle induced gamma-ray emission — useful for the analysis of light elements such as Li, B, F, Na, Mg, Al and Si in concentrations above $50\,ng/m^3$.
3. PESA: Particle elastic scattering analysis — useful for the analysis of hydrogen at all levels down to $20\,ng/m^3$.
4. RBS: Rutherford backscattering — useful for the analysis of very light elements such as carbon, nitrogen and oxygen in concentrations above a few hundred ng/m^3.

Each of these will be discussed; demonstrating their regions of applications, and individual strengths and weaknesses.

PARTICLE INDUCED X-RAY ANALYSIS (PIXE)

Particle induced X-ray emission or PIXE became popular as an analytical tool in nuclear physics laboratories in the mid 1970s. The earliest publications on PIXE included analyses of aerosol filter papers. This work continues today as evidence by the considerable work published in the International Conference series, PIXE and Its Analytical Applications (Vis, 1990). The physical processes involved in using light ion beams from accelerators to induced X-rays characteristic of the target being bombarded, have been adequately reviewed over the years (Johansson and Campbell, 1988) and Cohen and Clayton (1989) and will not be treated again in depth here.

The X-ray signals produced by ion bombardment of a target atom generate a unique fingerprint for that target atom. The X-ray energy or position of the peak in a spectrum determines the element present in the target, and the number of X-rays in this peak specifies the amount of that element in the target. The PIXE technique in aerosol analysis can be used to simultaneously determine over 20 different commonly occurring elements including: Al, Si P, S, Cl, K, Ca, Ti, V, Cr, Mn, Fe, Co, Ni, Cu, Zn, Br, Sr, Y, Zr, and Pb.

Typical PIXE spectra obtained from filters collected at Mascot, an inner suburb of Sydney Australia, and Cape Grim, an isolated global baseline station in north western Tasmania Australia, is shown in Figure 2. Both spectra were obtained from stretched Teflon® filters only 200 μg/cm^2 thick and exposed on the same day, May 31, 1995. Each of these PIXE spectrum was obtained using a 10 nA beam of 2.6 MeV protons from the 3 MV Van de Graaff accelerator at the Australian Nuclear Science and Technology Organisation (ANSTO). For 2.6 MeV proton beams the velocity of the ions striking the Teflon® filters was about 7% the speed of light and most protons pass straight through this thin Teflon® filter with very little energy loss (<1%). This is why the IBA methods are considered non-destructive for this type of aerosol analysis. They leave the delicate stretched Teflon® filters intact for further analysis by other techniques such as ion chromatography or neutron activation analysis. Proton irradiation times were typically only five minutes or less.

The number of peaks and hence the number of elements present in the two spectra are clearly different. Peaks corresponding to X-rays for elements from Al to Pb are readily seen. Each of these peaks not only identifies an element present in the aerosol, but can also be linked to a particular source. For example, Al and Si are associated with windblown soil, S with fossil fuel burning, Cl with sea spray, K with smoke and wood burning, Cr, Mn, Fe, Cu, and Zn with industrial processes, and Pb and Br with motor vehicles. Therefore proper analysis of spectra like those shown in Figure 2 can be powerful methods for source reconciliation whenever detailed source fingerprints are known. There is relatively little industry (Cr, Mn, Fe, Cu, and Zn), motor vehicle (Br and Pb), or windblown soil (Al and Si) aerosol contribution at Cape Grim compared with Mascot. Both sites have contributions from smoke (K), fossil fuel burning (S), and seaspray (Cl).

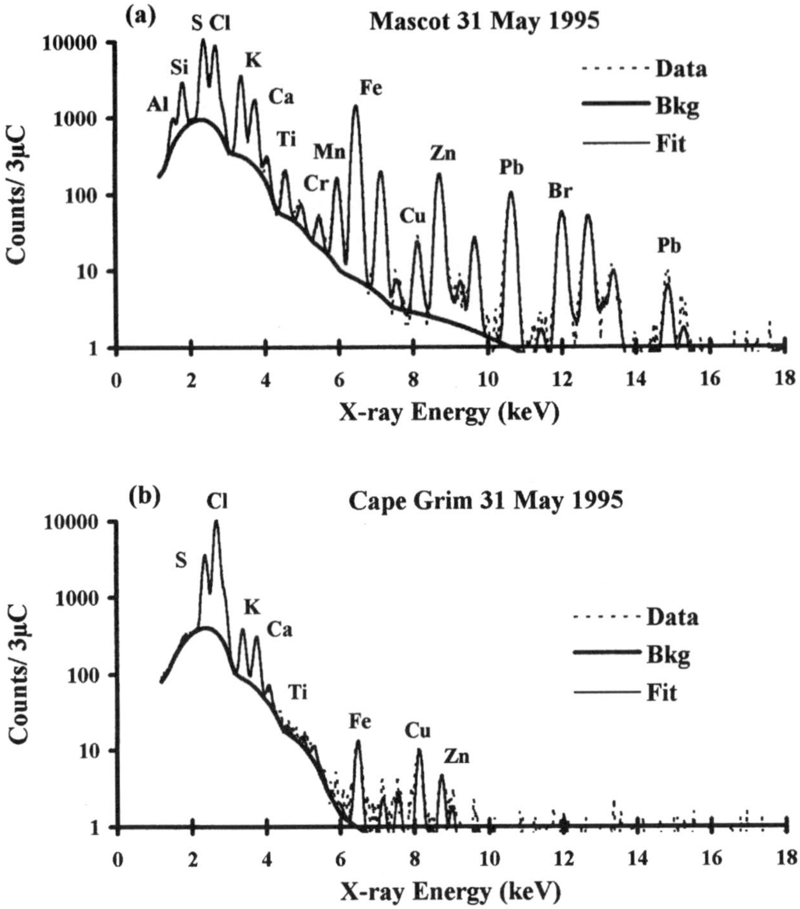

Figure 2. Typical PIXE spectra obtained for 2.6 MeV proton bombardment of Teflon® filters from (a) Mascot, an urban site in Sydney, Australia and (b) Cape Grim, a global baseline station in Tasmania, Australia.

The characteristic elemental peaks in Figure 2 sitting on the background vary in height over several orders of magnitude. They represent elemental concentrations down to a few parts per million by weight and clearly demonstrate the power of the PIXE technique for elemental analysis over several orders of magnitude in concentration. The dashed curve corresponds to the experimental data and the solid curve is a theoretical fit produced by the PIXAN computer code generated at ANSTO (Clayton, 1986). This code estimates the peak areas and calculates the elemental concentrations in the aerosol sample. The background underneath the peaks falls by two orders of magnitude between X-ray energies 2 and 5 keV, being essentially zero above 6 keV as shown in Figure 2. This lack of background in the region above X-ray energies of several keV, is the reason for the high sensitivity of PIXE in the Ca to Zn region of the periodic table as compared to other X-ray analytical techniques, such as X-ray fluorescence (XRF) (see Maenhaut, 1990). X-ray energies for common K, L, and M observed in PIXE spectra usually occur between 1 and 20 keV. The X-ray detection systems used have varying efficiencies between 1% and 100% with an optimal efficiency close to 100% for X-rays in the range of 10–20 keV (Cohen, 1980). Below 2 keV the X-ray detection efficiency drops off very rapidly as shown in the reduced backgrounds of Figure 2 in this region. Hence, the PIXE technique is generally not used for the detection of elements below aluminium.

PARTICLE INDUCED GAMMA RAY ANALYSIS (PIGME)

For sufficiently high ion beam energies the interaction of the ions with the nuclei of the target atoms may produce gamma-rays. These gamma-rays generally have energies from a 100 keV to several MeV. Analogous to X-ray production discussed above, the energy of the gamma-ray determines the element nuclei hit by the incoming ion beam. The number of gamma-rays detected is a function of the concentration of this element in the target. PIGME analysis is typically used to detect light elements like Li, Be, B, F, Na, Mg, and Al in a sample. The PIGME technique is ideally suited to complement the PIXE results for elements below silicon, where the X-ray detection efficiency for PIXE falls rapidly making measurements of light elements up to aluminium extremely difficult. This is one of the main reasons that PIGME and PIXE spectra are acquired simultaneously on the same filter sample (Cohen, 1994). Since PIGME originates from a nuclear process rather than an atomic reaction, it is significantly less sensitive for aerosol filter analyses than PIXE. For aerosol filters PIGME is generally most sensitive for fluorine and sodium analyses. The sodium originates from seaspray and windblown soil, and the fluorine is generally associated with industrial pollution from such processes as aluminium smelting.

A typical PIGME spectrum for a filter from the inner Sydney site of Mascot on February 22, 1995 is shown in Figure 3. As with the PIXE spectra, the sharp peaks are characteristic of the elements present and their areas are proportional to the elemental concentrations. Mascot is close to the coast therefore there is a reasonable amount of Na present in the aerosol (1540 ng/m^3 of air). Each element may have several identifying peaks. For example fluorine has gamma-ray peaks at 197 keV and three

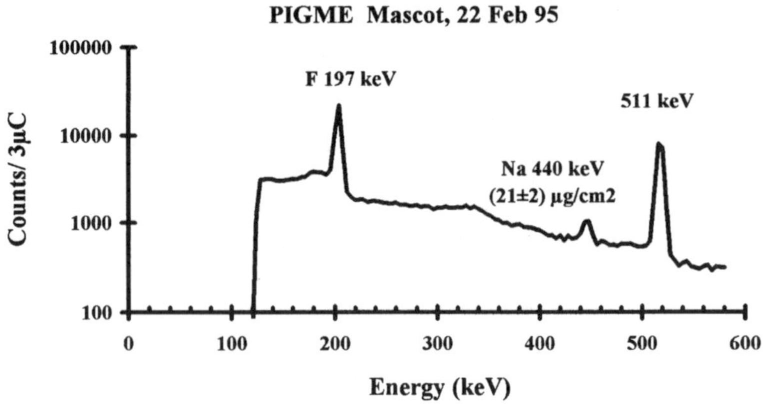

Figure 3. Typical PIGME spectrum obtained for 2.6 MeV protons on Teflon® filters.

other peaks between 5100 and 6130 keV, all of which could be used for analysis if required. The sodium gamma-ray peak at 440 keV is used for analysis of this element. It sits on a reasonably large gamma ray background but still supplies sufficient sensitivity to detect sodium above about 140 ng/m^3 of air for a 24 h average. Peaks from other light elements such as Li and Be, are generally not observed as their concentrations are mostly well below 100 ng/m^3. As is the case with Mg, Al, and Si the sensitivity of the PIGME technique is not sufficient to detect values below about a hundred ng/m^3 of air equivalent. The energies of some commonly occurring gamma-ray lines used for PIGME analysis are given in Table 1. A more

Table 1. Common gamma-ray line energies for light elements observed in PIGME analysis.

Element	Energy (keV)	Element	Energy (keV)
Li	477	Na	440
Be	3562		1634
B	429	Mg	417
F	110		1368
	197	Al	843
	1236		1013
	1349	Si	1778
	1357		
	6129		

comprehensive list of possible gamma-ray transitions for proton bombardment of most target atoms is given in Cohen and Clayton (1989).

PARTICLE ELASTIC SCATTERING (PESA)

Particle elastic scattering (PESA) involves an incoming ion beam elastically scattering off target nuclei and their energies being detected as a means of identifying selected chemical composition. For a proton beam traversing a thin target, such as an aerosol filter paper, this technique becomes particularly suited for hydrogen analysis if the filter itself does not contain hydrogen. The principles of the technique are schematically shown in Figure 4 where 2.6 MeV protons are incident normally onto a 220 μg/cm^2 thick (25 μm) Teflon® filter. If the detector is placed at small forward scattering angles (~30°), it will receive a range of scattered proton energies. The kinematics of the scattering process are such that for elements heavier than hydrogen, such as the carbon and fluorine in the Teflon® filter, protons of higher scattered energy are produced. The idealised spectrum is displayed at the bottom of Figure 4. Since hydrogen is lightest element, it has a unique signature, shown as the edges (a)–(b), at the lowest scattered proton energies. Other heavier elements produce the edges (c)–(d), at higher scattered proton energies, shown in Figure 4 and generally are not resolved.

A typical PESA spectrum for a Teflon® filter obtained at Mascot, Sydney on February 22, 1995 is shown in Figure 5. Note that Teflon® is

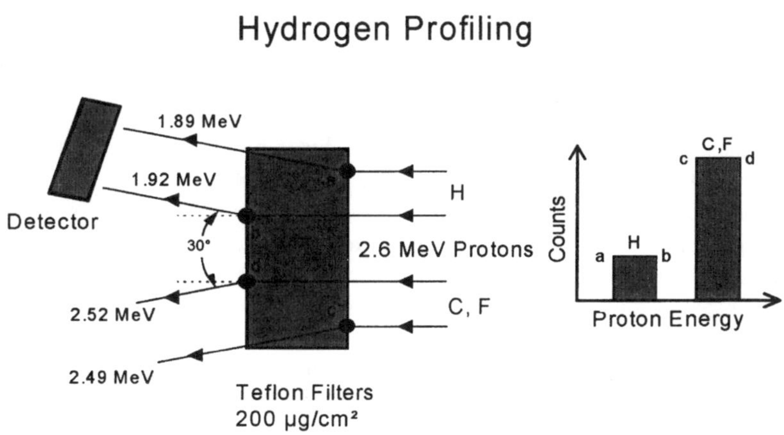

Figure 4. Schematic diagram of the principles behind total hydrogen measurements using proton elastic scattering from thin filters.

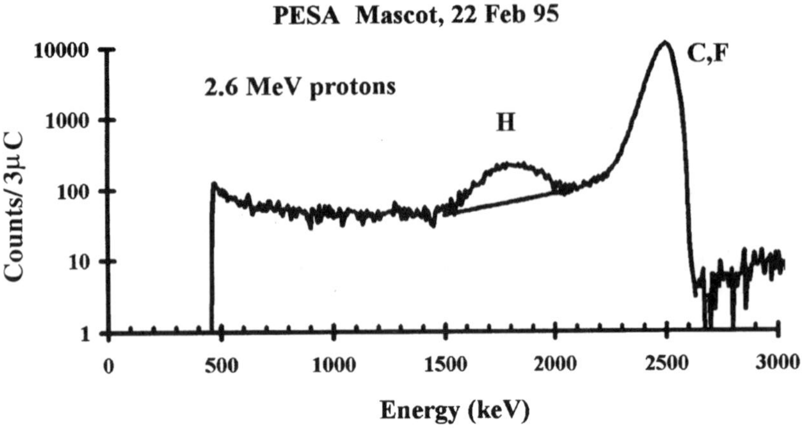

Figure 5. Typical PESA spectra for 2.6 MeV protons scattering off an exposed Teflon® filter from Mascot, in Sydney. The solid curve is the background if no hydrogen were present.

composed of carbon and fluorine and does not contain any hydrogen. The spectra is similar in shape to the schematic spectrum in Figure 4. Despite the presence of many elements heavier than carbon (Na, Al, Si, S, Ca, Fe, Zn, Pb, Br, etc.), the hydrogen peak in Figure 5 is still well resolved from these and corresponds to a total mass of hydrogen of 5.4 $\mu g/cm^2$ on the filter of 370 ng/m^3 of air. The area of the background subtracted hydrogen peak in Figure 5 is linearly proportional to the hydrogen concentration in the filter over the concentration range 0–200 $\mu g/cm^2$. This is calibrated by calculating the PESA system against known thickness Mylar films as shown in Figure 6. The constant of proportionality is high (1600 or more) showing that small concentrations of hydrogen produce large peak areas for these accelerator running conditions. This demonstrates the high sensitivity of this technique for total hydrogen analysis on filters.

The width of the hydrogen peak in Figure 5 is a strong function of the proton beam diameter (Cohen, 1993). The spectrum shown here was taken with an 10 mm diameter beam. One mm beams produce much narrower peaks with similar total areas. However, the hydrogen peak is still well separated from the peaks of heavier elements (such as carbon), and it is more important, for reasons previously discussed, to use the broader proton beams to analysis the filters. The PESA measurements and all IBA analyses are performed in vacuum, so that all free water vapour and volatile hydrocarbons are lost from the filter before analysis. Only the bound hydrogen (e.g. as ammonium ions) is detected.

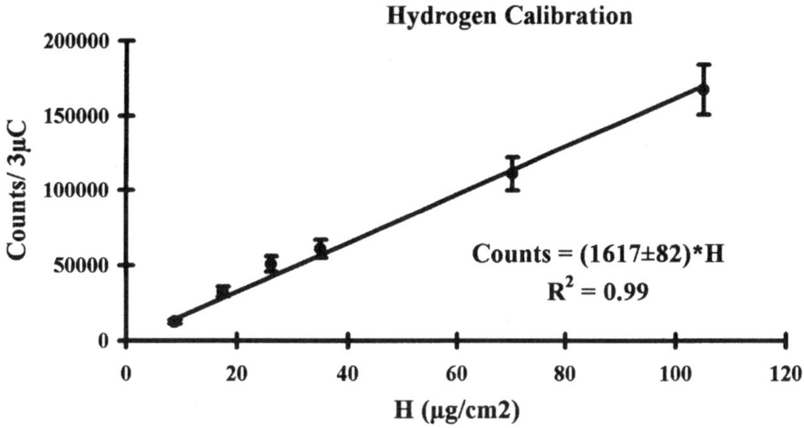

Figure 6. Plot of the area of the hydrogen peak area versus the hydrogen concentration for 2.6 MeV proton scattering off varying thickness of Mylar.

RUTHERFORD BACKSCATTERING (RBS)

For proton beam scattering off target nuclei heavier than hydrogen, the kinematics allows a small fraction of the beam to be elastically scattered at large backangles greater than 90°. This is called Rutherford backscattering and by placing a detector at backangles (angles around 160–170°), the energies of these scattered protons are measured and the mass of the target atoms struck calculated. For input ion beam energies below the target ion Coulomb barrier (several MeV) the processes are well characterised. This IBA technique of Rutherford backscattering can be used to measure the total carbon, nitrogen, and oxygen concentrations in atmospheric aerosol filters. A typical spectrum obtained from a Teflon® filter is shown in Figure 7.

Readily resolved characteristic peaks are obtained for the elements carbon, nitrogen, oxygen, and fluorine. The area of these peaks is related to the concentration of the element present. However, fluorine has several characteristic peaks that reflect several different types of nuclear reactions that can occur at the proton energies used. These other fluorine peaks lie under the nitrogen and oxygen peaks and interfere with them making the background subtraction under these peaks uncertain. This introduces systematic errors into the absolute concentration estimates for these elements. The smaller nitrogen peak, between carbon and oxygen is particularly affected by this problem. Carbon is also a major component of most filters. For example Teflon® is 24% by weight carbon. Therefore the carbon

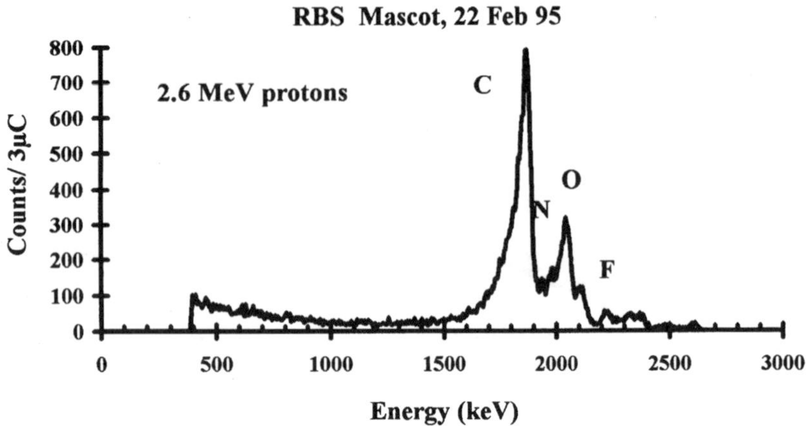

Figure 7. Typical RBS spectra obtained from 2.6 MeV protons on an exposed Teflon® filter.

contributions from the filter itself have to be subtracted out before an estimate of the total carbon content of the particle matter on the filter can be made. In the case of Teflon® filters this can be done by assuming that Teflon® is $(CF_2)_n$ and using the PIGME fluorine measurement to estimate the carbon content of the filter itself. This is then subtracted from the total carbon content measured by RBS.

Typically a thin 220 μg/cm^2 Teflon® filter might be loaded with 110 μg/cm^2 of total fine particle matter which on average was, 23% organic matter, 23% elemental carbon or soot, 23% ammonium sulphate, and 6% soil. The average total carbon and oxygen content of material on this filter would then be around 40% and 20% respectively of the total particle mass. This would mean the contributions to the total carbon content using RBS for the Teflon® filter and the fine particle mass are about one to one. The problems with carbon subtraction in the Teflon® filters and the interferences from the fluorine peaks make the absolute C and O concentration uncertain to at least $\pm 30\%$ and more for the smaller N peaks. RBS is generally not used to estimate the fluorine concentrations on the filters. This is better accomplished by using the PIGME analysis since the yields are more reliable and less dependent on the ion beam energy. The four IBA methods of PIXE, PIGME, PESA, and RBS described provide concentration estimates for elements from hydrogen to lead.

MINIMUM DETECTABLE LIMITS

The accelerator based ion beam analysis techniques described above are fast, non-destructive and very sensitive. In order to determine the minimum detectable limits for the range of elements measured, the background under each elemental peak in the typical plots shown in Figures 2, 3, 5, and 7 must be considered for each of the analysis techniques used. The minimum detectable limit (MDL) for a given element is defined as $3.29C \cdot \sqrt{B}$ where B is the background counts under the peak in the spectrum, and C is the conversion constant from counts to ng/m^3 for that element (Clayton, 1986). This is a conservative estimate of the MDL and takes into account inter-element interferences (Cohen and Clayton, 1989). The results of such a calculation for all four techniques and the trace elements measured by these IBA methods are seen in Figure 8. The MDL's shown were calculated assuming a typical run time of 300 s at a proton beam energy of 2.6 MeV, and a current of 10 nA on a thin Teflon® filters for a standard experimental arrangement (discussed in the next section). Converting the MDL's shown in Figure 8 from ng/m^3 of air sampled to μg/cm^2 of filter area is done by multiplying by approximately 0.013.

The PIXE technique utilises two characteristic X-ray lines, the K and the L series lines. The K lines cover light elements between Al and Zr (atomic numbers 13–40); and the L lines, the heavier elements, between Dy and Pb (atomic numbers 66–82) with MDL's lower than 10 ng/m^3. The

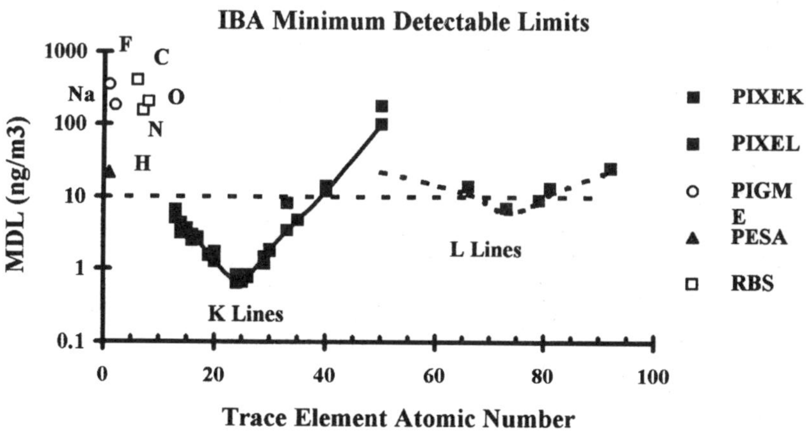

Figure 8. Typical MDL's for the four IBA techniques discussed. Calculations were made assuming 2.6 MeV protons, a beam current of 10 nA for 300 s, and standard experimental set-up.

PIGME technique is used for Na and F with MDL's of 180 and 350 ng/m^3 respectively. RBS sensitivities or C, N, and O lie between 100 and 400 ng/m^3 . Hydrogen from the PESA technique is measured with a sensitivity of 20 ng/m^3. Elements between Zr and Dy (atomic numbers 40–66), such as the rare-earth elements, have sensitivities between 10 and 100 ng/m^3, but these occur naturally at levels several orders of magnitude below this. Consequently they are not detected by these techniques in most practical cases. The average minimum detection limits of around 10 ng/m^3 for the majority of elements is more than sufficient for most components in atmospheric aerosols. It is several orders of magnitude below the typical fine mass loadings measured on filters and presently accepted health standards or goals, where they exist. An elemental concentration of about 1 ng/m^3 in air corresponds to between 50 and 100 ppm by weight for the average urban fine particulate atmospheric mass loading of between 10 and 20 $\mu g/m^3$. Alternatively, for an average ambient urban fine particle (PM-2.5) loading of 10 $\mu g/m^3$ (total mass per filter of around 300 μg), an elemental MDL of 1 ng/m^3 corresponds to only 30 ng of that element present on the filter.

The precision in each concentration is calculated as a function of statistical counting or accumulation errors, and experimental or measurement errors. The experimental errors tend to be relatively constant and are best estimated by regular measurements of known standards. Whereas errors due to statistical counting, depend on the concentration of a given element on a given day at a given site. High concentrations have lower associated errors. Typical experimental errors obtained for each analysis technique over a 30 month measurement period are shown in Table 2. This error is generally between 5% and 10%.

The error associated with each individual element concentration depends on the counting statistics for that element. Elements with concentrations near their MDL's will by definition have counting errors of around

Table 2. Typical experimental errors for the IBA techniques used in the Australian Aerosol Sampling Program (ASP) network over a 30 month period.

Technique	Error	Comments
Weight	3–5%	Depends on the mass loading
PIXE	3–3%	Measured on standards over 30 months
PIGME	9%	Measured on standards over 30 months
PESA	8–10%	Measured on standards over 30 months
RBS	4–9%	Measured on standards over 30 months

100%. For example, cobalt typically occurs at concentrations around 0.5 ng/m^3 which is comparable with its MDL, whereas fine sulphur has an average value of over 400 ng/m^3 which is 170 times its MDL. The statistical counting errors associated with these two elements were calculated to be 117% and 1% respectively. The statistical counting errors for each of the elements analysed by the IBA methods for 3μC charge irradiations are displayed in Table 3. It was calculated from the analysis of over 1000 typical filters acquired from the 25 site Australian network over a 12 month period and includes peak area estimation and background subtraction errors. The term RCM is the reconstructed mass discussed further below. It represents the sum of all measured elemental components for comparison with the total fine particle mass. Finally, the total error on each elemental measurement is obtained by assuming the statistical errors (Table 3) and the experimental errors (Table 2) are independent and adding the two in quadrature. This produces errors from a few percent to over 100% depending on the elements measured. This should be kept in mind when quoting elemental concentrations especially for elements like Co which occur at levels near the minimum detectable limit of the IBA system.

The four IBA accelerator based ion beam analysis techniques of PIXE, PIGME, PESA, and RBS are ideally suited for elemental analysis of aerosol particles on filters. They can each be performed simultaneously on the one aerosol sample, providing elemental concentrations on more than 35 different elements at a time for selected elements, from hydrogen to uranium. The techniques are non-destructive, fast, relatively inexpensive, and have high sensitivities especially for trace elements of common interest including: hydrogen, fluorine, sodium, aluminium to bromine, and

Table 3. Typical statistical counting errors associated with the elemental analysis of 1000 filters obtained from the Australian Aerosol Sampling Program (ASP) over 12 month period.

Element	Error %	Element	Error %	Element	Error %
Al	20	V	78	Pb	15
Si	5	Cr	61	H	2
P	40	Mn	27	F	5
S	1	Fe	5	Na	5
Cl	16	Cu	24	Total C	2
K	2	Co	117	Total N	17
Ca	6	Ni	47	Total O	6
Ti	32	Zn	14	RCM	6
		Br	16		

lead. Minimum detectable limits for these elements around a few ng/m^3 of air are obtainable in just a few minutes of accelerator running time. Sample throughputs of around 12 per hour or greater are possible. These are very powerful multi-elemental analysis techniques that have been used successfully world wide for some years now. The megavolt accelerators required to perform these techniques are commonplace with most developed and many developing countries possessing several of them. The last decade or more has seen a significant upsurge in the public's perception of the importance of pollution studies. Consequently there has been a corresponding increase in the use of these accelerators for aerosol analysis studies. Many ion beam analysis systems are fully automated and can analyse hundreds of samples a day and have handled thousands of samples a year. The usefulness of these fast multielemental analysis techniques is well established internationally. Clearly they will have a significant role to play globally, in future aerosol particle studies.

DATA COLLECTION, ANALYSIS AND INTERPRETATION

This section discusses the experimental system required to collect IBA data on aerosol filters, and the analysis and interpretation of this data. The principles given are generally applicable to most types of filters used in aerosol data collection. The emphasis is placed on Teflon® and Nuclepore type filters because they are considered thin and hence better suited to IBA methods. It was shown above that these methods require the irradiation of these filters by a high energy proton beam that passes directly through the filter. This is necessarily done in a vacuum chamber attached to an evacuated beamline along which these proton beams are transported (Cohen, 1992).

The typical frequency distribution of total mass for filters containing particles with aerodynamic diameters less than 2.5 μm (PM-2.5), for nearly 900 Teflon® filters collected in and around Sydney, Australia during 1995, is shown in Figure 9. The distribution is log normal, as expected, with an arithmetic average mass of around 10 μg/m^3 of air sampled. For the Teflon® filters used here and the 31 m^3 of air sampled, this corresponds to about 310 μg of material per filter or a thickness of 140 μg/cm^2. The Teflon® filter itself is around 220 μg/cm^2 in thickness, so that the average total amount of material presented to the proton beam was of the order of 360 μg/cm^2. The range of 3 MeV protons in this material is of the order of 15 mg/cm^2, hence the filter thickness, including the collected fine particle matter, is less than 3% of this range. The filters can then be considered as thin targets as far as the IBA analysis is concerned. This is an important

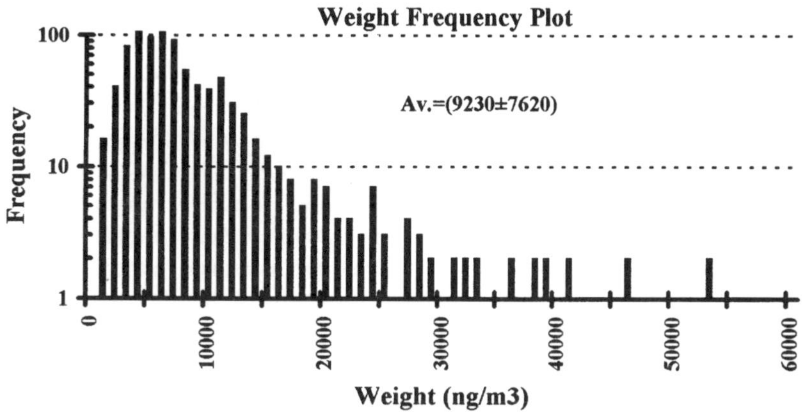

Figure 9. Frequency distribution of the mass obtained for 900 Teflon® filters from the Australian fine particle sampling network during 1995. The masses are shown in ng/m^3 of air sampled.

feature of the method and considerably simplifies the data analysis processes as is demonstrated in a later section on the determination of elemental concentrations. The fact that the filter is thin with respect to the proton beam also allows for the forward scattering process, PESA described above, to occur. PESA analysis can not be performed on filters that are too thick to allow the proton beam to penetrate or the forward scattered hydrogen atom to exit from the back surface of the filter. Typically loaded filters should be less than 500 g/cm^2 thick for these thin target methods to be applicable.

OPTIMAL EXPERIMENTAL DESIGN

Typical proton beam diameters vary from sub-millimetre to about 10 mm in diameter. It is important to sample as much of the filter area as possible in particle aerosol analysis. This can be accomplished by using large beams (10 mm diameter and above) or by scanning smaller beams across a representative area of the filter. Our experience has shown that the central 10–15 mm diameter portion of most filters in the diameter range of 25–50 mm is uniform in thickness, and representative of the overall characteristics of the particulate matter on the filter. It is therefore appropriate to just mount the entire filter, as collected from the sampling unit, directly in the analysis chamber and irradiate the central 10 mm or so with the proton beam. This is accomplished in a chamber system similar to the experimental set-up schematically shown in Figure 10. A cylindrical target chamber (20 cm in diameter) with a large 2 m wing extension perpendicular to the direction of the incoming beam, houses a meter long target stick that holds between 20 and 40 filters depending on their diameter. Each target is placed in the proton beam by a computer controlled stepping motor attached to the stick. A proton beam of energies around 3 MeV and currents of about 10 nA over a 10 mm diameter are typical for most filter analyses. It is considered important to keep the beam current densities generally well below 1 nA/mm^2 in order to avoid damage to the filters and subsequent evaporative losses due to local heating of the filter. A detection system for each of the four IBA methods is placed around the center of the chamber. The X-rays (PIXE) are detected at 135° backangle to the incoming proton beam. The gamma-rays (PIGME) are detected at the mirror image backangle to the PIXE detector. The backscattered particles for RBS are detected by a small (1 cm diameter) silicon surface barrier detector at 170° backangles and the forward scattered PESA detector is at a 30° downward forward angle to the incoming proton beam. All the detectors are inside the target chamber pointing at the center of the chamber at distance of between 50 and 100 mm from the filter. For optimum counting statistics, it is important to try to get each detector system as close as possible to the target filter in order to maximise the solid angle subtended. Practical limitations usually set the detector solid angles for these systems at a few millisteradians. A large solid angle is particularly important for

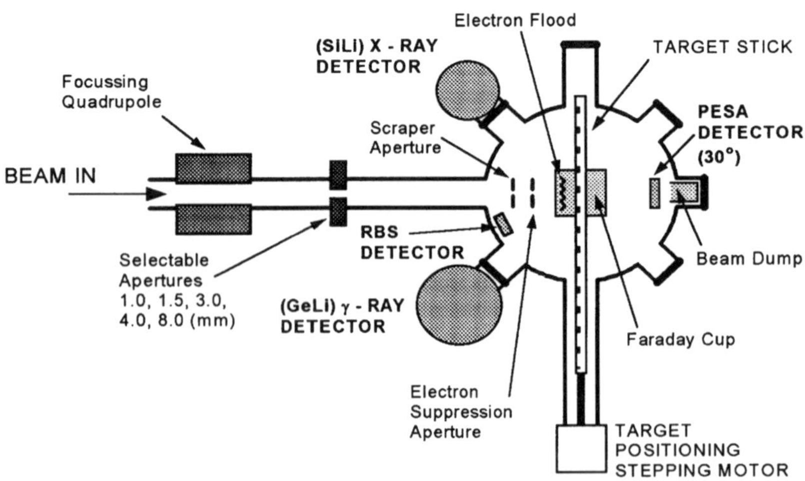

Figure 10. Schematic diagram of a typical IBA aerosol analysis chamber.

the PIGME detector since gamma-ray production rates are significantly lower than those for the other three methods.

BACKGROUND SUBTRACTION

The nature of ion beam analysis of thin aerosol filters produces two separate background subtraction problems. The first is similar to most analytical techniques in that a signal sits on a background produced by the physics of the method. This is clearly shown in Figures 2, 3, 5, and 7 for the four IBA methods discussed here. These backgrounds are easily stripped off automatically by standard peak fitting routines and analysis programs (Clayton, 1986; Cohen and Clayton, 1989) as shown by the solid background curves of Figures 2 and 5, for example. In IBA methods these backgrounds occur through energy loss, scattering and interactions of the ion beam as it passes through the filter material or from gamma-rays produced in the target scattering in the detection system. They are generally of little concern in this thin target application, are well understood, and easily subtracted out when peak areas and hence final elemental concentrations are estimated.

The second, sometimes more serious background subtraction problem, is produced by the ion beam passing directly through the filter medium and therefore effectively analyses both the deposit on the filter and the filter paper itself. It is then very important to consider on which filter medium the aerosol should be collected. This is particularly important for IBA analysis methods since they are multielemental and capable of analysing most elements from hydrogen to lead. For example, if fluorine was an important element to determine in air then it would be pointless to use a Teflon® filter. Teflon® is 76% by weight fluorine and the signal from the filter would swamp any possible fluorine signal from the deposit on the filter. Similarly, polycarbonate Nuclepore filters would not be much use if hydrogen was an element of interest in the filter deposits. It is not only the bulk composition of the filter that is important but also the presence of any key trace elements. For example, recent measurements by us have shown that some batches of 47 mm Nuclepore polycarbonate filters contain approximately 20 ng per filter of bromine. Under normal operating conditions this becomes a sizeable background subtraction when trying to estimate the total bromine present in the deposit on these filters and hence in the air. It is therefore essential to have a clean filter substrate with respect to the range of elements one wants to measure using IBA methods. The PIXE technique measures upwards of 20 different elements and maybe particularly sensitive to these background problems if selected elements are present in the filter matrix. It is important to measure a blank filter using each of the IBA methods before estimating concentrations from exposed filters.

A typical blank filter spectrum for a thin 25 mm diameter Teflon® $(CF_2)_n$ filter is shown in Figure 11. The background is the standard one expected from the secondary electron Bremsstrahlung. No characteristic peaks for elements from aluminium to lead were detected sitting above this background. This substrate is clearly a very clean filter medium for the PIXE method. A similar spectrum for a 47 mm diameter Nylon blank filter under PIXE analysis is shown in Figure 12. One can see the secondary electron Bremsstrahlung background. However, sitting on this are a few characteristic peaks for the elements Cl, Ca, Ti, Mn, Fe and Co. Equivalent concentrations per unit area of filter for these elements ranging from 10 ng/cm^2 for Fe to a over 5000 ng/cm^2 for Ti are seen in the figure. Clearly this is not a good filter to use if ambient Ti or Co concentrations are required, since the levels of these elements are generally lower in the atmosphere than those found in the filter medium itself.

Glass fibre filters or Teflon® coated glass filters, as commonly used in high volume samplers or TEOM samplers, are particularly bad substrates often containing a dozen or more key elements in high concentrations in the filter substrate. A PIXE spectrum for a standard blank TEOM glass fibre filter is shown in Figure 13. A range of elements from Al to Zr, including key elements found in atmospheric aerosols are present in the blank filter in concentrations well above those generally found in the atmosphere. In this particular filter blank the only elements able to be

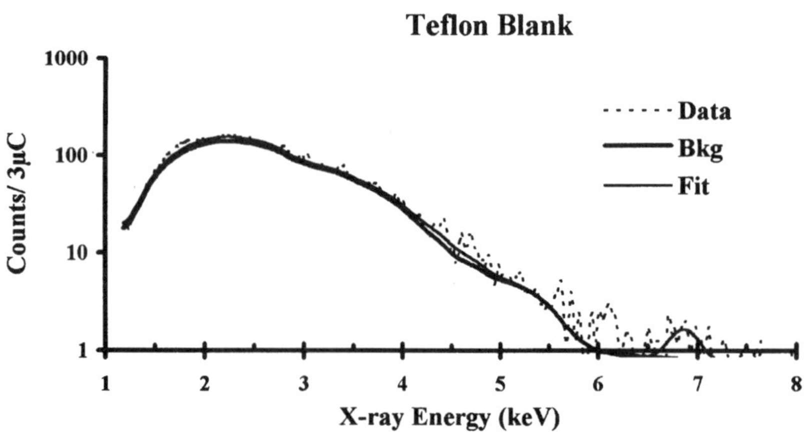

Figure 11. Typical PIXE spectrum for a thin blank Teflon® filter.

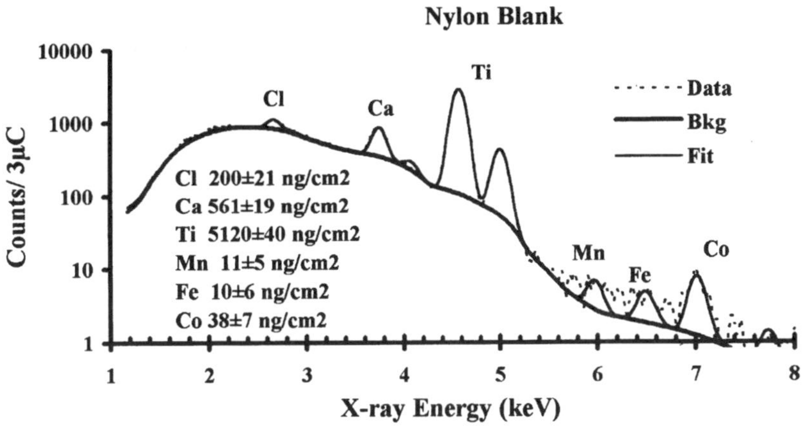

Figure 12. Typical PIXE spectrum for a thin blank 47 mm diameter Nylon filter.

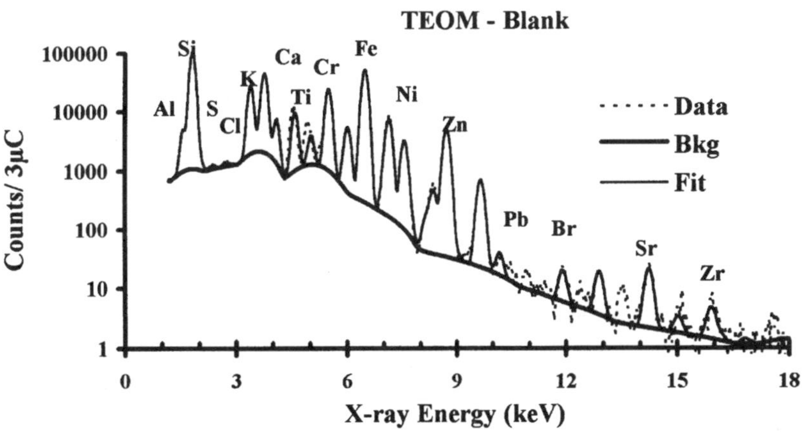

Figure 13. A typical PIXE spectrum for a blank TEOM glass fibre filter.

analysed for by the PIXE method were S and Cl. For most other elements from Al to Zr blank interferences were such that the analysis is unreliable. This makes these filters particularly unsuitable for IBA methods.

DETERMINATION OF ELEMENTAL CONCENTRATIONS

For the ion beam analysis techniques it has been shown qualitatively (for example, see Figure 6) that the area under a characteristic elemental peak is related to the concentration of that element present in the filter. Generally for the PIGME, PESA, and RBS methods the relationship between appropriate spectra peak areas and elemental concentration is determined by comparison with a reference standard. This technique provides the link from the measured peak areas to the determination of elemental concentrations.

The PIGME technique in filter analysis is practically used to measure sodium, fluorine, and sometimes aluminium concentrations. As mentioned it is a nuclear (not atomic) process using gamma-ray detection and therefore has a much lower yield than the other IBA techniques discussed here. Typically thin standards of known quantities of these elements (for example Micromatter Standards) are used to determine the linear relationship between counts in a peak and concentration. The gamma-ray energies are sufficiently high (>100 keV) so as to produce little self-absorption in the thin reference standards or in the relatively thin filters used. This means that no matrix corrections are needed for the conversion from reference standards to working filters for this technique. Typical yield conversion factors for selected elements for the ANSTO IBA aerosol analysis system are given in Table 4.

The PESA technique is only used to measure the total hydrogen content. Again the Mylar reference standards are thin (35 μg/cm^2 of hydrogen) as are the stretched Teflon® filters (220 μg/cm^2) used for atmospheric total hydrogen concentration determination. Note that the Teflon® filters are used exclusively when hydrogen needs to be determined, since this filter matrix contains little or no hydrogen itself, unlike the Nuclepore

Table 4. Typical conversion factors for PIGME, PESA and RBS techniques for the standard ANSTO aerosol analysis system using reference Mylar, Kapton and Micromatter standards. The proton energy was 2.6 MeV.

Element	Yield/ (μg/cm^2)/3 μC
PIGME (Micromatter Foils)	
Na	13 ± 2
F	76 ± 7
Al	1.2 ± 0.4
PESA (Mylar Foils)	
Hydrogen	620 ± 45
RBS (Mylar and Kapton Foils)	
Carbon	27 ± 3
Nitrogen	30 ± 4
Oxygen	25 ± 3

polycarbonate filters. For the standard ANSTO IBA system, typical yield conversion factors for hydrogen are around 650 counts/(μg/cm^2)/3 μC which is significantly higher than those quoted in Table 4 for the PIGME technique under the same conditions. This shows that the PESA detection method for hydrogen is very sensitive compared to the gamma-ray methods.

The RBS technique is used to estimate the carbon, nitrogen, and oxygen concentrations. Again Teflon® filters are used for this method because they are free from nitrogen and oxygen. They do however contain 24% by weight of carbon. The filter carbon content therefore needs to be subtracted out before a proper estimate of the carbon content of the particulate matter can be made. Calibration of the carbon, nitrogen, and oxygen analyses by RBS can be performed using standard thin Mylar foils for carbon and oxygen, and thin Kapton foils for nitrogen. The accuracy of these measurements should be around $\pm 6\%$ for carbon, $\pm 9\%$ for nitrogen, and $\pm 4\%$ for oxygen. This means that relative carbon, nitrogen, and oxygen measurements should be quite reliable even if systematic errors due to unreliable matrix subtraction problems may result in absolute concentrations with more than 30% error. Typical conversion factors for the RBS method for the standard ANSTO system are also given in Table 4.

Concentrations of over 20 different elements in a filter from aluminium to uranium can be measured by the PIXE system. It is time consuming to have to calibrate each of these elements as a function of peak area and concentration for a given matrix. The parameters for X-ray production by ion beam bombardment are sufficiently well known to predict the yields from both thin and thick targets from a given matrix within a few percent (Cohen and Clayton 1989). For a relatively thin target Cohen and Clayton (1989) give the following expression for the number of counts Y in an X-ray peak for an element bombarded by a proton of energy E MeV:

$$Y(E-\Delta E/2)=4.968\cdot 10^{-17}\, N\, \sigma(E-\Delta E/2) Q \Omega \varepsilon x \\ \exp[-(\mu/\rho)x \sec(\theta_o)]/\sin(\theta_i) \qquad (1)$$

Where:

ΔE — is the mean energy loss of the incoming ion at angle θ_i
N — is the number of target matrix atoms/cm^3
x — is the target thickness in μg/cm^2
Q — is the total ion charge hitting the target in μC
Ω — is the detection solid angle in steradians
ε — is the detector efficiency for the characteristic elemental X-ray
(μ/ρ) — is the mass attenuation coefficient (cm^2/g) for the characteristic X-ray emerging at angle θ_o to the target surface normal
ρ — is the target density in g/cm^3.

The X-ray production cross section σ (in barns, $10^{-24}\,cm^2$) is reasonably well known and polynomial fits for a range of different elements for K, L, and M shell X-ray production have been given by Cohen and Clayton (1989) for various proton energies.

Equation (1) is quite accurate as a thin target approximation for targets up to a few mg/cm^2 thick. Generally this equation is good to 10% or better for targets thinner than one quarter of the ion range in the matrix material. The full thick target equation (Cohen and Clayton, 1989) should be used for targets thicker than this. This equation integrates over the ion energy loss in penetrating the target as well as accounting for the absorption of the X-ray leaving the target at a given depth. This procedure is somewhat more complex than Equation (1) and generally can be avoided by using thin ($<1\,mg/cm^2$) filters.

Given the geometry of an IBA system (similar to the one shown in Figure 10) and a well characterised experimental set-up, it is possible to predict the peak areas of a PIXE spectrum similar to that shown in Figure 2 for a range of trace elements in a given matrix. Figure 14 is a theoretical plot for 2.6 MeV proton on a Teflon® filter target. The vertical axis is the number of K_α or L_α X-ray counts per ($\mu g/cm^2$) of trace element present in the Teflon® filter for every 3 μC of charge (10 nA for 300 s) passing through the filter. Two different X-ray line series, K and L lines, originating from the ion beam interacting with K or L shell target atom electron are used for the analysis. Typically K lines are used for elements below zirconium (X-ray energies between 1 and 16 keV energy) and L lines for elements above tungsten (X-ray energies above 8 keV). Using these two

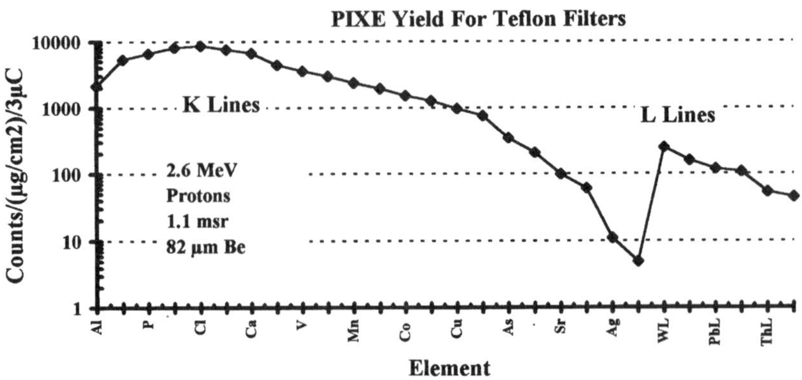

Figure 14. Yields for thin Teflon® filters for a typical PIXE system.

series most of the periodic table between Al and U can be covered with yields of 100 counts per peak or more for each μg/cm^2.

The theoretical curve of Figure 14 has several useful characteristics. The fall off in yield for elements lighter than chlorine is predominantly determined by the X-ray detector efficiency and any material placed between the filter and the detector crystal (ε in Equation (1)). The heavy element roll off above chlorine for K shell X-rays is determined by the X-ray detector crystal thickness and the fall off in X-ray production cross sections (σ in Equation (1)) with an increasing trace element atomic number. The height of the peak in the yield between sulphur and calcium is determined primarily by the detector solid angle Ω in Equation (1). A few reference standards can be used as follows: Al and Si to determine the low energy detection efficiency; Cl and Ca — the solid angle of detection; and Fe and Sr — the crystal thickness and to check the cross section roll off with the trace element atomic number. Having used these six reference standards to ascertain the shape of the curve in Figure 14 we can then use equation 1 to find the appropriate yields for any other element in the periodic table. Figure 15 is the measured to nominal ratio for the six thin Micromatter Standards between Al and Sr measured over a 30 month period. The standards were supplied as $\pm 5\%$ as shown by the dotted lines on the plot. The mean for these 30 measurements was (1.00 ± 0.03), showing that this type of PIXE calibration can be consistently repeated to $\pm 3\%$ over a long period of time if sufficient care is taken with experimental methods.

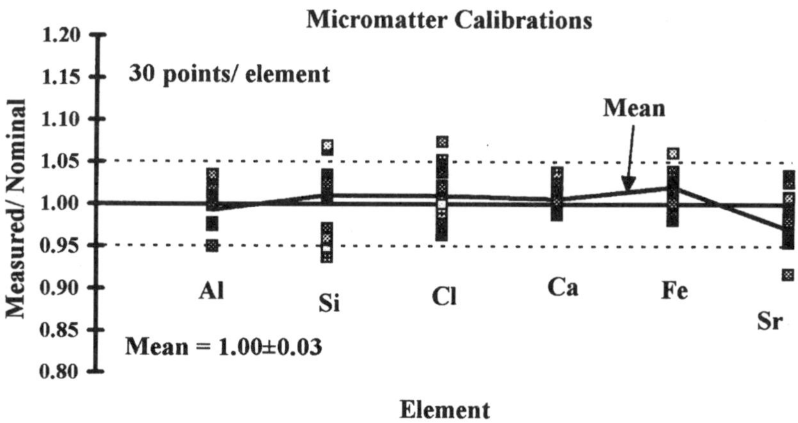

Figure 15. Measured to nominal PIXE concentrations for thin Micromatter foils of Al, Si, Cl, Ca, Fe and Sr over a 30 month period.

RECONSTRUCTED MASS

The four ion beam analysis techniques described here are capable of analysing over 20 different elements present in particulate matter on filters. These elements have been listed in Table 3. This is a large range of elements and if all their masses were added together we would expect it to be comparable with the total measured gravimetric mass of particulate material on the filter. The procedure of having all the analysed elemental constituents adding up to the measured gravimetric mass is called mass closure and is important in source fingerprinting and reconciliation's techniques discussed later.

It is interesting to combine some of these many measured elements to try and estimate concentrations of compounds of known chemical composition. For example, the major constituent of seaspray is NaCl. Since both these elements are measured, thus it should be possible to put together a composite variable that is representative of the seaspray component of our fine aerosol. Windblown soils can be expected to be composed of oxides of the elements Al, Si, Ca, Ti, and Fe. These are some of the major elements present in this matrix. Again a composite variable can be constructed from these elements that is representative of fine soil components of aerosols. Other combinations of pure elements that represent signatures for interesting aerosol components can be derived.

The Aerosol Group from the University of California, Davis has been very successful in using these composite variables to describe the aerosol measurements from their IMPROVE fine particle sampling network across North America (Cahill, 1990; Malm *et al.*, 1994). Following the work of Malm *et al.* (1994) it is possible to define five composite variables based on elements measured to investigate how well these IBA methods produce mass closure for fine atmospheric particles. Here we discuss the following five composite variables, namely, sulphate, organic matter, seasalt, soil, and smoke. These five variables together with an independent elemental carbon measurement by laser integrated plate or other methods (Malm *et al.*, 1994) produce good mass closure for fine particle aerosol measurements.

Sulphate

Sulphur primarily enters the atmosphere as SO_2 gas and is converted to sulphuric acid (H_2SO_4), which is then neutralised by ammonia gas to form ammonium bisulphate NH_4HSO_4 or ammonium sulphate $(NH_4)_2SO_4$ depending on the degree of neutralisation. The rate of this transformation depends on the amount of ammonia gas and water vapour in the air. Using the atomic weights of each of these chemical species we can estimate the sulphate ion concentration in its various forms from the elemental sulphur concentration. This is shown in Table 5. The definitions have been tested by comparison with independent SO_4^{2-} measurements using ion chromatography and sulphur measurements using PIXE. The agreement is excellent with the (SO_4^{2-}/S) ratio being near 3.0 with a correlation coefficient of $R^2 > 0.8$. This is discussed further in next section.

Seasalt

Weast (1977) gives the major components of seasalt as sodium (32% by weight), chlorine (58%) with minor components of magnesium (4%), sulphur (3%), calcium (1%), potassium (1%) and bromine (0.25%). On average, sodium chloride represents 90% of the dried weight of sea water. Therefore it is useful as an estimator of the seaspray component in fine particles. The IBA analysis methods measure the total sodium and chlorine concentrations separately. However we need to be aware of other possible sources of these elements, as well as check that they appear in the correct proportions to be associated with seasalt.

Studies by Cohen (1995) have shown that the average fine particle sodium and chlorine contents were around 5% of the total fine particle mass. This was similar to the soil content across the network (5–10%). There is some sodium in soils, about 1% (Weast, 1977). This small correction to the total sodium measured for the sodium present in soil is of little consequence and the main source of sodium in fine particles is seasalt. Chlorine, however, has several other sources including motor vehicles, and it is well known that volatile compounds like chlorine and bromine can be lost from the filters particularly in the presence of acidic aerosols.

The molar weight of Na = 23 g, Cl = 35.5 g, and NaCl = 58.5 g. Therefore the sodium chlorine component of seasalt can be estimated in several ways. Sodium chloride equals 2.54 times Na concentration, or sodium plus the chlorine concentration, or even 0.61 times Cl concentration. To obtain

Table 5. The estimate sulphate, ammonium sulphate, ammonium bisulphate and sulphuric acid concentrations from the elemental sulphur concentrations.

Form	Definition	Neutralisation	Comments
SO_4^{2-}	$= 3.000 * S$	—	Sulphate ion
$(NH_4)_2SO_4$	$= 4.125 * S$	100%	remote, urban sites
$(NH_4)HSO_4$	$= 3.594 * S$	50%	urban, industrial sites
H_2SO_4	$= 3.063 * S$	0%	heavy industrial sites and Arctic or Antarctic (no NH_3)

the best estimate we show a plot of $2.54 * Na$ versus $Na + Cl$ monthly average concentrations for selected sites across the Australian, Sydney network for the two year period from January 1992 to December 1993. Figure 16a is a plot for the coastal sites of the network and Figure 16b is for the inland sites. The solid lines on each plot represent the expected line for sodium chloride where the [(Na + Cl)/2.54Na] ratio should be unity. The coastal sites in Figure 16a gave the ratio $[(Na+Cl)/2.54Na] = (0.98 \pm 38)$ consistent with sodium or chloride which is a reasonable estimator of seasalt at these sites. However the inland sites showed a ratio of only (0.59 ± 0.20), consistent with significant chlorine loss demonstrating that the chlorine element was not a good estimator of seasalt at these sites. The results in Figures 16a and b show that elemental sodium was an excellent surrogate for sodium chloride and that:

$$\text{Seasalt} = 2.54 * \text{Na} \qquad (2)$$

This is a good estimate of the total fine particle seasalt component of an aerosol.

Soil

Windblown soil is composed mainly of the oxides of Mg, Al, Si, Ca, Ti, and Fe with obviously many other trace elements. Weast (1977) gives the average composition of sandstone and sedimentary rocks and the summation of the five major oxides of Al, Si, Ca, Ti, and Fe accounts for more than 85% of the total composition. The major exclusions are MgO, Na_2O, K_2O, and water. In order to put together a composite variable that represents soil in aerosols, define fine soil as the summation of these five oxides with a multiplicative factor of 1.16 included to compensate for the exclusion of the other three oxides and water (Malm *et al.*, 1994). This exclusion factor of 1.16 is calculated from the expected abundance of these oxides in sedimentary soils (Weast, 1977). That is:

$$\text{Soil} = 2.20 * \text{Al} + 2.49 * \text{Si} + 1.63 * \text{Ca} + 1.94 * \text{Ti} + 2.42 * \text{Fe} \qquad (3)$$

This equation also assumes that the two common oxides of iron Fe_2O_3 and FeO occur in equal proportions. The factor of 2.42 for iron also includes the estimate for K_2O in soil through the $(K/Fe) = 0.6$ ratio for sedimentary soils. The elemental components of fine soil represented in equation 3 have been confirmed by independent work of Malm *et al.* (1994) and the references therein.

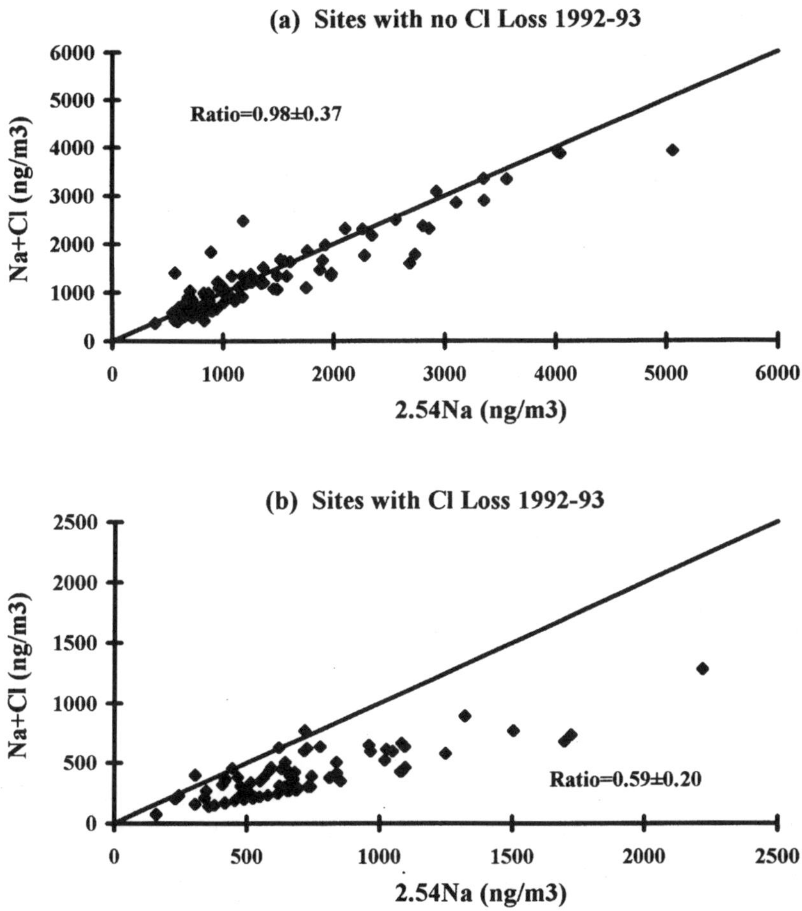

Figure 16. A plot of 2.54 Na vs Na + Cl for the monthly data (a) from coastal sites within the fine particle Australian network, (b) from inland sites within the fine particle Australian network.

Smoke

Fine potassium is an accepted indicator for smoke from wood burning and bush fires. In order to obtain a reliable smoke estimator we need to first subtract the fine potassium associated with soil. Weast (1977) gives a ratio of $(K/Fe) = 0.6$ for sedimentary soils. An estimate for smoke can be obtained from the measured total fine potassium with the potassium associated with soil removed. Smoke is given by:

$$\text{Smoke} = (K - 0.6 * Fe) \tag{4}$$

This parameter has the potential to be negative in non smoky, or industrial areas where the Fe content of the aerosol far exceeds the expected Fe content of natural soil in those areas. This parameter has been tested at sites in Australia, that have actually been shut down by smoke from bush fires. It was found to be a very sensitive indicator of smoke when fine potassium levels exceed those expected from natural soil. It is a good parameter for monitoring wood burning in metropolitan areas during the winter times. It can also be increased by factors of two–six between summer and winter seasons at sites affected by the domestic burning of wood.

Organic Matter

Organic matter is generally a significant component of fine aerosols. Cahill *et al.* (1989) have successfully measured organic matter by two methods using chemically inert Teflon® filters from 34 sites from their US network during 1982–1987. The two techniques were the remaining mass method, based on the difference between the measured mass and the sum of all measured components; and the non-sulphate hydrogen method. The IBA PESA technique for measuring the total hydrogen content on filters has good precision and high accuracy and is well suited to the non-sulphate hydrogen method of estimating organic matter. This technique assumes that:

1. The total hydrogen on the filter, as measured by PESA, is composed mainly of hydrogen from organic material and hydrogen from ammonium sulphate (assumes the sulphate ion is fully neutralised).
2. Hydrogen in water is removed in vacuum during analysis.
3. Hydrogen in ammonium nitrates is not well retained by Teflon® filters and is therefore negligible.
4. The hydrogen content of soils is also negligible (Weast, 1977).

The average organic matter is assumed to be composed of 9% H, 71% C and 20% O by weight (Malm *et al.*, 1994). This was deduced from independent measurements of organic matter from carbon by thermal optical reflectance. Cohen (1995) analysed the total hydrogen, carbon, and oxygen content of hundreds of Australian Teflon® filters from 25 sites during January to June 1993. After removing known contributions of oxygen in soil (56%), oxygen in $(NH_4)_2SO_4$ (48%), and hydrogen in $(NH_4)_2SO_4$ (6%), the best estimate of the composition of organic matter is (12 ± 10)% H, (68 ± 58)% C and (20 ± 18)% O. The errors in this estimate were quite large, consequently the recent values (Malm *et al.*, 1994) of 9% H, 71% C, and 20% O were chosen until more precise ones can be obtained. Using these assumptions one can define organic matter as calculated from hydrogen as:

$$\text{Organic matter} = 11\ (H - 0.250 * S) - \text{aerosols in form } (NH_4)_2SO_4 \quad (5a)$$

$$= 11\ (H - 0.156 * S) - \text{aerosols in form } NH_4HSO_4 \quad (5b)$$

$$= 11\ (H - 0.063 * S) - \text{aerosols in form } H_2SO_4 \quad (5c)$$

where: the H and S concentrations were measured in ng/m^3 giving the organic matter in ng/m^3. Organic matter for a fully neutralised aerosol, ammonium bisulphate aerosol, and acidic sulphuric acid aerosol is presented in Equations (5a), (5b), and (5c) respectively.

RECONSTRUCTED MASS (RCM)

The sum of all the above composite variables should provide a reasonable estimate of the total fine mass for comparison with the measured gravimetric mass of fine particles on Teflon® filters. To test this we defined the reconstructed mass (RCM) as the sum of these five composite variables plus the independently measured elemental carbon concentration, namely:

$$\begin{aligned} \text{RCM} = {} & \text{Ammonium Sulphate} + \text{Seasalt} + \text{Soil} \\ & + \text{Smoke} + \text{Organic Matter} + \text{Elemental Carbon} \end{aligned} \quad (6)$$

Contributions for most of the major elemental aerosol components measured in fine mass except P (0.1%), V (<0.05%), Cr (<0.05%), Mn (0.1%), Co (<0.05%), Ni (<0.05%), Cu (<0.05%), Zn (0.2%), Br (0.4%), and Pb (1%) are found in Equation (6). The sum of these excluded elements contributes on average less than about 2% to the total fine particle mass (Cohen, 1995). RCM mass estimate does not include nitrates, a significant fraction of which are volatilised from the Teflon® filter during collection (Malm *et al.*, 1994) and are thus not measured in gravimetric mass either. The accuracy of Equation (6) can be tested by comparing RCM with the measured gravimetric weight of the filters. Such a comparison is shown in Figure 17a for all sites in the Australian network for 1995. A least squares fit to the data shown as a solid line shows that, on average RCM accounted for $(92 \pm 9)\%$ of the gravimetric mass. This is an excellent result when one considers that the vacuum analysis reduces the fine mass by 5–8% and 2% of trace elements are not included in the definition of RCM. Clearly the composite variables define by RCM in Equation (6) are a good estimate of the total gravimetric mass on our Teflon® filters and consequently produce reasonable mass closure.

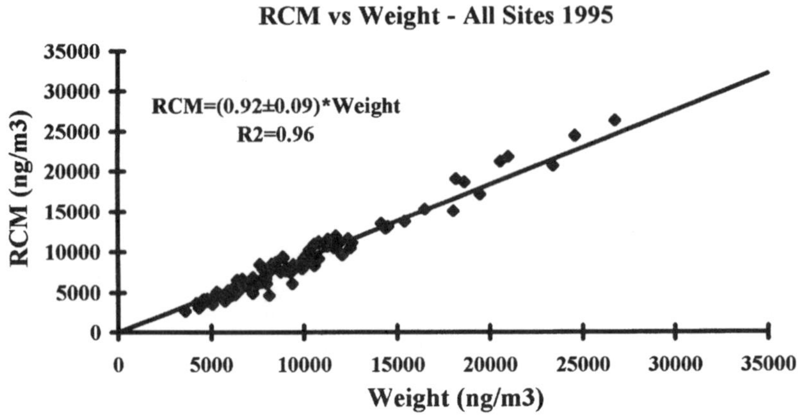

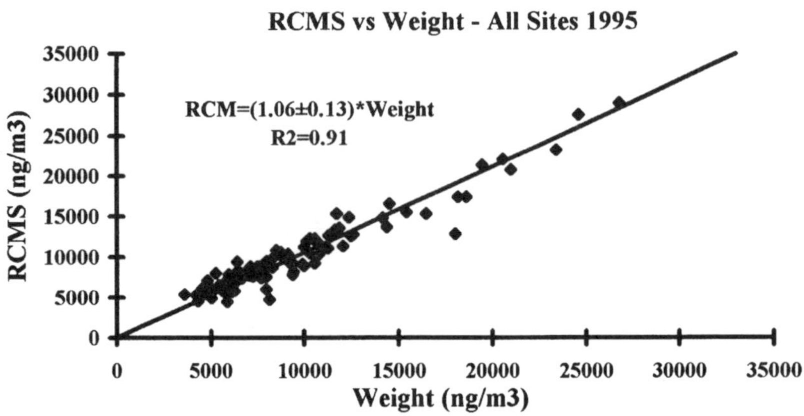

Figure 17. A plot of the reconstructed mass RCM as defined (a) by Equation (6) versus the gravimetric mass for all sites in the Australian fine particle network in 1995. The filter medium was Teflon® (b) by the sum of all elemental concentrations measured by IBA methods versus the gravimetric mass for all sites in the Australian fine particle network in 1995. The filter medium was Teflon®.

Another possible definition of the reconstructed mass is the sum of all individual elements, H, C, N, O, Na, Al, Si, P, S, Cl ,K, Ca, Ti, V, Cr, Mn, Fe, Co, Cu, Ni, Zn, Br, and Pb measured by our IBA techniques. This sum RCMS and is not as accurate or precise as the RCM defined above. The addition of so many small numbers to produce a larger one gives rise to larger errors on the final number. Furthermore, the major components of this summation of total carbon, nitrogen, and oxygen concentrations assume no chemical form for these elements and they are not precisely determined by the RBS technique. The total carbon C as measured by RBS methods will contain carbon from organic matter, elemental or sooty carbon. As well, the carbon in the filter matrix that must be subtracted out. This subtraction of carbon in the filter matrix is often difficult to do precisely and unless one can use filters with no carbon content, this method is not recommended. The plot in Figure 17b is similar to the one in Figure 17a except that this data uses the sum of all elemental concentrations (RCMS) as measured by the IBA techniques for the reconstructed mass estimate. A good correlation with the gravimetric mass is found in the RCMS data (as is the case with the RCM data). However, RCMS data has a larger spread, $\pm 13\%$, and tends to over estimate both RCM by 14% and gravimetric mass by 6%. Since we expect 5–8% water vapour to be lost from the filters in vacuum analysis (Cohen, 1995), RCMS is an overestimate of the reconstructed mass; but unlike RCM, it does contain measured components for nitrogen and oxygen. In fact, it is encouraging that the gravimetric mass is spanned so closely by the two reconstructed mass estimates RCM and RCMS.

COMPARISON WITH OTHER METHODS

Comparison of both sampling and filter analysis methods against other techniques and known reference standards, is important for good quality assurance and control in fine particle analysis. Sampling methods can be compared by running different samplers at the same site for the same sampling period and comparing results. This is an excellent way to pick up system artefacts and to obtain a feel for the true spread in experimental data. In order to compare sampling systems, it is important to analyse filters from different systems in the same way using the same techniques. Data from a stacked filter system is compared with data collected by a cyclone sampling system in Figures 18a and b. Both samplers collected PM-2.5 fine particles over a 24 hours period at the same site. A direct comparison of the gravimetric masses over a twelve month period during 1995 is shown in Figure18a. The stacked filter system that measured both PM-2.5 and 2.5–10 μm size particles on two different filters consistently produced 15% lower PM-2.5 estimates than cyclone system for mass loadings over the range from 1 to 16 $\mu g/m^3$. The standard deviation of the least squares fit to the data was $\pm 18\%$ so that this difference was not significant at this level. However the data does show that for comparisons of this type spreads of $\pm 20\%$ in the fine mass for different samplers at the same site, is not uncommon. Figure 18b is a plot of the sulphur concentrations

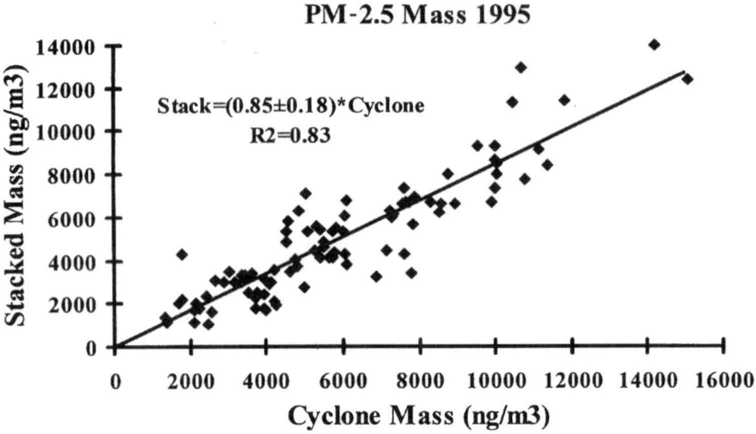

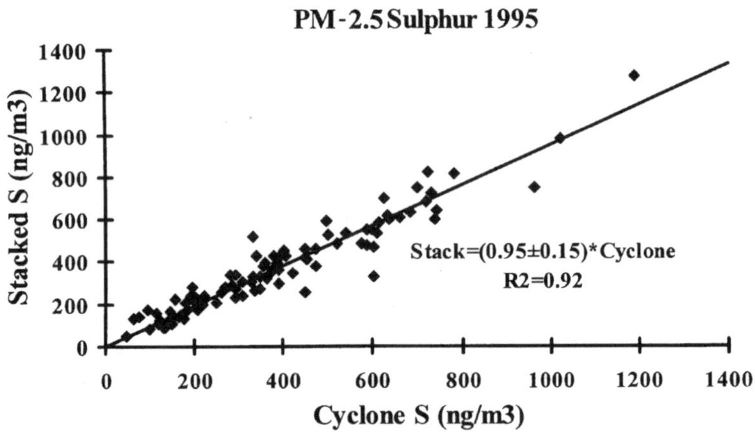

Figure 18. (a) Comparison of gravimetric mass over a 12 month period for a stacked filter system and a cyclone system measuring PM-2.5 particles at the same site for the same sampling times. (b) Comparison of PIXE sulphur concentrations, over a 12 month period, for a stacked filter system and a cyclone system measuring PM-2.5 particles at the same site for the same sampling times.

as measured by PIXE on the same filters and samplers as used in the data depicted in Figure 18a. This shows that the sulphur concentration for PM-2.5 particles as measured by the stacked filter system and the cyclone system agree within $(95 \pm 15)\%$. Comparisons of this type provide great confidence in the use of these two sampling systems as well as the consistency of the PIXE method for sulphur analysis.

A good method to check the quality control of analysis techniques is to run comparisons on known reference standards for as many different elements as possible. This has been performed on these IBA techniques over the years with great success. A comparison on thin Micromatter foils of known elemental concentrations taken over a 30 month period using the PIXE technique is found in Figure 15. The data illustrates that with careful experimental techniques $\pm 3\%$ precision over extended periods of time is possible, $\pm 5\%$ is attainable by most groups with some experience, and $\pm 10\%$ achievable by most using these thin aerosol samples. More comparisons for a range of elements employing IBA techniques with known references standards made at our laboratories over the years is found in Figure 19. With appropriate attention to detail, precision of less than $\pm 10\%$ is achievable even on thick reference standards like those shown in Figure 19.

Several comparisons of concentrations obtained by IBA techniques with those using other methods such as ion chromatography (IC), neutron activation (NAA), and inductively coupled plasma mass spectrometry (ICP-MS) have been made (Cohen, 1994; 1995). At the $\pm 5\%$ level these comparisons were most favourable. The results of a comparative study of the analysis of IAEA dust, fly ash, and soil samples using NAA, ICP-MS, PIXE and PIGME methods are shown in Figure 20. The plot contains 80 points spanning six decades of concentration from parts per million to a 100% and covers more than 20 different elements. The ratio of the fit for NAA and ICP-MS to PIXE and PIGME on the same samples is (0.975 ± 0.10)

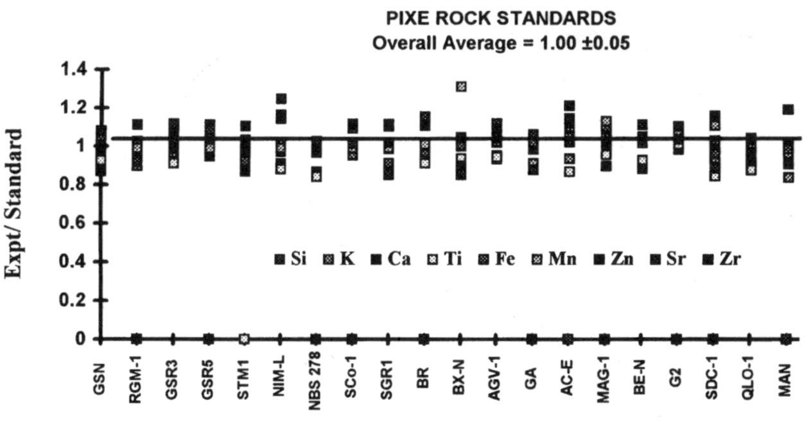

Figure 19. Comparison of measured to nominal concentrations for a range of known reference standards using IBA methods of analysis.

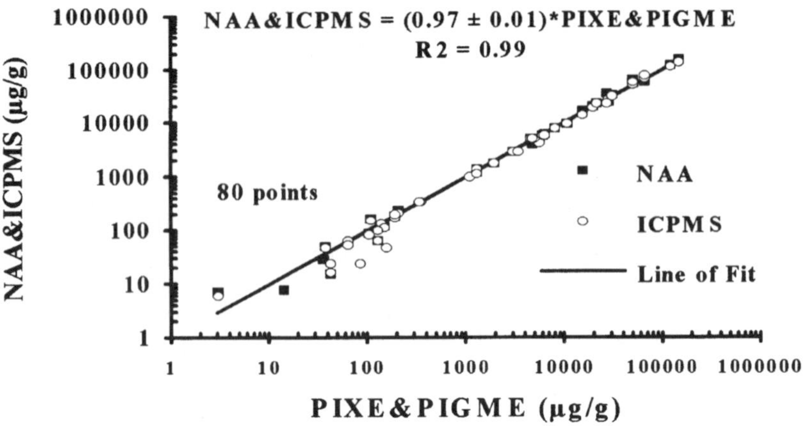

Figure 20. A comparison of PIXE and PIXE techniques with NAA and ICP-MS on three different IAEA reference samples: dust, soil, and fly ash.

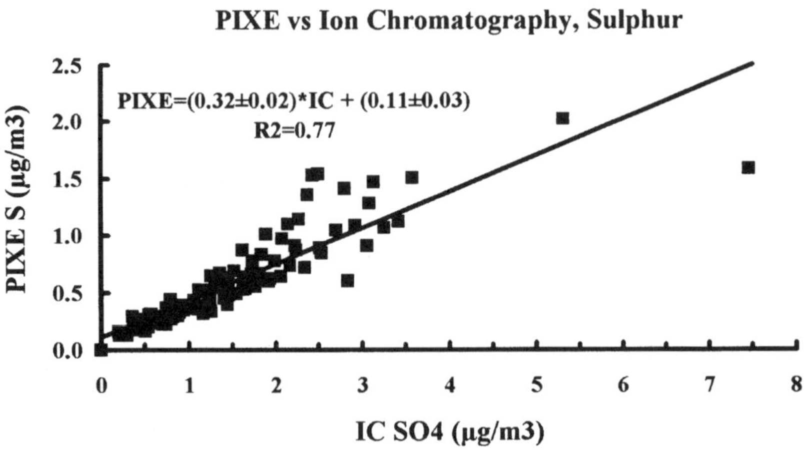

Figure 21. A comparison of the PIXE technique for sulphur with ion chromatography for the sulphate ion.

with a coefficient of least squares fit $R^2 = 0.99$. This was an excellent agreement since the analyses were all carried out in different laboratories. A comparison of sulphur concentrations by PIXE, with the sulphate ion concentration as measured by IC for 102 filters collected at the same site and sampled on the same day is also shown in Figure 21.

Since PIXE is a non-destructive method, unlike IC, it allows for further analyses to be completed afterwards. Sulphate ion concentrations (SO_4^{2-}) are expected to differ from sulphur concentrations by a factor of three. The gradient of the least squares fit to the data shows this to be the case. This is again an excellent result for two totally independent measurement techniques. Similar results have been obtained for chloride ions (Cl^-) by IC and chlorine by PIXE, sodium by NAA and PIGME and chlorine, manganese, bromine, vanadium, and potassium by NAA compared with PIXE for these elements (Cohen, 1995). An example of the precision that can be obtained for careful comparisons between totally different techniques on a sample for an individual element is found in Figure 22. It shows the comparison between chlorine using PIXE and NAA on over a hundred different samples. The concentrations ranged from 0 to 3 μg/m^3 of chlorine and as indicated in the plot, the correlation was high with agreement between PIXE and NAA to 4%. This along with other published evidence over the past few years provides great confidence in IBA techniques for accurate and precise multi-elemental analysis of fine particle filters. These methods have considerable advantages for aerosol analysis, however there are some drawbacks as with all techniques. The common advantages and disadvantages for IBA techniques for aerosol filter analysis are compared in Table 6.

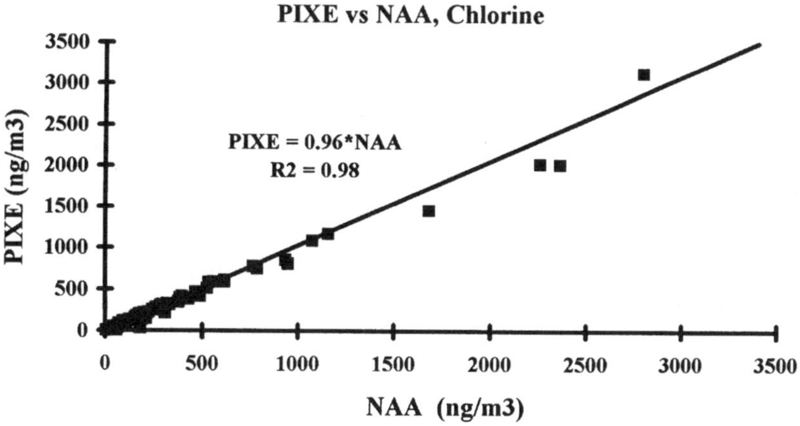

Figure 22. Comparison of chlorine measured by PIXE and NAA on over 100 Teflon® filters.

Table 6. Comparison of the advantages and disadvantages of IBA methods for aerosol filter analysis.

Advantages	Disadvantages
1. Non-destructive (low beam currents)	1. No chemical information
2. Minimal sample preparation	2. Requires homogeneous solid samples
3. Quantitative ($\pm$3–5% if careful)	3. Can not easily do liquid and gaseous samples.
4. Multielemental (H to U)	4. Some particle size effects for low atomic number materials.
5. Fast, 15 min irradiation times	5. Not cost effective for a few samples
6. Sensitive (μg/g to 100% for many elements)	6. Requires access to an ion beam accelerator (3 MV)
7. Can handle small sample sizes ($<$1 mg)	
8. Cost effective for hundreds of samples	

TYPICAL RESULTS FROM A LARGE AREA NETWORK

The power of the IBA methods discussed above will be demonstrated by presenting some typical results obtained for a large area fine particle sampling network, known as the Aerosol Sampling Program (ASP), centerd in Sydney, Australia. Sydney is the most populated city in Australia and is situated 34S, 151E on the eastern seaboard of the Australian coastline in the State of New South Wales. It is a coastal city with a moderate climate. The population of New South Wales is about four million people with more than three million people living within a 150 km radius of the center of Sydney. Approximately 100 km to the north and 75 km to the south are the major industrial cities of Newcastle and Wollongong respectively, with industries such as iron and aluminium smelting, sulphuric acid plants, lead and zinc production and fertiliser works. There are seven major coal fired power stations which burn over 20 M tonnes of coal a year, and supply the city of Sydney with power. These are located predominantly to the north and west of Sydney as shown in Figure 23. The major Sydney basins contain the city center and are bounded by the coast and a mountain range some 100 km to the west. This together with large periods of sunshine, extensive use of leaded petrol for cars, industry to north and south, and the burning of coal containing sulphur, results in a few high pollution days per year with poor visibility, photochemical smog, ozone, sulphur dioxide, and nitrogen dioxide episodes.

In the middle of 1991 a collaboration was established to sample fine particles using cyclone samplers covering an area up to 200 km from the center of Sydney, in order to determine the elemental composition of these particles using IBA methods. The siting of the sampling units within this

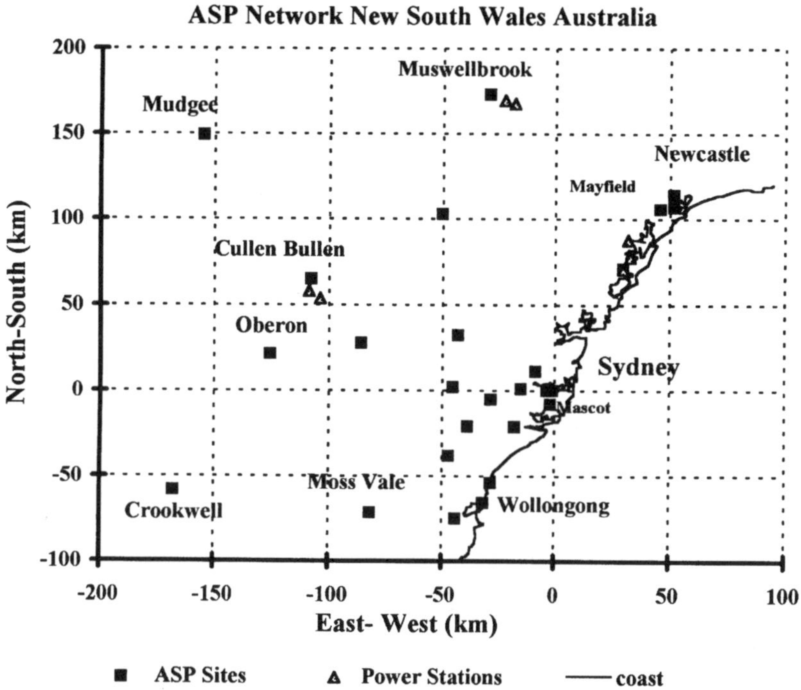

Figure 23. Location map of the Sydney, Australia aerosol sampling network sites in 1993.

sampling area is shown in Figure 23. The sites selected included, at one extreme, the most urbanised and industrialised areas in eastern Australia, and at the other, rural sites adjoining large areas of national parks and virgin bushland. Samples of fine particles with diameters $<2.5\,\mu m$ were collected on stretched 25 mm diameter Teflon® filters over a 24 h period, every Sunday and Wednesday at each sampling site. These filters were about $(220 \pm 20)\,\mu g/cm^2$ thick and typically contained a total fine particle mass on the filters for analysis of between 50 and $500\,\mu g/cm^2$. This converted to a mass per cubic metre (ng/m^3) of air sampled of between $3.5\,\mu g/m^3$ and $35\,\mu g/m^3$. The sampling network provided up to 300 filters per month for analysis by the four IBA techniques discussed above. To date over 12,000 filters have been analysed by these IBA methods for more than 20 elements, each from this ASP network and other related sampling sites. This makes for a very powerful multi-elemental database that can be used for fine particle characterisation, the study of regional and seasonal variations, and fingerprinting of fine particle sources in the Sydney region.

Average Aerosol Composition

When so many different elements are measured in a given filter analysis by IBA methods it is possible to estimate the average fine particle composition on the filter fairly accurately. In Table 7 the analysed data for a representative filter collected at the Mascot site (see Figure 23) is presented. This site is in the inner city near Sydney's main international airport and dominated by motor vehicles with some light industry in the surrounding areas. Elemental concentrations in μg/cm^2 of filter area, the percentage of each element relative to the gravimetric mass, and minimum detectable limit (MDL) for each element for this filter are shown. The errors quoted are total error estimates and represent the counting statistics as well as instrumental errors for each element. The mass quoted was obtained from the gravimetric mass and RCM is the reconstructed mass as discussed earlier. For this example, the RCM was (106 ± 7)% of the gravimetric mass. typically for Teflon® filters RCM ranges from 70% to 110% with an average for all filters of less than 100% as shown in Figure 17a.

Table 7. Elemental concentrations for a typical fine particle filter (24 h) from the inner city site of Mascot.

Species	Conc. (μg/cm^2)	Mass %	MDL (μg/cm^2)	Species	Conc. (μg/cm^2)	Mass %	MDL (μg/cm^2)
Mass	196 ± 10	100	2.0	Cr	0.02 ± 0.01	0.01	0.02
H	5.4 ± 0.5	2.7	0.3	Mn	0.7 ± 0.2	0.3	0.02
C	103 ± 6	52	11	Fe	4.2 ± 0.2	2.2	0.03
N	10.0 ± 0.9	5.2	2	Cu	0.19 ± 0.05	0.9	0.03
O	26 ± 1	13	2	Co	0.02 ± 0.02	0.01	0.07
Na	21 ± 2	11	2	Ni	—	—	0.04
Al	1.0 ± 0.2	0.5	0.15	Zn	1.1 ± 0.2	0.5	0.03
Si	2.3 ± 0.1	1.1	0.07	Br	0.6 ± 0.1	0.3	0.06
P	0.07 ± 0.03	0.04	0.07	Pb	2.0 ± 0.3	1.0	0.2
S	10.5 ± 0.3	5.3	0.06	RCM	208 ± 15	106	
Cl	16 ± 3	8.0	0.06				
K	1.38 ± 0.05	0.7	0.03	Seasalt	52 ± 9	27	2
Ca	1.7 ± 0.1	0.8	0.03	Soil	21 ± 3	11	0.1
Ti	0.14 ± 0.04	0.07	0.03	Organics	31 ± 3	16	0.3
V	0.04 ± 0.03	0.02	0.03				

The numbers quoted in Table 7 can be converted to approximated values of ng/m^3 of air sampled by multiplying by 72. This table contains 25 or more different chemical species. Using the arguments provided above, an estimate of the major percentage compositions can be made by selecting just five composite variables: ammonium sulphate, seasalt, soil, smoke, and organic matter. The average annual percentage contributions of these variables together with lead for the Sydney fine particulate aerosol for 1992 and 1994–95 are given in Table 8. The two compositions show that the average annual aerosol composition has remained similar over this four year period. The large ranges for these parameters is explained by the large seasonal fluctuations experienced and is discussed further below.

Organic matter and elemental carbon or soot account for nearly half of the total fine particle matter in the Sydney aerosol (Table 8). The next major constituent is ammonium sulphate. Note that nitrates are not listed in Table 8 because all the data is derived by Teflon® filters and nitrates are not well retained on these filters. It is expected that nitrates together with about 1% of trace elements such as V, Cr, Cu, Zn, etc., and water vapour account for the difference between the average RCM and the gravimetric mass quoted in the table.

Seasonal Variations in Aerosol Compostion

It is possible to follow seasonal variations not only in the fine particle loadings, but also in different elemental concentrations for sampling sites

Table 8. Average annual percentage contribution of major composite variables for the ASP network for 1992 and 1994–95. Elemental carbon was measured independently using laser integrated plate methods (Cohen, 1995).

Composite Variable	Jan–Dec 1992		Jul 94–Jun 95	
	Average(%)	Range(%)	Average(%)	Range(%)
Ammonium Sulphate	23	15–35	21	12–33
Seasalt	7	2–15	10	1–23
Soil	6	4–18	10	6–15
Smoke	10	0–40	13	0–50
Organic matter	23	10–40	23	6–38
Elemental carbon	22	13–35	29	18–39
Lead	1	0–3	0.9	0–3
Mass (μg/m^3)	8	5–15	10	7–17
RCM (%)	81	68–95	92	83–96

that have been in operation for many years. This can provide insights into possible chemical processes that may occur as well as corroborating different meteorological conditions that occur seasonally. The average monthly fine particle mass for the Mascot site in Sydney for the period January 1992 to December 1995 is found in Figure 24. There are clear maxima in the winter and minima in the summer. This is common for Sydney fine particle aerosols. The weather conditions are calmer and more conducive to temperature inversion layers in the winter producing higher fine particle masses. In the summer the temperature is warmer, there is more ultraviolet light, and the conditions for converting sulphur dioxide from a gas to particulate sulphate ions and hence detectable sulphur on filters is more favourable. This is seen in Figure 25 where the average monthly sulphur concentrations for the two year period from 1992–93 for the Muswellbrook site are shown. Seasonal variation in the total mass and sulphur are clearly cyclic and opposite in nature illustrating how different external parameters play a vital role in fine particle concentrations and their elemental compositions. Following these seasonal and long term variations for a considerable time period, for particular elements of interest, can prove very useful to air pollution managers and regulators. In Sydney we have had a concerted campaign over the last few years to reduce the lead in the atmosphere originating from motor vehicles due to the use of leaded petrol.

Fine particulate matter at the Mascot site in Sydney is dominated by motor vehicles. During the winter months motor vehicles may contribute

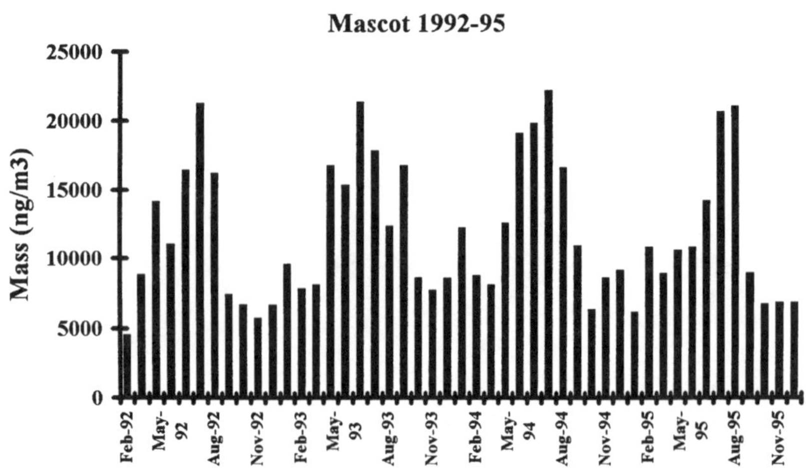

Figure 24. Average monthly fine particle mass for the Mascot site in Sydney for the period January 1992 to December 1995.

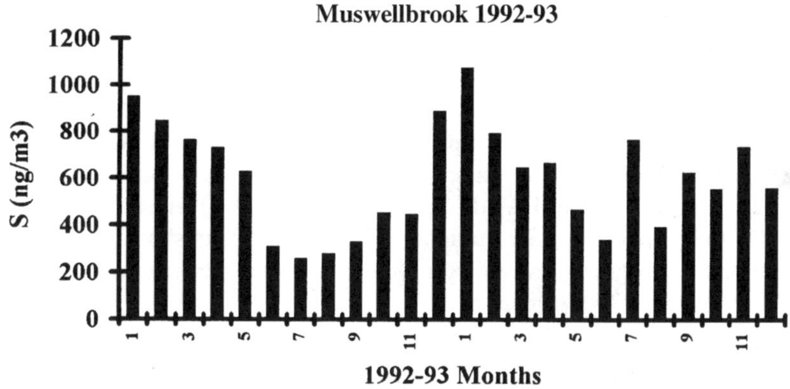

Figure 25. Average monthly sulphur concentrations at the Muswellbrook site for 1992–93.

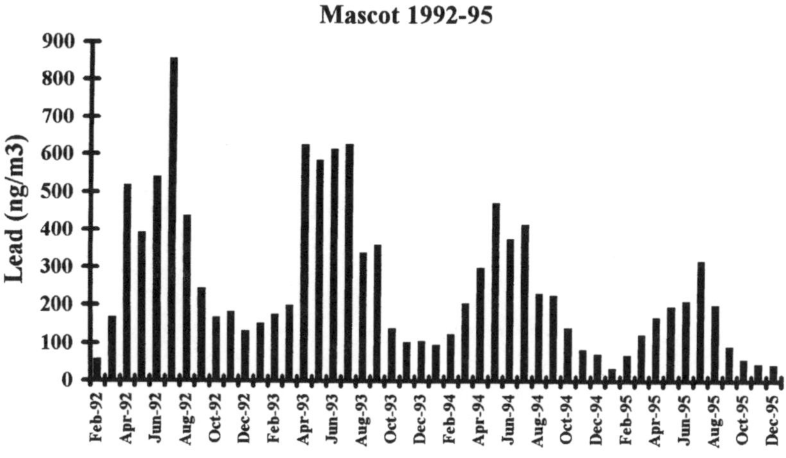

Figure 26. Average monthly fine particle lead concentrations at Mascot during 1992 to 1995 inclusive.

over 80% to the fine mass. This drops to 30% in the summer. The seasonal variations in lead concentrations at the Mascot site are shown in Figure 26 for the same period as the one in Figure 24. It further illustrates a significant drop in the average monthly lead concentrations over this period. The peak winter lead concentrations in 1992 were around 600–800 ng/m^3. In 1995 this had dropped to 300–400 ng/m^3. Similar summer minima drops are also seen in the plot. These reduced levels of lead in air were very encouraging since the government had reduced the lead content in petrol from 0.4 to 0.2 g/L and also introduced pricing penalties on lead petrol during this period. Clearly, these measures were working.

Other seasonal trends can be followed in this manner. For example, seasalt as defined in Equation (2), and calculated from the PIGME sodium concentration, can be shown to peak in the summer months when the more favoured afternoon seas breezes bring large amounts of seaspray from the coast to regions further inland (Gimsenius *et al.*, 1996). By following the PIXE fine potassium concentrations it is also possible to show the relative importance of domestic wood burning during the winter in several suburban areas in the Sydney region.

Regional Variations in Aerosol Composition

A large area network of many sampling sites, such as the one in Figure 23, enables one to study regional differences in fine particle mass as well as composition. The average annual fine mass for each of 24 sites in the Sydney ASP network during 1993, in a three dimensional display is shown in Figure 27. The network covers an area of 60,000 square kilometres, being 300 km along the Australian coast from Wollongong, through Sydney to Newcastle, and up to 200 km inland. The figure shows a relatively even cover of fine particles across the network with rural sites (200 km inland) like Crookwell, Oberon, and Mudgee having lower mass concentrations than coastal inner city or industrial sites in Wollongong,

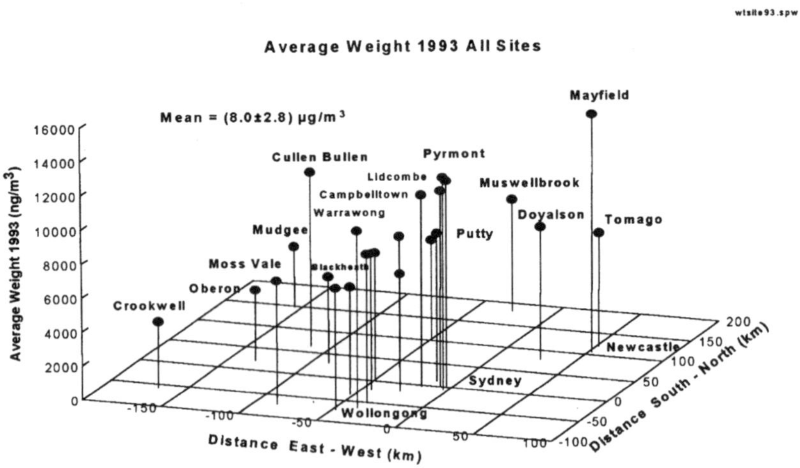

Figure 27. The annual average fine particle mass for the 24 sites in the ASP network in Sydney, Australia during 1993.

Sydney, and Newcastle. To explain the differences in fine particle mass between each of these sites in the network we need to look more closely at local industry, meteorological conditions, and possible interregional transport of pollution into these sites. The possibility of looking at individual elemental concentrations over this region helps to do this with a greater certainty.

Similar three dimensional plots for the average sulphur concentrations during December 1992 and lead concentrations during July 1992 are shown in Figures 28 and 29 respectively. The two plots show quite different regional distributions with sulphur being relatively wide spread across the network from Muswellbrook in the north to Moss Vale in the south and west as far as Crookwell, Oberon and Mudgee. The average sulphur concentration over the region during December 1992 was $(0.55 \pm 0.13)\ \mu g/m^3$, and the majority of this was in sulphate form. If all of this sulphate ion was fully neutralised to ammonium sulphate, then an estimate of the ammonium sulphate across the region can be obtained by multiplying the sulphur levels given in Figure 28 by 4.125 as given in Table 6. Care should be taken in extending this argument too far. Although there is strong evidence to show that all sulphur occurred as sulphate on the filters, there is no evidence to demonstrate that all of this was fully neutralised to the ammonium sulphate form. Some of the over forms of sulphate given in Table 6 may be also relevant.

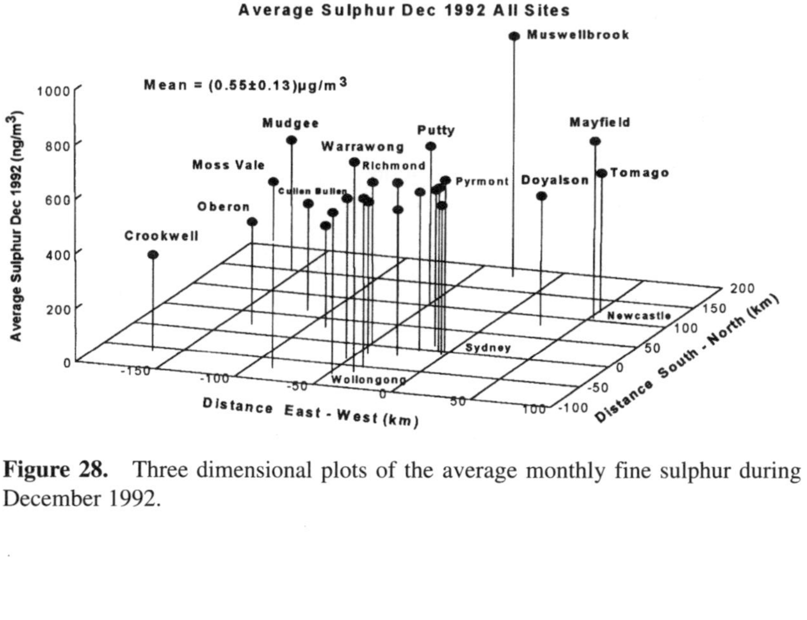

Figure 28. Three dimensional plots of the average monthly fine sulphur during December 1992.

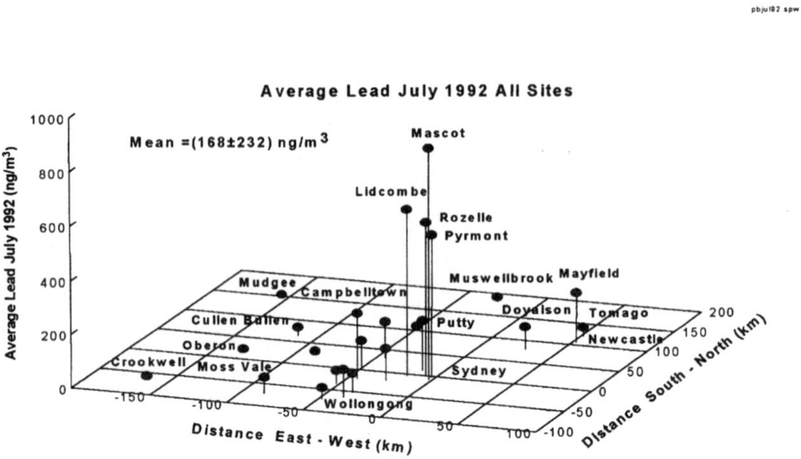

Figure 29. Three dimensional plots of the average monthly fine lead during July 1992.

In contrast to sulphur, lead concentrations during July 1992 given in Figure 29, demonstrate marked regional variations across the network. Inner Sydney sites at Lidcombe, Mascot, Rozelle, and Pyrmont clearly dominate the distribution. The industrial sites at Wollongong, 80 km to the south of Sydney, and Newcastle, 150 km to the north of Sydney, had fine lead concentrations typically three to five times lower than the inner Sydney sites. Correlations of lead with bromine over the network region showed that lead concentrations were clearly dominated by contributions from motor vehicles (Cohen *et al.*, 1995). In the discussion of Figure 26 the seasonal variations of lead at the inner Sydney site of Mascot and how it was dominated by emissions from motor vehicles was explained. If this type of data is combine with information similar to the regional data shown in Figure 29, then a fairly complete picture of lead from motor vehicles in the Sydney region can be made.

The ASP network sites shown in Figure 23 can be grouped into bins which reflect their distance from the center of Sydney to the coast line as defined by the cities of Wollongong, Sydney, and Newcastle. Then various elemental concentrations can be plotted as a function of distance from these source points. Figures 30 and 31 are such plots: lead for distance from the center of Sydney, and seasalt, as defined by Equation (2), for the distance from the coast. The plots show the relative fall off of these elemental concentrations and hence of the fine particle sources associated with them as one moves away from the source. They demonstrate the

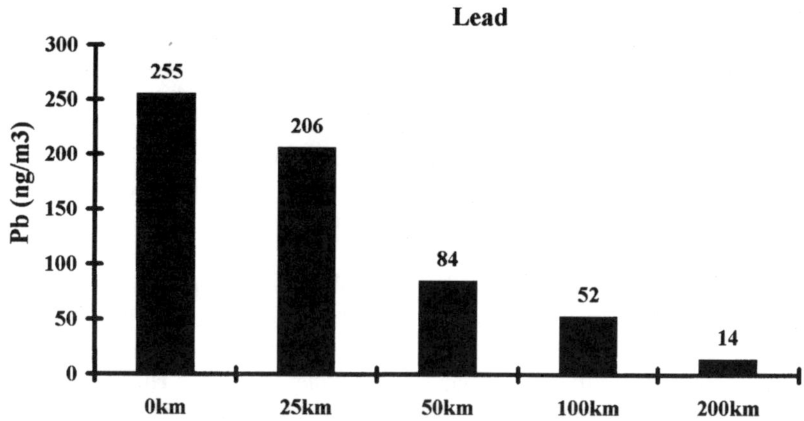

Figure 30. Plot of average lead concentrations versus distance from the center of Sydney during 1992–93.

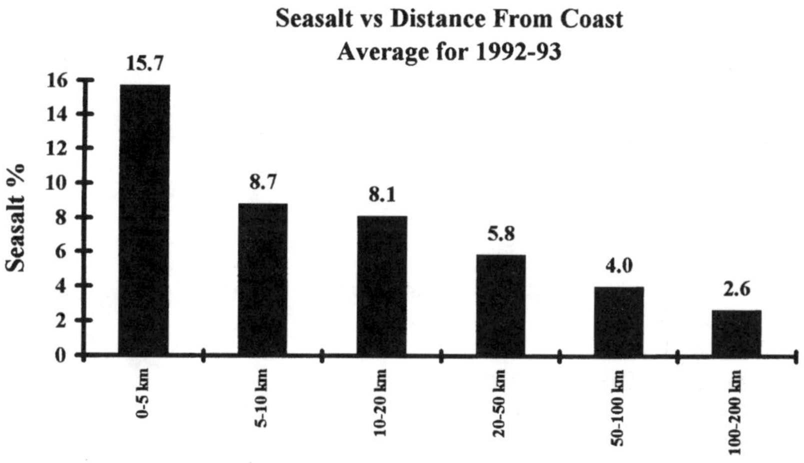

Figure 31. A plot of average percentage seasalt versus distance from the coast for 1992–93.

capability of fine particles to have a local effect with lead that is really confined to the Sydney basin (20 km), or a regional effect as with seasalt that extends hundreds of kilometres inland. They also demonstrate the need of analytical techniques to provide a range of elemental analysis from 10 to several thousands of ng/m^3 of air sampled on a filter sample that may only contain 100 µg of sample. The accelerator based ion beam analysis techniques as discussed here are capable of doing just that.

Elemental Correlations in Aerosols

Demonstrating that one element correlates strongly with another is a powerful way of identifying possible sources of both natural and anthropogenic pollution. A good example of this was shown in Figures 16a and b where the correlation between the elements sodium and chloride were used to define a seasalt source based on NaCl. They also show that this expected inter-element correlation breaks down when acidic aerosols produce loss of chlorine from the filter.

A big advantage of the multi-elemental IBA techniques is that, with upwards of 20 different elements to chose from, many more of these correlations can be tested. For example, motor vehicles are a major contributor of fine particle pollution in the inner Sydney regions. The correlation between lead and bromine, both of which are additives to Sydney petrol in known proportions can be tested (Cohen, 1996). Figure 32 is a plot of the lead bromine correlation for the Mascot site for the 48 months from January 1992 to December 1995. It shows a strong correlation ($R^2 = 0.98$) between lead and bromine with a measured gradient of (2.02 ± 0.28) and an offset of (46 ± 40) ng/m^3. The gradient is very close to that for Sydney petrol where a lead to bromine ratio of (Pb/Br) = 2.1 is expected, and shows that the bulk of the lead in the air around Mascot is associated the petrol combustion. A similar lead to bromine plot for the industrial site at Mayfield, in Newcastle north of Sydney is found in Figure 33. In this situation the expected ratio of (Pb/Br) = 2.1 is shown as the solid line, and the

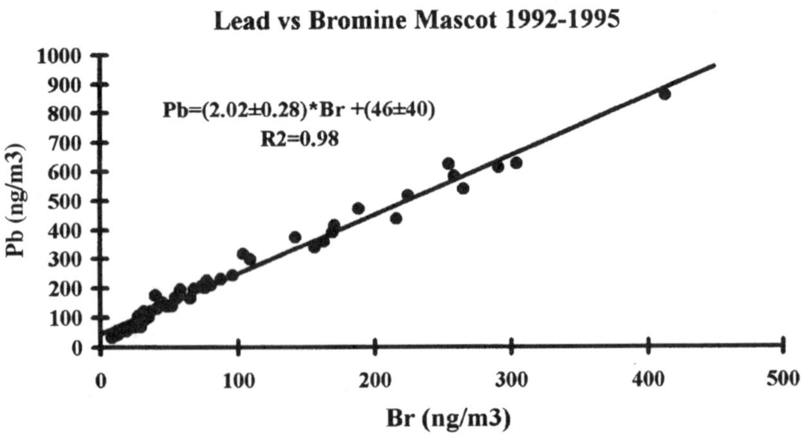

Figure 32. Plot of lead versus bromine for 48 months from January 1992 to December 1995 for the Mascot site which is dominated by motor vehicles.

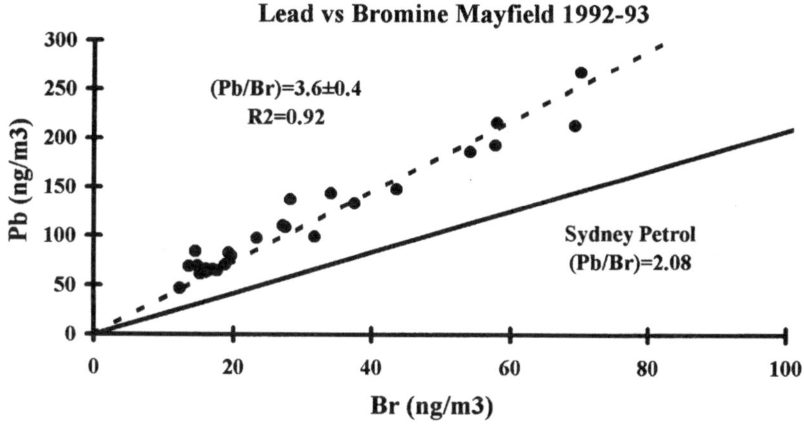

Figure 33. Average monthly lead versus bromine concentrations for the industrial site of Mayfield in Newcastle north of Sydney for the period 1992–95 inclusive.

bulk of the data lies above this line and is still strongly correlated. The data does not lie on this solid line for two possible reasons: the presence of excess lead not associated with motor vehicles, or the loss of bromine due to the presence of acidic aerosols. Either or both of these reasons are possible for this site since it is an industrial one close to large metal manufactures, and has high sulphate levels relative to other ASP sites in the network.

Similar correlation plots can be formulated for composite variables (defined in the section on reconstructed mass), and various elements to test for sources of fine particles. For example, a plot of soil, as defined by Equation (3), against silicon and at a rural site not effected by other anthropogenic sources will give the average silicon content of airborne soil at the site. The correlation between soil and silicon at this site is very strong, with the gradient of the plot giving the average silicon content of the soil as 24%. This percentage of silicon compares favourably with the Mason (see Weast, 1977) value for crustal material of 28% silicon. Similarly the titanium correlates well with soil giving an average content of 0.7% that again is comparable with the Mason value of 0.44%. Figures 34a and b are such plots for silicon and titanium, known components of soil, for the rural site of Oberon hundreds of kilometres from the major coastal city centers. A similar correlation plot for soil and silicon for the heavy industrial site of Mayfield in Newcastle is given in Figure 35.

All the data lies above the 24% silicon line and is not well correlated. This demonstrates components of soil not associated with the rural type

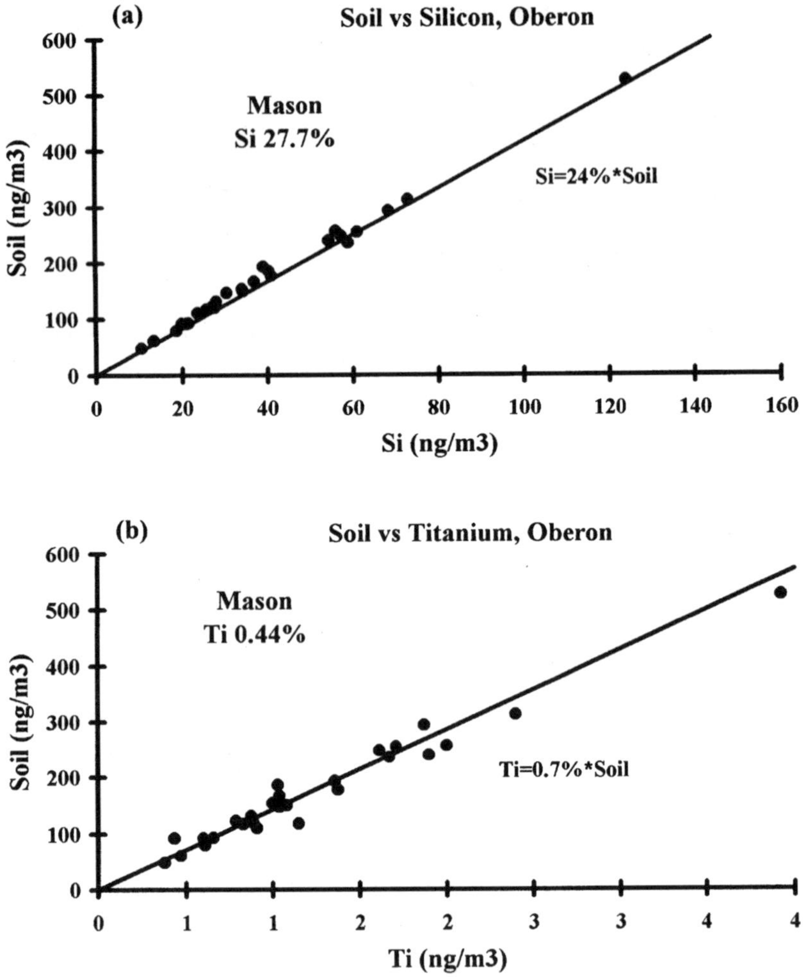

Figure 34. Plot of soil versus: (a) silicon concentration for the rural site of Oberon in the Australian ASP network; (b) titanium concentration for the rural site of Oberon in the Australian ASP network.

soil. A closer inspection of the data shows that the excess soil is produced by an additional iron component in this soil consistent with the large amounts of steel production in the vicinity. Plots like Figure 35 can therefore be used to estimate the amounts of natural and anthropogenic soil that make up the total soil content at that site. For example, if one drops a

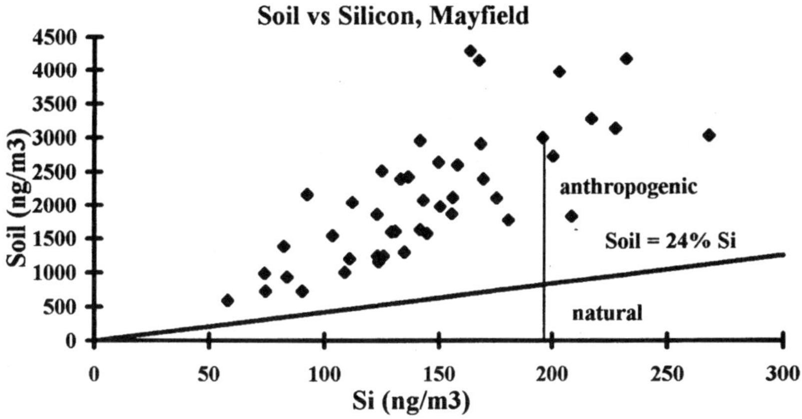

Figure 35. Correlation plot of soil versus silicon for heavy industrial site of Mayfield in Newcastle.

vertical line down from a given data point in Figure 35, then the amount of soil from 0 to the 24% silicon line may be an estimate of the natural component. The remainder from the line to the point itself is an estimate of the anthropogenic component.

Source Fingerprinting of Aerosols

In order to obtain a chemical fingerprint of a given fine particle source, two things must be determined: which elements characterise that fingerprint, and the relative proportions of each of those elements that provide the signature for that fingerprint at a sampling site. The fingerprint for several sources can then be used in Chemical Mass Balance (CMB) programs (Watson *et al.*, 1990) to determine the relative contributions of each of them to the total ambient particle mass. The major advantage of accelerator based IBA methods is that they provide multi-elemental fingerprints that give enough elements to uniquely determine a given source fingerprint.

The multivariate analysis method of Principal Component Analysis (PCA) followed by varimax orthogonal rotation may be used to determine the appropriate elements for each chemical fingerprint (Hopke, 1991). Having determined the appropriate elements, events such as favourable wind information and sampling site selection are employed to optimise the source receptor parameters being considered. The measured ambient relative elemental concentrations together with these parameters can then be used to estimate each individual fingerprint or profile for use throughout a network of sampling sites. For this process to work well, there must be a sufficient number of sites in a network that have been operating for a long enough period of time so that representative data can be used to identify sufficient fingerprints that fully specify the atmospheric aerosol being sampled. The Australian ASP network defined earlier and shown in Figure 23, has sufficient sites that have been operating for several years, and is an ideal network to show the type of fingerprint that can be obtained. These fingerprints can be put to in chemical mass balance codes.

For this ASP network six primary fine particle fingerprints were identified by PCA techniques to explain the total mass loadings on the filters (Cohen *et al.*, 1995). These were: motor vehicles, coal combustion, industry, smoke, soil, and seaspray since their elemental makeups identified them as the major contributors to these particular fingerprints. Their relative elemental concentrations are shown in Figures 36a–f and the elements used in each of the fingerprints are given in Table 9. There was considerable overlap between elements for each fingerprint. For example, hydrogen was present in motor vehicles, coal combustion, and smoke; and sulphur was present in motor vehicles, coal combustion, seaspray, and industry. However, the ratio in which all of these elements occurred in a given fingerprint was found to be sufficiently unique to identify the presence of a given fingerprint. The six fingerprints shown in Figure 36, together with the concentrations of each elemental species, were then used as input for the USEPA CMB7 programs (Watson *et al.*, 1991) to estimate their relative and absolute contributions at each site at a given time. The results of such a calculation for the Mascot site in Sydney for the months of July and December in 1994 are shown in Table 10. The total fine mass was higher in the winter than in the summer due to calmer conditions and

Table 9. The elements associated with six fine particle source fingerprints used in the Australian ASP network.

Fingerprint	Elements
Motor Vehicles	H, C, N, O, Na, Al, Si, S, Cl, Fe, Zn, Br, Pb, Elt. C
Coal Combustion	H, C, N, O, Na, Al, Si, P, S, K, Ca, Fe, Elt. C
Smoke	H, C, N, O, Cl, K, Ca, Elt. C
Soil	O, Al, Si, K, Ca, Ti, Mn, Fe
Seaspray	Na, S, Cl, K, Ca
Industry	H, C, N, O, P, S, V, Cr, Cu, Pb, Elt. C

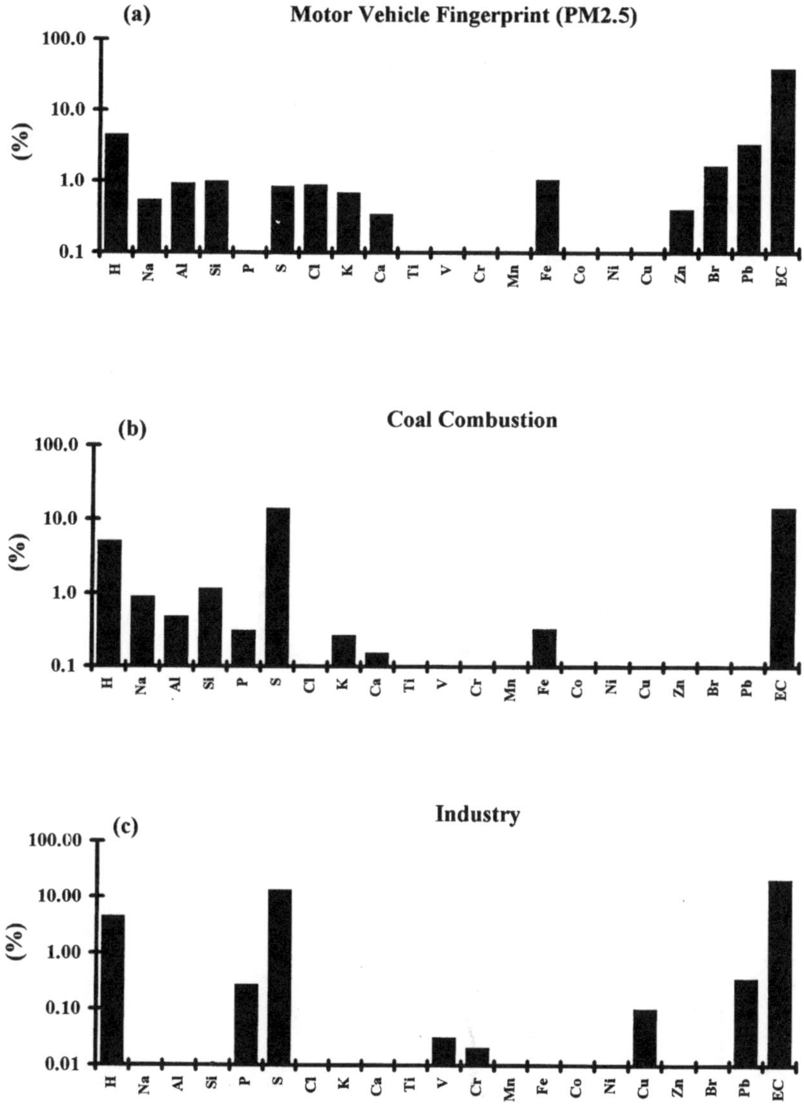

Figure 36(a–c). (continued)

the presence of inversion layers. The results indicate that motor vehicles were the major source of fine particles in the winter, but that local industry was the cause in the summer. The seaspray component for this coastal site was higher in the summer than the winter; because afternoon sea breezes

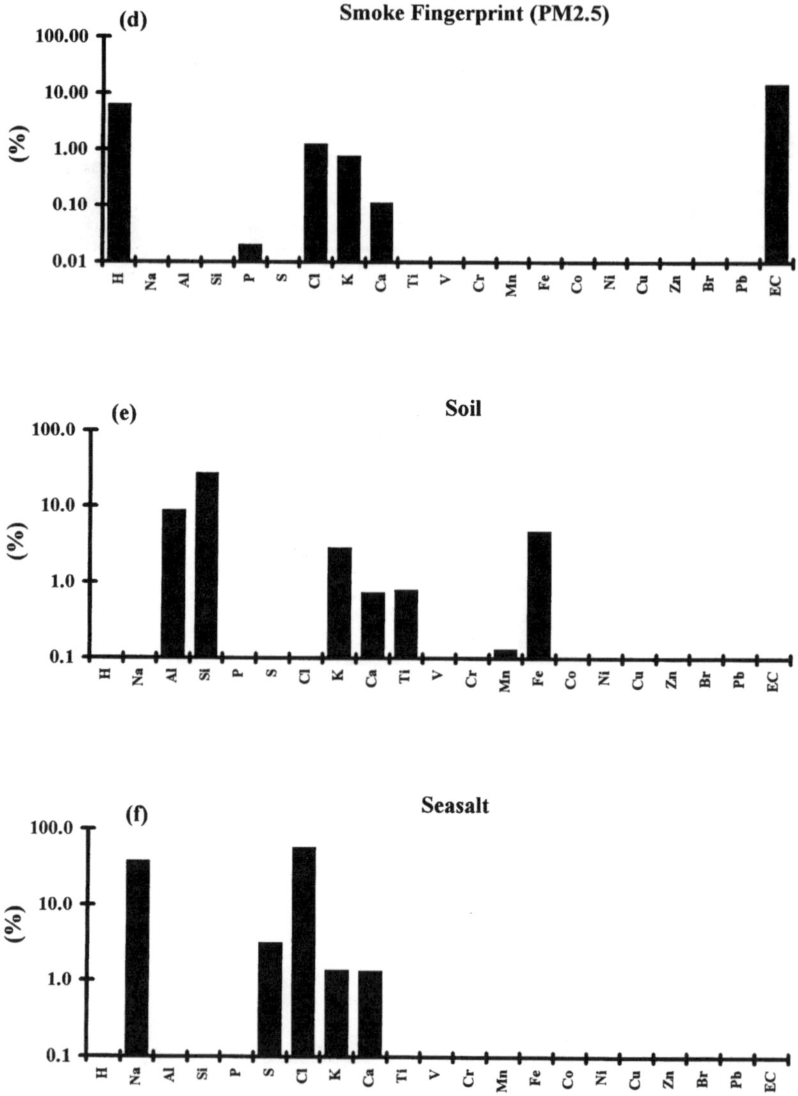

Figure 36. Elemental fingerprints for (a) motor vehicles, (b) coal combustion, (c) smoke, (d) soil, (e) seasalt, and (f) industry for the Australian ASP network.

from the north and east were dominant in summer, while winds were mainly from the south and west in the winter. These methods provide powerful techniques for studying the regional and seasonal trends in fine particle sources across a sampling network.

Table 10. The percentage fingerprint contribution at Mascot for fine particles (PM-2.5) for 1994. The US EPA Chemical Mass Balance code CMB7 was used to calculate these contributions using local source fingerprints given in Figure 36.

Fingerprint	Fingerprint Contribution (%) for Mascot Site		
	Winter July 1994	Summer Dec 1994	Average Whole 1994
Motor Vehicles	79 ± 6	24 ± 5	59 ± 18
Soil	—	2 ± 1	2 ± 2
Seaspray	4.0 ± 0.9	24 ± 2	9 ± 8
Industry	17 ± 3	51 ± 6	31 ± 12
Mass $\mu g/cm^2$	300 ± 30	130 ± 7	180 ± 70
Mass $\mu g/m^3$	22 ± 2	9.1 ± 0.5	13 ± 5

Averaged over all the ASP network sites as shown in Figure 23, for a two year sampling period the fingerprints for motor vehicles, industry, and coal combustion accounted for over 70% of the fine particle mass. Soil and seaspray contributions were generally small, between 3% and 5% of the total fine particle mass for all sites. This was not unexpected as both these particulate components were generated by mechanical methods and were therefore expected to dominate the coarser mass components. The use of these six fingerprints in the CMB codes enabled us to estimate their seasonal variations. Typical data for selected fingerprints for the 12 month period from January to December 1994 are shown in Figure 37. Most

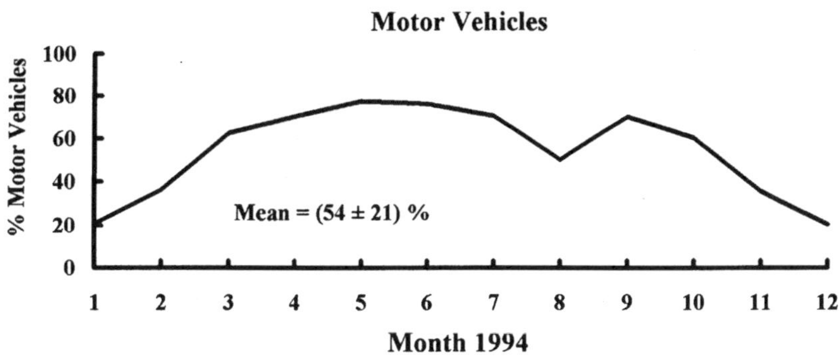

Figure 37. (continued)

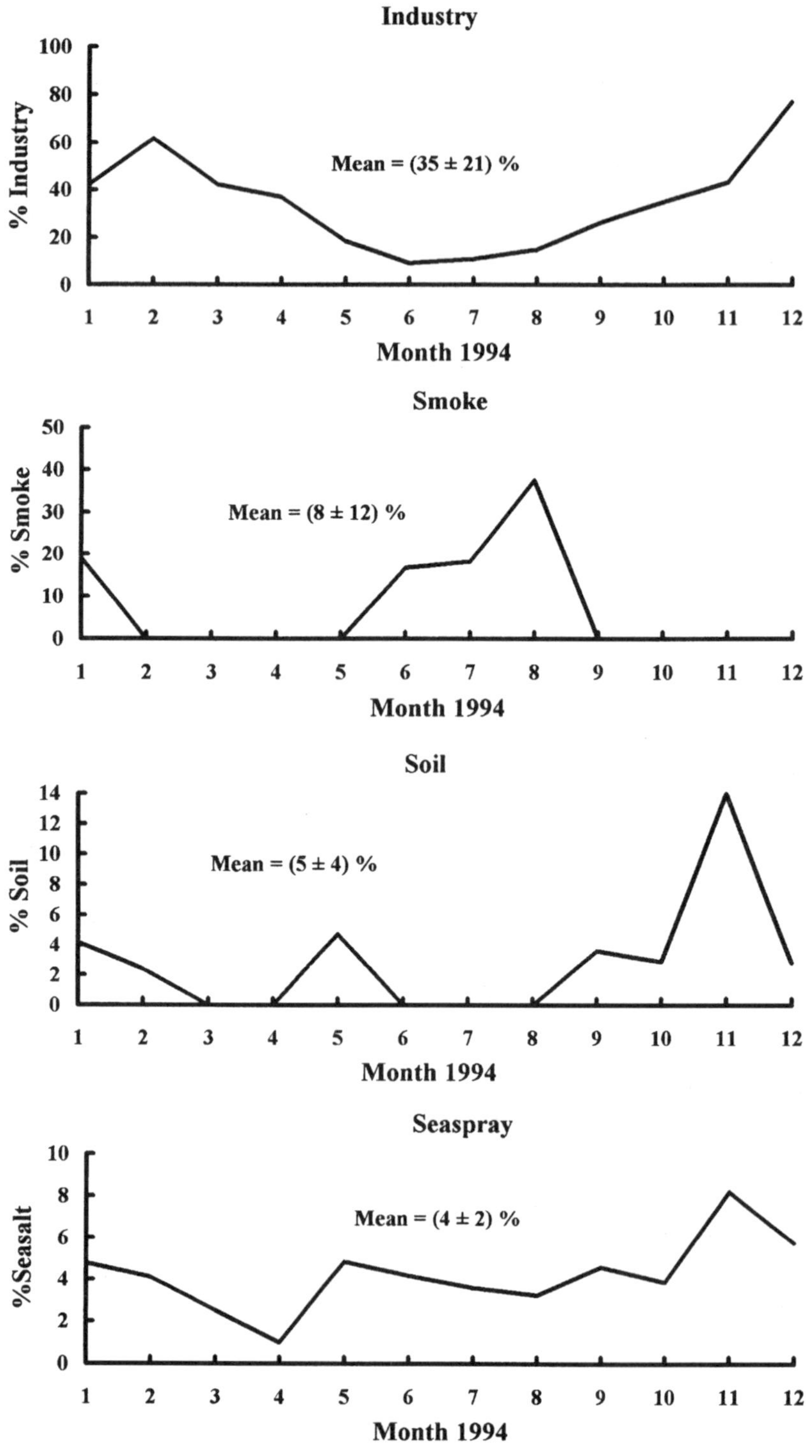

Figure 37. Average fingerprint contributions for selected fingerprints shown in Figure 36 for the ASP network in Australia during 1994.

fingerprints showed significant seasonal variations. For example, fine particles associated with wood and vegetation burning on average accounted for 12% of the total fine particle mass across the network. However there were very strong seasonal variations that accounted for the high standard deviation associated with the mean smoke mass as seen in Figure 37. Contributions from motor vehicles and industry, which includes coal combustion, in the major cities of Wollongong, Sydney, and Newcastle, dominated with 31% and 27% of the total fine particle mass respectively. Contributions from coal combustion dominated in the rural areas away from the coast, with an average of 14% of the fine particle mass.

CONCLUSION

In summary, accelerator based ion beam analysis (IBA) techniques, such as PIXE, PIGME, PESA, and RBS provide a very powerful, routine, multi-elemental analysis methods for fine particle aerosol studies. They provide elemental information on all commonly occurring elements in atmospheric particulate matter including: H, C, N, O, F, Na, Mg, Al, Si, P, S, Cl, K, Ca, Ti, V, Cr, Mn, Fe, Cu, Ni, Zn, As, Sr, Y, Zr, Br, and Pb. The precision and accuracy of these IBA methods are now well established at around the $\pm 5\%$ level for carefully controlled measurement of most major elements. There have been many round robins and concentration comparisons performed with other analysis methods. Favourable results were obtained. One of the main advantages of these techniques is their ability to provide all of this data on relatively small sample sizes, usually only a few hundred micrograms in weight. The ability to furnish elemental concentration information on all of the most significant elements from hydrogen to lead, means that statistical techniques such as PCA and CMB programs can be applied to the large databases generated. This in turn provides further information on source fingerprints and hence possible sources of fine particles in our atmosphere.

ACKNOWLEDGMENTS

The financial support of the Energy Research and Development Corporation and the NSW Environmental Trust during some phases of this work is gratefully acknowledged. The assistance of Macquarie University, the University of NSW, Pacific Power and the NSW EPA is appreciated; as is the analyses for comparison with the PIXE and PIGME results provided by J. Fardy, D. Hill and L. Dale. Thank you to David Garton and the 3 MV accelerator staff for their support, and Jarrod Martin for considerable help with the software programs that produced much of the data shown in the figures.

REFERENCES

Cahill, T.A., R.A. Eldred, N. Motallebi and W.C. Malm, *Aerosol Sci. and Technol.*, **10** (1989) 421.

Cahill, T.A., *Nucl. Instrum. Meth. Phys. Res.*, **B49** (1990) 345.

Cahill, T.A., Particle induced X-ray emission, in *Metals Materials Characterisation Handbook*, 9th edition, Vol. 10, American Society for Metals (1992) pp. 102–108.

Clayton, E., *PIXAN, The Lucas Heights PIXE Analysis Computer Package*, Australian Nuclear Science and Technology Organisation, AAEC/M113 November (1986).

Cohen, D.D., *Nucl. Instrum. and Meth.*, **178** (1980) 481.

Cohen, D.D., *Clean Air*, **26** (1992) 113.

Cohen, D.D., *Nucl. Instrum. Meth. Phys. Res.*, **B79** (1993) 385.

Cohen, D.D., *Proceedings of the 9th Pacific Basin Nuclear Conference*, Sydney, Australia, May 1–6, **1** (1994) pp. 357–364.

Cohen, D.D., *Energy Research and Development Corporation*, Study Report No. 259, June (1995) Canberra, Australia.

Cohen D.D., *Proceedings of 13th Int. Clean Air Conf.*, Adelaide, Australia, September 22–25 (1996) pp. 6–12.

Cohen D.D. and E. Clayton, PIXE, in *Ion Beams for Materials Analysis*, Academic Press, New York (1989), Eds. J.R. Bird and J.S. Williams.

Cohen, D.D., G.M. Bailey, D. Garton, E. Stelcer, P. Crisp, R. Rothwell, J. Banks and R. Hyde, *Proceedings of the Air Toxic Conference*, Sydney, August 10–11, **2** (1994) pp. 125–140.

Cohen, D.D., R. Kondepudi, R. Hyde and P. Crisp, *NSW Environmental Trust Grant Report* No. 1993/RD/G02, March (1995) Sydney, Australia.

Gimsenius, A., D.D. Cohen and H. Whitlow, *Proceedings of 13th Int. Clean Air Conf.*, Adelaide, Australia, September 22–25 (1996) pp.1–5.

Hopke, P.K., *Receptor Modelling for Air Quality Management*, Elsevier Science, Amsterdam (1991).

Johansson, S.A.E. and J.L. Campbell, *PIXE — A Novel Technique for Elemental Analysis*, Wiley, New York (1988).

Johansson, S.A.E. and T.B. Johansson, *Nucl. Instrum. and Meth.*, **137** (1976) 473.

Maenhaut, W., *Nucl. Instrum. and Meth. Phys. Res.*, **B49** (1990) 518.

Malm, W.C., J.F. Sisler, D. Huffman, R.A. Eldred, T.A. Cahill, *J. Geophys. Res.* **99** (1994) 1347.

Vis, R. (Ed.), *Proceedings of 5th Int. Conf. on PIXE and Its Analytical Applications,* in *Nucl. Instrum. and Meth. Phys. Res.* **B49** (1990).

Watson, J.G., N.F. Robinson, J.C. Chow, R.C. Henry, B. Kim, Q.T. Nguyen, E.L. Meyer and T.G. Pace, *Receptor Model Technical Series*, Volume 3, CMB7 User's Manual, US Environmental Protection Agency, Report EPA-450/4-90-004, January (1990).

Weast, R.C., *CRC Handbook of Chemistry and Physics*, CRC Press (1977) 57th ed., Ohio, USA, pp. 203.

5 TRACE ELEMENT DETERMINATION OF AIRBORNE PARTICLES BY NEUTRON ACTIVATION ANALYSIS

S. LANDSBERGER

Nuclear Engineering Teaching Lab
University of Texas, Pickle Research Campus, R-9000
Austin, Texas 78712 USA

INTRODUCTION

During the 1960s neutron activation analysis (NAA) and X-ray fluorescence were at the historical forefront of multielemental determination in atmospheric studies. These analytical methods were major contributing factors to the initial development of source-receptor modeling. Without these multilemental data, mathematical models could never be so rigorously employed. Later on other multielemental techniques, including particle induced X-ray emission and inductively coupled plasma, were accepted as alternative and complimentary methods. But NAA still remains unique, primarily for its non-destructive characteristic and high sensitivity for a wide range of trace elements found in the atmosphere: urban, rural, and remote geographical areas of the world. The methods and techniques of NAA are now very well understood and employed in a wide variety of environmental samples. Furthermore, for airborne particulate matter, NAA can determine virtually all elements with the notable exceptions of carbon, oxygen, nitrogen, fluorine, and lead. The sensitivity for sulphur is poor and is seldom reported in NAA studies.

NAA is analytically unique in that there are various means of increasing the sensitivity of the elements in question. The signal to noise ratio, and ultimately detection limits can be affected by the following:

- changes in neutron flux
- choice of neutron energy (thermal or epithermal)
- variations in irradiation time
- decay and counting time
- choice of isotope used in the analysis
- choice of counting geometry position
- implementation of low-level counting techniques such as: lead shielding, Compton suppression, and gamma–gamma coincidence
- choice of filter used in air sampling

Automated peak deconvolution programs must also be properly benchmarked and rigorously tested to ensure reliable results. Similar measures for deconvolution programs are required for other nuclear, atomic, and chemical methods.

Since the middle 1980s there have been several books and review articles on NAA methods as well as general radiochemistry (Alfassi, 1989; 1994; Bode, 1996; Dams, 1981; Das and Van der Sloot 1989; Das *et al.*, 1983; Ehman and Vance, 1989; 1990; Ertmann and Petri, 1986; First International Workshop, 1981; Heydorn, 1984; 1988; IAEA 1987a, b; Kolthoff *et al.*, 1986; Landsberger, 1994a; Parry, 1990; and Tolgyessy and Klehr, 1987; 1989). Special mention should be given to the in depth reviews on nuclear and radiochemical analysis compiled by Ehmann and Yates (1986; 1988) and Ehman, *et al.* (1990, 1992). Neutron activation cross-section data have also been by compiled the IAEA (1987c) and Mughabghab *et al.* (1984). Specific articles related to NAA or other nuclear/atomic methods for atmospheric chemistry studies have also appeared (Alian and Sansoni, 1985; 1988; 1989; Dams, 1985; 1992; Krivan *et al.*, 1990; Landsberger, 1992; Landsberger and Bielgalski, 1995; Lodge, 1989; Maenhaut, 1989; 1992; Obrusnik, 1990; Perry and Young, 1977; Smitt and Segebade, 1985). Discussions of gamma-ray spectrometry techniques (Debertin and Helmer, 1988; Gilmore and Hemmingway, 1995; Knoll, 1989) and aspects of operating a NAA laboratory (IAEA, 1990) contain valuable information relevant to electronic set-up, calibration, low-level counting, and overall quality assurance.

The salient features of NAA for elemental characterization of airborne particles are discussed in this chapter. Not included are: detailed aspects of NAA theory; radiochemical methods; prompt-gamma; charge-particle or fast neutron activation analysis method (Csikai, 1987); cyclic activation analysis (DeSilva and Chatt, 1982; Spyrou and Kerr, 1979); deadtime corrections; loss-free counting (Westpahl, 1996); or detection limits, accuracy and precision (Currie, 1968; Heydorn, 1980). Descriptions for neutron sources, interaction of radiation with matter, nuclear detection instrumentation, and radiation protection (often overlooked in many review articles) can be found in Chemical Analysis by Nuclear Methods (Alfassi, 1994) and all the references therein.

BRIEF OVERVIEW OF NEUTRON ACTIVATION ANALYSIS

When a neutron collides with a nucleus, the following reactions may occur: elastic scattering (n, n), inelastic scattering (n, n'), radiative capture (n, γ), charge-particle reaction (n, α), or (n, p), (n, 2n), and fission (n, f). The most useful reaction in NAA is radiative capture, which is denoted as:

$$n + {}^{A}Z \longrightarrow {}^{A+1}Z^{*} \longrightarrow {}^{A+1}Z + \gamma \qquad (1)$$

where:

${}^{A}Z$ — target nucleus,
${}^{A+1}Z^{*}$ — compound nucleus which is usually in an excited state.

It can de-excite with the emission of a gamma-ray called a prompt-gamma. In normal NAA this gamma-ray is not used in the analysis. Instead delayed gamma from the product ${}^{A+1}Z$ is employed. This means that the product ${}^{A+1}Z$ is required to be radioactive, emitting at least one gamma-ray photon. It can be detected by a high resolution gamma-ray detector for both energy and intensity. If the product ${}^{A+1}Z$ is a stable isotope, it can not be detected. For example, while magnesium in nature has three stable isotopes ${}^{24}Mg$, ${}^{25}Mg$, and ${}^{26}Mg$, only the last one with an abundance of 11% can be used in NAA employing the ${}^{26}Mg(n, \gamma){}^{27}Mg$ reaction. Neutron-induced reactions of ${}^{24}Mg$ and ${}^{25}Mg$ give the stable isotopes ${}^{25}Mg$ and ${}^{26}Mg$, respectively.

Since stable isotopes of an element can undergo different reactions depending on the neutron energy, it is possible that the same product ${}^{A+1}Z$ can be produced via different nuclear reactions. For example, ${}^{28}Al$ is produced by the following three different reactions:

$$^{27}Al + n \longrightarrow {}^{28}Al + \gamma \qquad (2)$$

$$^{28}Si + n \longrightarrow {}^{28}Al + p \qquad (3)$$

$$^{31}P + n \longrightarrow {}^{28}Al + \alpha \qquad (4)$$

The first reaction (Equation (2)) is the expected reaction in thermal NAA, while Equations (3) and (4) are referred to as primary nuclear interference reactions. In NAA of airborne particles there are very few nuclear interference reactions at the level of elements found in the atmosphere. The gamma-ray spectra of neutron activated samples are far less complicated than those generated by geological samples. The two noted nuclear interferences exceptions for neutrons sources with an epithermal component are: $^{27}Al(n,\gamma)^{28}Al$ and $^{28}Si(n,p)^{28}Al$; and $^{26}Mg(n,\gamma)^{27}Mg$ and $^{27}Al(n,p)^{27}Mg$. Typically they can be measured and taken into account. Spectral interferences are those in which gamma-rays from one type of reaction give almost the same energy of emitted gamma-ray or have an overlapping peak. In air particle measurements there are several specific interferences and these are discussed later in the chapter.

Epithermal NAA

Many reactors have irradiation positions in which there is an epithermal component. In the past decades, it has been shown that epithermal NAA is advantageous to the determination of radionuclides that can be produced by resonance neutrons. This technique has been judiciously reviewed and used by many researchers particularly for geological and biological samples (Alfassi, 1985; Chisela *et al.*, 1987; Ehmann and Vance, 1989; Ehmann *et al.*, 1980; Glascock *et al.*, 1985; Landsberger, 1994a; Lavi *et al.*, 1989; Parry, 1982; Rowe and Steinnes, 1977; Tobler *et al.*, 1990). Elements such as sodium, chlorine, aluminum, and manganese produce radionuclides that can dominate the spectrum in short- or medium-lived NAA. Using a neutron filter such as cadmium or boron, the activity of the radionuclides ^{27}Al, ^{38}Cl, ^{56}Mn (short-lived), and ^{24}Na (short- and medium-lived) can be substantially reduced, allowing other elements of interest to be determined with much greater precision. Advantage factors for certain elements that have a high I_0/σ_{th} (neutron radiative capture resonance integral (I_0)/thermal cross-section (σ_{th})) have been generalized for counting statistics and detection limits (Brune, 1969), and placement of samples closer to the detector due to much lower activities for epithermal NAA (Tian, 1986; Tian and Ehmann, 1984). In the past there have been very few studies employing epithermal NAA for airborne particles (Dams *et al.*, 1975; Janssens *et al.*, 1975). More recently research has been conducted using epithermal NAA with and without Compton suppression to drastically reduce the background for epithermal irradiated air filters from the Arctic and the Great Lakes (Biegalski and Landsberger, 1994; Landsberger, 1988; Landsberger and Wu, 1993; Landsberger *et al.*, 1991; 1997a) and for environmental tobacco smoke (Landsberger and Wu, 1995; Landsberger *et al.*, 1993). Several illustrations depicting the enhanced signal to noise ratios for iodine, indium, arsenic, antimony, and nickel using epithermal NAA and/or Compton suppression are shown in Figures 1–5. Compton supression is ideally used for radionuclides that emit one or predominately one gamma-ray. A complete description of Compton supression techniques and its use in environmental studies is given elsewhere (Landsberger, 1994b; Landsberger and Peshev, 1996).

Values for the thermal neutron capture cross-section and the resonance integral are given in several references by IAEA (1987c) and Mughabghab *et al.* (1984). Elements important to airborne particles having relatively high ratios of I_0/σ_{th} include: antimony, arsenic, barium, cadmium, indium, iodine, molybdenum, samarium, selenium, silver, tin, tungsten, uranium, and zinc (Table 1). These activation products are either short- or medium-lived nuclides. Usually for airborne particles there is little background interference from the production of long-lived radionuclides, thus with the exception of nickel, the use of long epithermal neutrons irradiations is of little value. Two other important types of analysis that employ epithermal neutrons using the neutron-charged particle reactions are $^{29}Si(n,p)^{29}Al$ and $^{58}Ni(n,p)^{58}Co$. The former may be important to properly determine the silicon contribution to the production of the $^{27}Si(n,p)^{27}Al$ reaction. Iron can also be determined using the $^{54}Fe(n,p)^{54}Mn$ reaction.

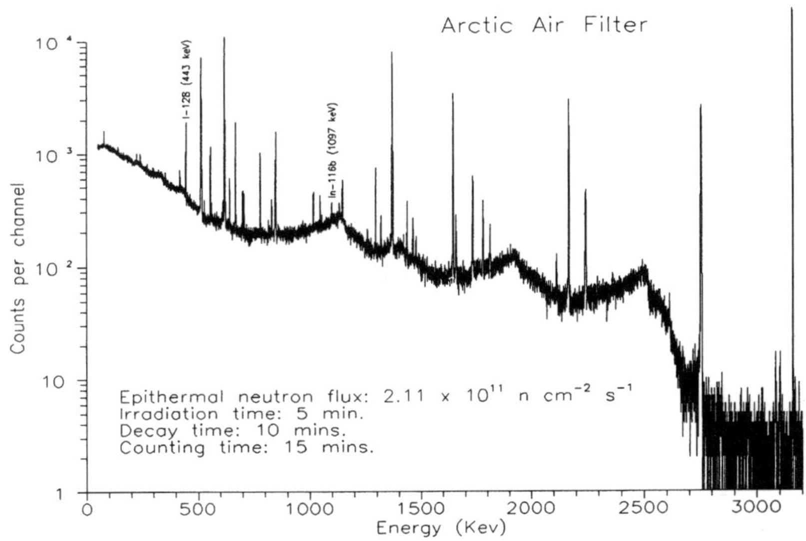

Figure 1. Gamma-ray spectrum of an Arctic air filter irradiated with epithermal neutrons for short-lived NAA.

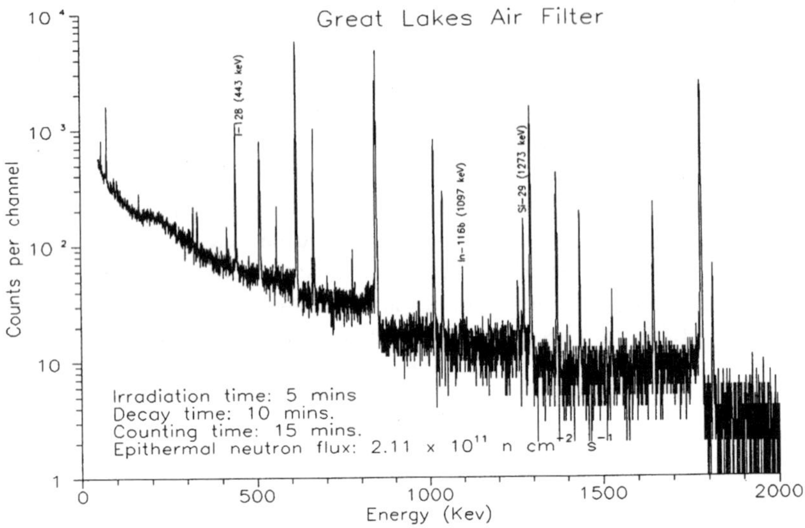

Figure 2. Gamma-ray spectrum of a Great Lakes air filter irradiated with epithermal neutrons and counted in a Compton suppression system for short-lived NAA.

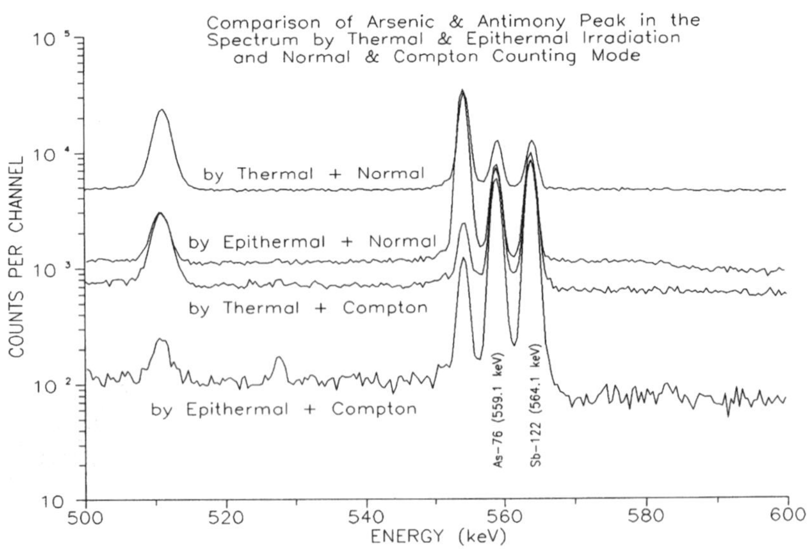

Figure 3. Gamma-ray spectra of Arctic air filters irradiated with thermal or epithermal neutrons with and without Compton suppression for medium-lived NAA.

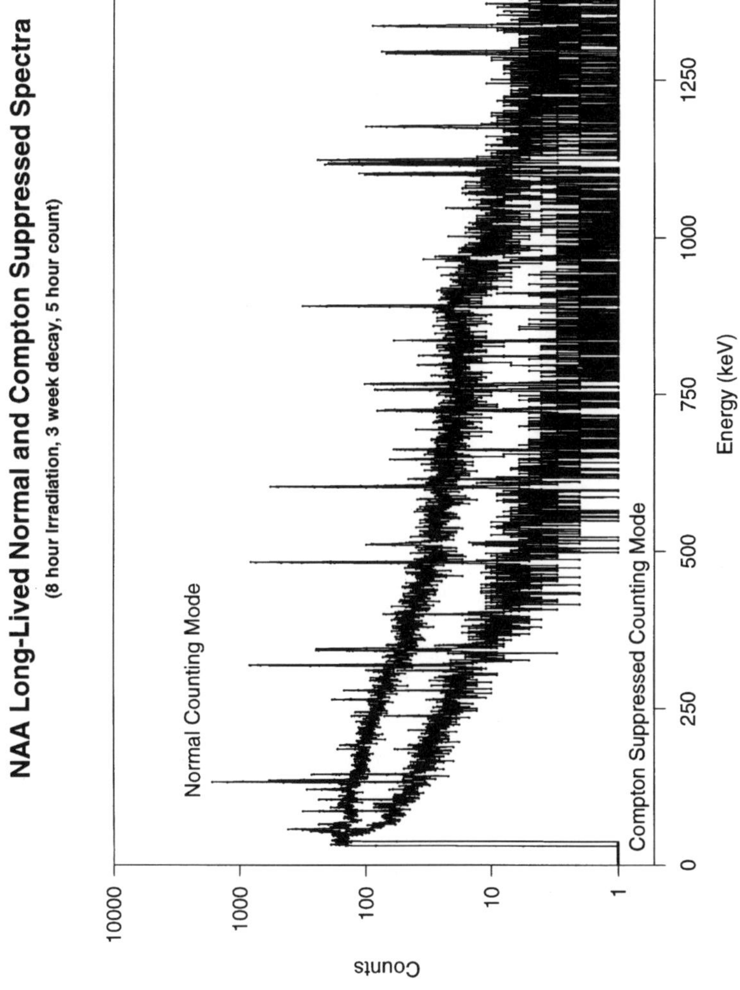

Figure 4. Comparison of gamma-ray spectra of a Great Lakes air filter with and without Compton suppression for long-lived NAA.

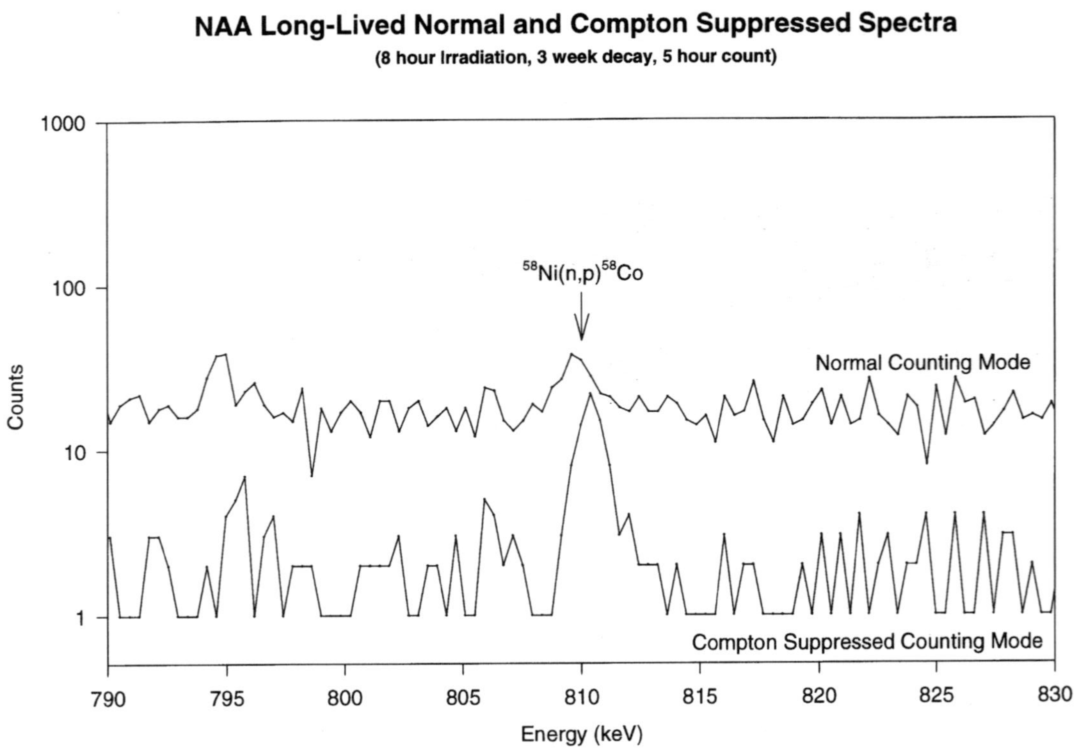

Figure 5. Comparison of normal and Compton suppressed gamma-ray spectra for the determination of nickel through the ^{58}Ni(n, p)^{58}Co using the 810.8 keV photon.

Table 1. Nuclear data for resonance integral reactions.

Element	Nuclear reaction	I_0/σ_{th}	γ-energy (keV)	Half-life
Short-lived				
Ag	$^{109}Ag(n, \gamma)^{110}Ag$	1380	657.8	24.6 s
Ba	$^{138}Ba(n, \gamma)^{139}Ba$	3	165.9	1.39 h
I	$^{127}I(n, \gamma)^{128}I$	24	442.3	25.0 min
In	$^{115}In(n, \gamma)^{116m}In$	16	1097.3	54.2 min
Se	$^{77}Se(n, n')^{77m}Se$	—	161.7	17.4 s
Si	$^{124}Si(n, p)^{29}Al$	—	1273.0	6.5 min
Sn	$^{124}Sn(n, \gamma)^{125}Sn$	8	332.4	9.5 min
U	$^{238}U(n, \gamma)^{239}U$	102	74.6	23.5 min
Medium- and long-lived				
As	$^{75}As(n, \gamma)^{76}As$	15	559.1	26.3 h
Cd	$^{114}Cd(n, \gamma)^{115}Cd/^{115m}In$	66	336.3	53.4 h
Fe	$^{54}Fe(n, p)^{54}Mn$	—	834.3	312 d
Mo	$^{98}Mo(n, \gamma)^{99}Mo$	56	140.0	2.75 d
Ni	$^{58}Ni(n, p)^{58}Co$	—	810.8	70.9 d
Sb	$^{121}Sb(n, \gamma)^{122}Sb$	33	564.0	64.8 h
Sm	$^{152}Sm(n, \gamma)^{153}Sm$	14	103.2	46.3 h
W	$^{186}W(n, \gamma)^{187}W$	490	685.8	23.9 h
Zn	$^{68}Zn(n, \gamma)^{69m}Zn$	3.3	438.6	13.8 h

NEUTRON ACTIVATION ANALYSIS CALCULATIONS

NAA is based on the measurement of characteristic gamma-rays from a radionuclide formed by a specific neutron reaction. The corresponding radioactivity is governed by the usual radioactive decay law. If other factors are considered in the measurement, the final equation is (Tolgyessy and Klehr, 1987):

$$A = \sigma \Phi (m/M)\, N_A SDC\theta P_\gamma \xi \tag{5}$$

where:

A — measured activity (Bq) from the product of a expected reaction,
σ — activation cross-section of the reaction (cm^2),
Φ — activating flux (neutrons/cm^2 s),
m — amount of the element determined (g),
M — atomic weight of the element to be determined (g/mole),
N_A — Avogadro constant (6.022×10^{23} molecules/mole),

S — $[1-\exp(-\lambda t_i)]$ saturation factor (λ is decay constant of the radioactive product, t_i is duration of the irradiation time),
D — $\exp(-\lambda t_d)$ decay factor (t_d is duration of the decay time),
C — $[1-\exp(-\lambda t_c)]$ correction factor for nuclide decay during the counting time (t_c is duration of the counting time),
θ — relative natural isotopic abundance of the activated isotope,
$P\gamma$ — probability of emission of a photon with energy E,
ξ — detector efficiency for the measured radiation energy.

The above equation can be greatly simplified by using the comparator method. If equal weights or known concentrations of both samples and standards have the same irradiation, decay and counting times then Equation (5) becomes:

$$C_{\mathrm{sam}} = C_{\mathrm{std}} * (A_{\mathrm{sam}}/A_{\mathrm{std}}) \tag{6}$$

where:

C_{sam} — unknown concentration of the element in the sample,
C_{std} — known concentration of the element in the standard,
A_{sam} — activity of the sample,
A_{std} — activity of the standard.

In this case, the unknown concentration of the element in the sample can be obtained by comparing the activities of the gamma-ray peaks. The terms D and C in Equation (5) usually have to be used as well since they often vary. Normalization of the weights between standards and unknowns must be done often. The overall equation is then given as follows:

$$C_{\mathrm{sam}} = C_{\mathrm{std}} * (A_{\mathrm{sam}}/A_{\mathrm{std}}) * (S_{\mathrm{std}}/S_{\mathrm{sam}}) * (D_{\mathrm{std}}/D_{\mathrm{sam}}) * (C_{\mathrm{std}}/C_{\mathrm{sam}}) * (W_{\mathrm{std}}/W_{\mathrm{sam}}) * (F_{\mathrm{std}}/F_{\mathrm{sam}}) \tag{7}$$

where $W_{\mathrm{std}}/W_{\mathrm{sam}}$ and $F_{\mathrm{std}}/F_{\mathrm{sam}}$ are the weights and neutron fluences of the sample and standard, respectively.

NEUTRON SOURCES

There are several neutron sources that can be used in NAA including a particle accelerator (Csikai, 1987) or artificial isotopes, such as plutonium–beryllium (Pu(Be)). The most common source is from a fission reactor due to its high neutron flux. All the descriptions in this chapter are for reactor neutrons using thermal or epithermal neutrons between 10^{11} and 10^{14} n cm^2 s^{-1}. Olmez (1981) has detailed the determination of aluminum, silicon, and iron for air particles using fast neutrons, but generally this technique is of limited use for samples having a low mass.

ANALYTICAL LABORATORY PROCEDURES

Irradiation Facilities

Although very fast pneumatic transfer facilities employing ultra short-lived radionuclides can be used in biological and geological studies, they are probably of limited value in the investigations of air particulate matter. It is known that lead can be determined using the $^{207}Pb(n, n')^{207m}Pb$ reaction with its characteristic 0.8 s half-life. Since ^{207}Pb is radiogenic (its isotopic composition varies in nature to a significant degree), serious errors can arise in the determination of lead in environmental samples by this method. The cost benefit of such a system is not justified, since almost all the elements that can be determined by ultra short-lived isotopes can also be identified by conventional thermal and epithermal NAA. A far more cost effective approach is the installation of a dedicated cyclic activation facility to improve the sensitivity for the determination of a variety of radionuclides with half-lives less than one minute (e.g. ^{74m}Se and ^{110}Ag).

Preparing Filters for Irradiation

1. Sample preparation prior to analysis

This is completed in an inactive area and preferably one that is as clean as possible with respect to trace element contamination. A so-called "clean room" is not absolutely essential, but it is useful to have a clean work-bench (e.g. laminar flow hood) and a chemical-type fumehood for operations resulting in acid fumes, e.g. during cleaning of tools and sample vials. If quartz encapsulation of the samples is being used, then a suitable facility for sealing the ampoules will be needed.

2. Sample preparation following activation

This working area includes a chemical-type fumehood for operations (e.g. opening of vials) that may lead to the escape of radioactive gases or particles. An area for storage of radioactive samples with adequate floor loading strength so that lead bricks can be placed for radiation shielding is also required. The area also contains the terminus of the pneumatic transfer (if one is available) for analyses with short-lived activation products. Special consideration should be given to health physics practices particularly for procedures that require transfer of activated samples.

Irradiation and Counting Procedures

Depending on neutron flux density, mass of air particulate matter, and efficiency of the germanium detector, irradiation times may vary; but generalized schemes can be adhered to. Typically two irradiations are performed to determine short-lived radionuclides (one to several minutes) and medium- and long-lived radionuclides (several hours–one day). These neutron irradiations are done for the same sample. If any epithermal NAA is required then an additional irradiation is necessary. For activation analysis to produce short-lived radionuclides, irradiation times from one to ten minutes are used. The radionuclides produced are shown in Table 2. A decay period of at least five minutes is needed to let the usually high ^{28}Al activity decrease by two

Table 2. Properties of reactions producing short-lived radionuclides.

Element	Isotope	Half-life	Gamma-ray energies (keV)
Ag	^{110}Ag	24.6 s	657.8
Al	^{28}Al	2.24 min	1778.9
Ba	^{139}Ba	83.2 min	165.9
Br	^{80}Br	17.7 min	616.2
Br	^{80m}Br	4.42 h	37.1
Ca	^{49}Ca	8.7 min	3084.4
Cl	^{38}Cl	37.3 min	1642.4, 2167.5
Co	^{60m}Co	10.48 min	58.6
Cu	^{66}Cu	5.1 min	1039.4
I	^{128}I	25.0 min	442.3
In	^{116m}In	54.2 min	416.9, 1097.3
K	^{42}K	12.36 h	1524.7
Mg	^{27}Mg	9.45 min	843.8, 1014.4
Mn	^{56}Mn	2.58 h	846.7, 1810.7, 2212
Na	^{24}Na	15.0 h	1368.6, 2754.1
Se	^{77m}Se	17.4 s	161.7
Sb	^{122m}Sb	4.15 min	61.5
Si[a]	^{29}Al	6.6 min	1273.0
Sn	^{125m}Sn	9 min	332.4
Sr	^{87m}Sr	2.81 h	388.4
Ti	^{51}Ti	5.8 min	320.1
U	^{239}U	23.5 min	74.6
V	^{52}V	3.76 min	1434.1

[a]Si is detected by the $^{29}Si(n,p)^{29}Al$ reaction.

half-lives. High backgrounds due to the Compton effect are caused by this high activity. Counting periods from 5 to 15 min are usually employed to determine the rest of the elements. To improve detection limits for barium, iodine, indium, potassium, and strontium, longer decay (30–60 min) and counting (30–60 min) periods may be used. Detection limits for these elements and silicon are significantly improved by using epithermal neutrons.

For the determination of medium- and long-lived activities, irradiations of several hours are needed to sufficiently activate the samples in order to achieve good counting statistics. The other major limiting factor for good detection limits is the high background resulting from ^{24}Na and ^{82}Br. Improved detection limits for arsenic, antimony, cadmium, molybdenum, tungsten, and uranium can be achieved by epithermal NAA since the Compton background due to ^{24}Na is significantly decreased. This procedure would be at the expense of not having enough activity for the long-lived isotopes (since the epithermal flux density can be one order of magnitude less than thermal flux density). If filters can be cut in half, with one portion being used for medium-lived nuclides and the other for long-lived activation products, then epithermal NAA may be advantageous. Elements that can be determined by medium-lived radionuclides are shown in Table 3, while those for long-lived isotopes are shown in Table 4.

Table 3. Properties of reactions producing medium-lived radionuclides.

Element	Isotope	Half-life	Gamma-ray energies (keV)
As	^{76}As	26.3 h	559.1
Au	^{198}Au	2.7 d	411.8
Br	^{82}Br	35.3 h	554.3, 776.5
Cu	^{64}Cu	12 h	511.0
Cd[a]	^{115m}In	53.5 h	336.3, 527.9
Hg	^{197}Hg	64.1 h	77.4
K	^{42}K	12.36 h	1524.7
La	^{140}La	40.23 h	328.8, 487.0, 1596.2
Mo	^{99}Mo	66.02 h	140.5
Na	^{24}Na	15.02 h	1368.6, 2754.1
Sb	^{122}Sb	2.72 d	564.0
Sm	^{153}Sm	46.7 h	103.2
U[b]	^{239}Np	2.35 d	277.7
W	^{187}W	23.9 h	685.8
Zn	^{69m}Zn	13.8 h	438.6

[a]Cd is determined from the $^{114}Cd(n, \gamma\beta)^{115}Cd \rightarrow {}^{115m}In$ reaction.
[b]U is determined from the $^{238}U(n, \gamma\beta)^{239}U \rightarrow {}^{239}Np$ reaction.

Table 4. Properties of reactions producing long-lived radionuclides.

Element	Isotope	Half-life	Gamma-ray energies (keV)
Ag	^{110m}Ag	249.8 d	657.8
Ce	^{141}Ce	32.5 d	145.4
Cr	^{51}Cr	27.72 d	320.0
Cs	^{134}Cs	2.06 y	795.8
Co	^{60}Co	5.27 y	1173.2, 1332.4
Eu	^{152}Eu	13.4 y	1408.0
Fe	^{59}Fe	44.5 d	1099.2, 1291.6
Fe[a]	^{54}Mn	312 d	834.5
Hf	^{181}Hf	42.4 d	482.2
Hg	^{203}Hg	46.6 d	279.2
Ni[b]	^{58}Co	70.9 d	810.8
Rb	^{86}Rb	18.7 d	1076.6
Sb	^{124}Sb	60.2 d	1691.0
Sc	^{46}Sc	83.8 d	889.3, 1120.5
Se	^{75}Se	119.8 d	136.0, 264.7, 400.7
Sn	^{113}Sn	114.4 d	391.7
Sr	^{85}Sr	64.84 d	514.0
Ta	^{182}Ta	115.0 d	1221.4
Th[c]	^{233}Pa	27.0 d	311.9
Zn	^{65}Zn	243.8 d	1115.5
Zr	^{95}Zr	64.0 d	756.7

[a]Fe is determined using the $^{54}Fe(n, p)^{54}Mn$ reaction.
[b]Ni is determined using the $^{58}Ni(n, p)^{58}Co$ reaction.
[c]Th is determined using the $^{232}Th(n, \gamma\beta)^{233}Th \rightarrow {}^{233}Pa$ reaction.

A detailed list of the salient features of the elements that can be determined in air filters, including their spectral and nuclear interferences previously reported is found in Landsberger (1992). For the determination of very short-lived radionuclides such as: ^{77m}Se, ^{110}Ag and ^{20}F, Teflon® filters can not be used since fluorine is a major component of the substrate. This results in a very radioactive sample for a period of about one minute until the short-lived ^{19}F decays away.

GAMMA-RAY SPECTROMETRY

Most of the measurements in NAA are performed using high resolution (FWHM of 2.0 keV or less for the 1332 keV ^{60}Co peak) gamma-ray spectrometers with a high purity germanium coaxial detector. It is advantageous to have a high efficiency (25% or greater) detector for counting. With present day technology detectors, efficiencies greater than 100% are available. In particular, counting times can be reduced significantly for the longer-lived radionuclides with even superior peak to background ratios. To maximize the amount of information obtainable from radioactive samples, detectors are also available that are capable of processing both low and high energy gamma-rays. Very little work has been performed in low energy analysis for air particulate matter (Janssens *et al.*, 1975; Nakanishi and Sansoni, 1977). In principle, this method should not be nearly as complicated for the analysis of air filters as for geological samples. The self-attenuation of the gamma-rays is minimized since the mass of the sample is usually small and close matrix matching with standards is possible. The detector is usually surrounded by a cadmium- and copper-lined lead shielding with a sample holder for reproducible positioning of the samples. An automatic sample changer is very useful for many types of routine measurements.

CALIBRATION

There are several ways to calibrate the NAA system for elemental analysis. The first involves the use of liquid standards that are prepared either from commercially available solutions or basic chemicals. If basic chemicals are chosen, extreme care should be given to the stoichiometry of the chemical including any hydrated forms. Inorganic chemicals recommended for use in calibration have been reviewed by Moody *et al.* (1988). Becker (1986) and Greenberg (1994) have detailed the use of primary standards in NAA and accuracy in standards preparation, respectively. All chemicals require oven drying to remove the moisture content. Liquid standards can also be pipetted into acid washed vials and heat sealed. Both polyethylene and quartz vials can be utilized for the encapsulation of liquid samples. If polyethylene vials are chosen, liquid samples should be doubly encapsulated and heat sealed for irradiation due to inherent leakage. The heat sealed vial needs to be placed in a larger vial, also heat sealed. After irradiation, all liquid standards are transferred to labelled inert vials for counting. In some reactors there are no facilities available for liquid irradiations, or the neutron flux is so high it causes evaporation and extensive leaking as a result of thermal heating.

In the comparator method, calibration can also be done by pipetting a known amount of one or more elements onto a filter paper. This method is commonly used but several factors must be taken into consideration. The filter paper should contain only minimal concentrations of contaminants when compared to the amount pipetted (hence the need for a trace element characterization of the filter paper before its use). When combining two or more standard solutions of two elements it has to be ascertained that the solution of one element does not contain trace or minor quantities of the second one. For instance, it is conceivable that commercially bought or laboratory prepared compounds may contain unspecified quantities of other elements. Filters need to be placed in acid washed vials for irradiation, then transferred to labeled inert vials prior to counting. Some types of filter paper may not be transferable owing to their condition following irradiation, particularly for high neutron flux conditions. Filter papers can also be prepared in the form a pellet, but care must be given to prevent contamination from the apparatus used for the procedure. The advantage of pelletization is that an exact geometry can be reproduced for irradiation and counting procedures.

Calibration can also be done using the comparator method with reference materials. In general, reference materials should not be used for the calibration procedure but rather for testing the analytical accuracy of the methods and techniques. Many geological or biological reference materials have trace and minor elements certified with accuracy errors up to $\pm 20\%$ (and in some cases even more) or not certified with known elemental concentrations. This means that the final derived elemental concentrations in the actual samples will not have accuracies better than the reported errors of the individual elemental concentration in the reference material. The existence of spectral interferences may also limit the application of certified reference materials made from geological material. For instance, the 136.0 and 264.7 keV gamma-rays coming from the $^{74}Se(n,\gamma)^{75}Se$ activation product are susceptible to substantial interferences from the same photopeaks arising from the $^{180}Hf(n,\gamma)^{181}Hf$ and $^{181}Ta(n,\gamma)^{182}Ta$ reactions, respectively. Other elements, such as nickel, exhibit poor sensitivities in thermal NAA and therefore should not be used in any calibration procedure.

Calibration can also be achieved using the k_0-method. Essentially, it is an absolute technique in that the amount of material is calculated from the measured counts by means of Equation (5). The nuclear constants of each nuclide is combined together to give the k_0 constant. Since there are some uncertainties in the nuclear constants, the k_0 values are determined experimentally. Several activation groups world-wide are pursuing this method (Blaauw, 1995; DeCorte *et al.*, 1987; 1993; Heydorn and Damsgaard, 1994). Epithermal neutron flux density distributions, rigidly standardized counting procedures, and coincidence counting effects need to be considered.

ACCURACY

There are several factors which can severely limit the accuracy of NAA including:

1. Changes in Neutron Flux Profile

Neutron flux monitors are always used for long irradiations but are seldom used for short-lived NAA. The relative neutron flux variation of air filter samples irradiated for one minute in an identical position in the reactor is shown in Figure 6. Flux monitors irradiated with air filter samples for one hour in a vertical batch with epithermal neutrons is illustrated in Figure 7. Each short-lived NAA sample was irradiated with a molybdenum wire, while for medium-lived NAA a wire was placed at each level. As seen from the figures, in order to obtain a much higher accuracy, flux monitors are crucial. It is often wrongly assumed that flux wires are only needed for longer irradiations. Some reactors like the SLOWPOKE have an inherent flux variation of only $\pm 2\%$. For such reactors a wire may be of minimal benefit, but for the majority of research reactors the stochastic processes governing flux variations may be significantly larger.

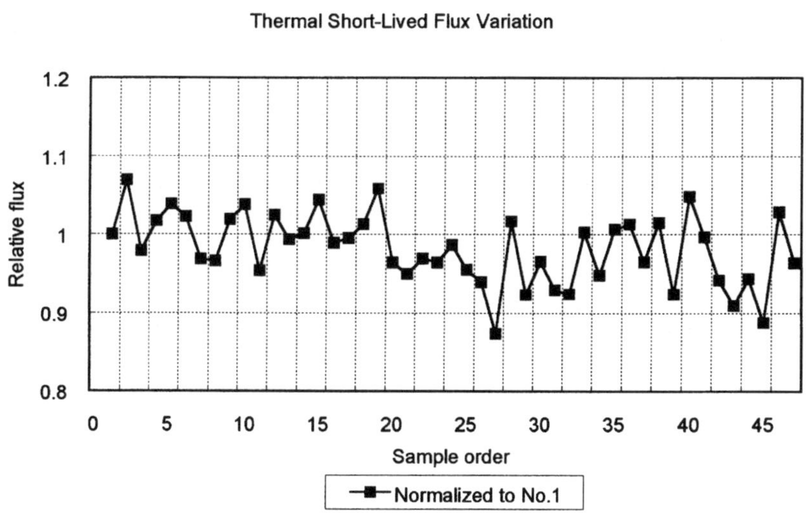

Figure 6. Relative neutron flux in short-lived NAA in the identical irradiation position.

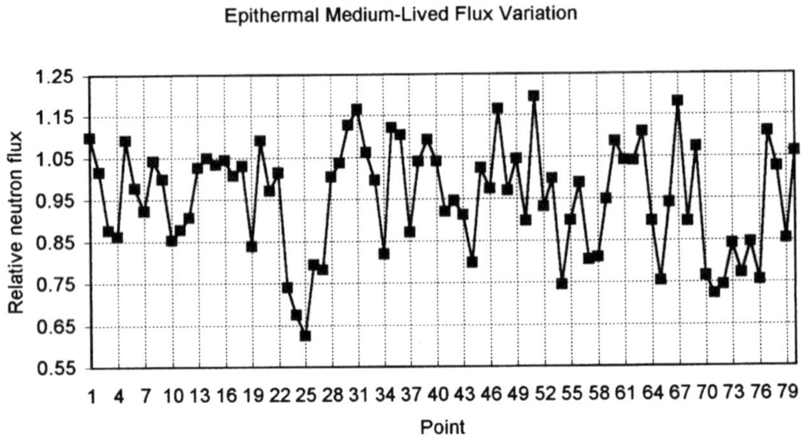

Figure 7. Relative neutron flux in medium-lived epithermal NAA for stacked samples.

2. Irradiation, Decay and Counting

Exact times for short irradiations is crucial. Each short-lived system must be characterized for reproducability. While control of short irradiation times can be easily monitored, often it is reactor operations' staff who monitor long irradiation times. A good communication between the staff and NAA researcher must exist so that exact times of irradiation are monitored. For instance a two minute mistake in an one hour irradiation will add a 3% error to the final result.

3. Inaccurate Placement of Samples in Front of Detectors Including Volume Variations Between Standards and Samples

Radiation from the sample to the detector decreases as $1/r^2$ (Figure 8). This is particularly significant when counting at close distances from the detector. Sample positioning is crucial and every effort should be made to retain a nearly identical geometry between the samples and standards. The change in measured activity for a europium standard diluted with different amounts of water is also shown in Figure 8. The closer the sample is counted in front of the detector, the more variation in the net counts relative to the volume size.

Due to their size and structure, most air filters must be folded or pelletized to fit into counting vials. Once encapsulated, some of the filter's particulate material may be off center. Automatic turning of the sample in

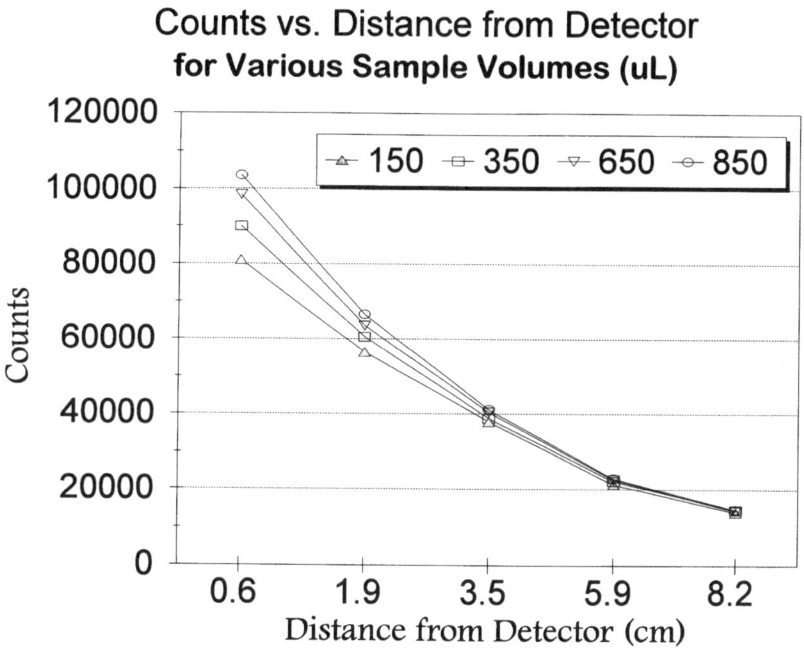

Figure 8. Relative counts vs distance from detector for various sample volumes for 0.15, 0.35, 0.65, and 0.85 ml.

front of the detector will ensure better reproducibility of the counting. A quality control check can be performed by turning the sample 180° and seeing if the new calculated concentrations are significantly different. The activity changes of an irradiated soil sample counted at 0°, 90°, 180°, and 270° is shown in Figure 9.

4. Unknown Changes in Relative Natural Abundances

The three elements which can be of the greatest concern for natural isotopic shifts are Pb, B, and S. Lead and boron are never reported in airborne particles, while sulphur has a very poor sensitivity and is seldom reported.

5. Inaccurate Correction for Deadtime Losses and Pulse Pile-up either by Software or Hardware Methods

High rate gamma spectroscopy using a loss free counter or the pulser method has been recently reported in several studies (Korun and Martinic,

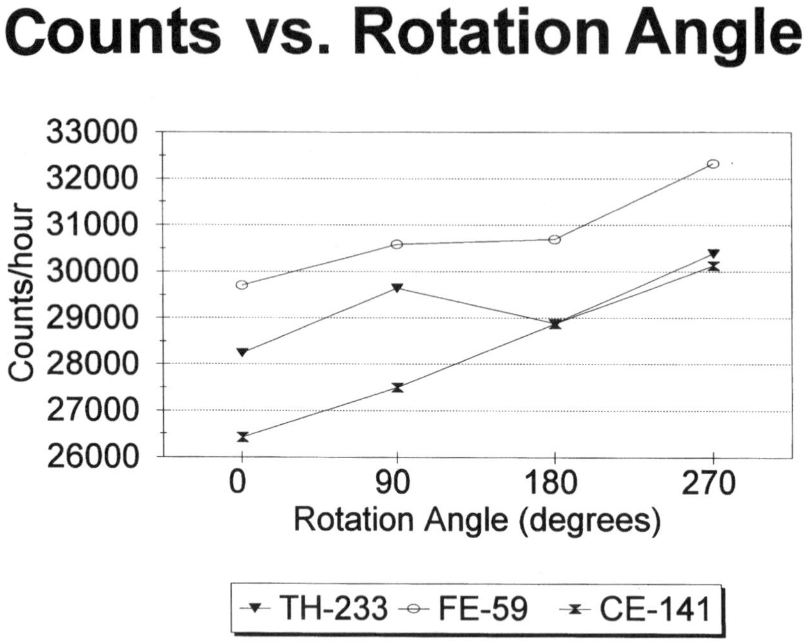

Figure 9. Variation of activity for a soil sample counted in the same position and rotated at different angles.

1996; Lindstrom *et al.*, 1996; Westphal, 1996) while few have described on the use of loss free counting on real samples. Short-lived NAA of Arctic aerosol samples using a loss free counter is reported in Biegalski *et al.* (1996). The real advantage of the loss free counters for airborne particles is for ^{77m}Se ($t_{1/2}$ = 17.4 s) and ^{110}Ag ($t_{1/2}$ = 24.6 s). A more detailed description of these methods is found in the above references.

6. Poor Deconvolution of Overlapping Peaks by Computer Programs

Spectrum evaluation can be performed by peak fitting routines available from commercial companies (e.g. CANBERRA, EG&G, APTEC) or from other laboratories such as HYPERMET (Fazekas *et al.*, 1996), SAMPO (Aarnio *et al.*, 1992) or CINA (Nelson, 1987). There are many other in house programs developed by various NAA laboratories. These types of software are compatible with PC computers. Great care must be given to low intensity peaks that may have crucial environmental significance, but may not be peak fitted very well due to poor counting statistics. A critical comparison for such photopeaks between sophisticated programs and manual fitting must be made. Christensen and Heydorn (1987) proposed a quality assurance procedure in the determination of overlapping peak areas. In general there are few spectral interferences in the analysis of aerosols, and they can easily be accounted for. In fact the vast majority of photopeaks can be evaluated manually by taking backgrounds on both sides of the peak or computer programs can be written to do this automatically (Landsberger *et al.*, 1994).

7. Sample Blanks and Vial Contamination

Investigations in airborne particulate matter accuracy and sensitivity can be severely compromised by an inherent trace element contamination from filters and counting vials. Before any sampling is undertaken, a detailed analysis of the chosen filter type must be completed. At least five to ten filters should be analyzed in order to determine trace concentrations and variability of the results. A large standard deviation may be unacceptable if the element of environmental interest has concentrations in the same range as the blank element. Another consideration in choosing filters is the presence of elements in high amounts and those having large activation cross-sections. For instance, high Compton backgrounds in short-lived activation analysis attributed to aluminum, $^{27}Al(n,\gamma)^{28}Al$; chlorine, $^{37}Cl(n,\gamma)^{38}Cl$; manganese, $^{55}Mn(n,\gamma)^{56}Mn$; and sodium, $^{23}Na(n,\gamma)^{24}Na$ reactions result in poor analytical detection limits for elements such as vanadium and copper. Bromine, $^{81}Br(n,\gamma)^{82}Br$, and sodium, $^{23}Na(n,\gamma)^{24}Na$, photopeaks have the same effects in long irradiations on the determination of some of the radionuclides of medium half-life (e.g. ^{76}As, ^{122}Sb, and ^{42}K). Many ^{82}Br peaks usually seen in such spectra contribute to the high background resulting from the Compton effect. It is important for each new set of filters acquired that blank determinations be made since variability in trace elements and concentrations are not uncommon for different filter batches produced. Berg and Royset (1993) present a comprehensive table of blank values of 30 different elements in 19 different filter types from Gelman, Millipore, Nuclepore, and Whatman.

Trace contaminants can arise in field blanks as a result of improper handling, unclean bags or containers used in transportation, or from the sampling apparatus itself. Field blanks and real samples should also include filter samples taken up- and downwind from major and minor pollution sources. This is especially true for samples taken from pristine environments where contributions from specific local pollution sources may give rise to abnormally high elemental concentration levels.

Analytical problems can arise from trace constituents in vials. Trace elements such as Al, Br, Cl, Mn, Na, Se, V, Zn etc., are present in many types of irradiation containers (Alian and Sansoni, 1985). These concentrations are usually negligible when analyzing geological material, but are important when analyzing air filters. Other possibilities include placing filters in ultra thin precleaned polyethylene bags. It is advisable for short-lived activation studies that the filters be removed from their irradiation containers and placed into clearly labelled inert vials for counting. This procedure also holds true for medium- and long-lived NAA. During longer irradiations, several types of filters (e.g. Whatman) may become very flaky. Their transfer to inert vials is not only difficult but may result in loss of sample material. One way to overcome this difficulty is to press the Whatman filters into a pellet. Great care is required so that no external contamination arises from the handling the filter or the presser itself. Other types of filters (e.g. Nuclepore) may be much more resistant to irradiation damage making the transfer to inert vials possible.

If filters can not be removed from vials, an evaluation of the trace contamination in irradiation containers is needed. Irradiation vials require soaking a minimum of 24 h in diluted concentrated HNO_3, followed by three rinses with distilled and/or deionized water, and then thoroughly air- or oven-dried. Cleaned vials are stored in a sealed polyethylene bag or container until use. For long-lived irradiations, ultrapure quartz may also be utilized, but precaution is needed when torches are used to heat seal the vials, for safety reasons, to prevent ignition of the sample, or add to the contamination.

8. Volatilization of Elements During Irradiation

Mercury is the element most susceptible to volatilization as a result of heating from the reactor. As with other environmental specimens, air filter samples should be encapsulated in glass to ensure mercury containment. Although there has been little reported in the literature, bromine may also be potentially volatilized especially when in the form of organic compounds.

9. Unforeseen Nuclear or Spectral Interferences Specific to Airborne Particles

There are fewer nuclear or spectral interferences for NAA of airborne particles as compared to geological specimens. Some are as follows:

(*a*) *Aluminum*

The interfering $^{28}Si(n, p)^{28}Al$ reaction must be taken into account when using the $^{27}Al(n, \gamma)^{28}Al$ reaction. The interfering $^{31}P(n, \alpha)^{28}Al$ reaction, so important in biological samples due to the presence phosphorous, is negligible in particulate matter.

(*b*) *Antimony*

Antimony is usually determined by the 564.0 keV gamma-ray belonging to the $^{121}Sb(n, \gamma)^{122}Sb$ reaction. There is a small but significant spectral interference arising from ^{76}As. A gamma-ray of energy 562.8 keV is characteristic of the ^{76}As decay scheme. It is approximately 2.5% of the abundance of the dominant 559.1 keV photopeak. If the arsenic concentration is much higher than that of antimony, the interference may be very important. The interference factor should be calculated using single elemental standards. Since ^{122}Sb has a half-life almost three times longer than that of ^{76}As, it can take a long for the ^{76}As to decay to negligible amounts. Spectral resolution from the 559.1 keV gamma-ray of ^{76}As is also important.

(*c*) *Bromine*

When using the 616.2 keV photopeak from the $^{79}Br(n, \gamma)^{80}Br$ reaction, a good deconvolution is needed from the 620.4 keV gamma-ray arising from the double escape peak belonging to the 1642.4 keV photopeak of ^{38}Cl, and the possible 619.0 keV photopeak coming from the medium-lived radionuclide, ^{82}Br. There is also an internal transition of ^{80m}Br that has a half-life 4.42 h and "feeds" the 616.2 keV photopeak. If identical irradiation and counting times are not applied to the samples and standards, corrections may be needed (Biegalski and Landsberger, 1994). The very sensitive 37.1 keV photopeak of the internal transition may be used. The overlapping peaks and self-attenuation of the low energy gamma-ray need to be considered.

(*d*) *Cadmium*

Cadmium is best determined using the $^{114}Cd(n, \gamma)^{115}Cd \rightarrow {}^{115m}In$ reaction with the 336.3 keV gamma-ray characteristic of ^{115m}In. A decay time of at least 35 h is required so that equilibrium can be achieved between ^{115}Cd and ^{115m}In and the ^{115m}In half-life of 4.49 h, that may have originally come from any indium in the sample, and will decay to negligible amounts. The 336.3 keV gamma-ray is then representative of the 53.5 half-life of ^{115}Cd.

(*e*) *Magnesium and manganese*

The reactions used to determine magnesium and manganese are $^{26}Mg(n,\gamma)^{27}Mg$ (843.8 keV) and $^{55}Mn(n,\gamma)^{56}Mn$ (846.7 keV), respectively. A good peak deconvolution program is needed to separate these two overlapping peaks. Otherwise the less sensitive 1810.7 or 2112 photopeaks of ^{56}Mn can be used. This may not always be feasible for low concentrations of manganese. An alternative procedure is to wait until the 843.8 keV gamma-ray ($t_{1/2}=9.45$ min) has mostly decayed away in an hour. The 846.7 keV gamma-ray ($t_{1/2}=2.58$ h) is then practically interference free (Landsberger, 1988). Magnesium which is also interfered by the $^{27}Al(n,p)^{27}Mg$ reaction can be measured using a pure aluminum (magnesium free) standard.

(*f*) *Mercury*

Mercury can be detected using the low-energy 77.4 keV photopeak from the $^{196}Hg(n,\gamma)^{197}Hg$. Spectral interferences may arise from the many other low energy gamma-rays or high energy X-rays in the vicinity of the photopeak. Mercury is also determined by the $^{202}Hg(n,\gamma)^{203}Hg$ (279.6 keV) reaction and is interfered by the $^{74}Se(n,\gamma)^{75}Se$ (279.2 keV) reaction. Usually, mercury is not detected in particulate matter since it is mainly in the gas phase and special samplers are required. However, since selenium can also be present in the gas phase it requires consideration. A subtraction of the interferences can be made by normalizing to the 400.7 keV photopeak of ^{75}Se. Normalization to the 136.0 keV or the 264.1 keV gamma-rays should not be done because spectral interferences already exist from other gamma-rays (see selenium).

(*g*) *Molybdenum*

Mo is determined using the 140.5 keV photopeak from $^{98}Mo(n,\gamma)^{99}Mo$ reaction. Usually a decay time of five to six days is needed to optimize the sensitivity for measurement of ^{99}Mo. The uranium fission reaction, $^{238}U(n,f)$ strongly interferes with this determination, therefore a careful evaluation of uranium in a sample is needed. There is a significant decrease in the fission interference and improvement of detection limits using epithermal neutrons (Landsberger, 1986).

(*h*) *Selenium*

Selenium is usually determined using $^{74}Se(n,\gamma)^{75}Se$ (136.0 and 264.1 keV) reaction in long-lived NAA. The interfering reactions include: $^{180}Hf(n,\gamma)^{181}Hf$

(136 keV) and $^{181}Ta(n, \gamma)^{182}Ta$ (264 keV). In air measurements the interferences arising from hafnium and tantalum are quite small and usually the 264.1 keV gamma-ray is used because it is situated on a lower background. The less sensitive 400.7 keV interference free photopeak can also be utilized.

(*i*) *Uranium*

Uranium can be detected using the 74.6 keV peak belonging to $^{238}U(n, \gamma)^{239}U$ reaction for short-lived NAA. Excitation of lead bricks (used for shielding) by gamma-rays interacting as a result of the photoelectric effect will produce photopeaks corresponding to the lead K X-rays at 72.8 keV and 74.9 keV. Even the use of cadmium- and copper-lined lead bricks may not totally eliminate this problem.

(*j*) *Rare-earth elements*

Rare-earth elements in airborne particles can occasionally be used for source signatures from various industries. Care is needed to account for proper uranium fission interferences for cerium, lanthanum, and neodymium (Landsberger, 1986) and spectral interferences for samarium (Landsberger and Simsons, 1987).

10. Human errors

As in all analytical chemistry procedures there are many areas where errors can occur. Some are obvious and others not. Besides the usual occurrences: standards preparation (e.g. wrong dilution factor), and mislabeling samples, there are additional errors in incorrect readings of irradiation, decay, and counting times. Of particular importance is the need to carefully document the sample counting position. When irradiating air filters that have high amounts of sodium, chlorine, or even manganese for short-lived NAA, changing the sample position may be needed for a fraction of the samples due to high deadtimes. The presence of high amounts of sodium and bromine may also lead to high deadtimes in medium-lived NAA, and changes in sample position are sometimes necessary.

DETECTION LIMITS

There has been a plethora of work done on detection limits (Gilbert and Kinnison, 1981; Helsel, 1990; Long and Winefordner, 1983; Rao *et al.*, 1991;

Slymen *et al.*, 1994). The classical definitions of detection limits for radioactivity have been defined by Currie (1968). For many sets of samples there will always be some elemental concentrations that are below the analytical detection limit. In addition certain statistical programs will only operate with a completely filled data set. One way to overcome this deficiency is to assign a value between zero and the detection limit as calculated by a random number generator.

Concentration of Blank Above the Analytical Detection Limit

A statistical treatment of Whatman filter field blanks was utilized in an Arctic study (Zhang, 1996) where one-eighth of the filter was used in the analysis. If more than 50% of elemental contents of the blank filters in a year are above the analytical detection limit, then the mean and standard deviation of the elemental concentrations are calculated based on the number of these blank filters. The elemental concentrations of the blank filter that are greater than the mean plus two standard deviation are deleted. The mean and standard deviation of the elemental concentrations are recalculated and the critical value for testing sample concentration is:

$$A = B + 3\sigma_b \tag{8}$$

where:

A — critical value of the sample being analyzed,
B — mean of the elemental concentration of the blank filters,
σ_b — standard deviation of the elemental concentration of the blank filters.

Finally, the air concentrations are obtained from the following:

$$\begin{aligned} &\text{If: } X_{\text{sample}} > A \\ &\text{Then: } C = 8\,(X_{\text{sample}} - B)/V \\ &\text{Else: } C < 8 \times 3\sigma_b/V \end{aligned} \tag{9}$$

where:

X_{sample} — elemental concentration of the sample,
C — air elemental concentration,
V — volume through whole filter.

The number 8 comes from the fact that 1/8 of the filter was used.

Concentration of Blank Below the Analytical Detection Limit

If more than 50% of the elemental concentration of the blank filters in a year is below the analytical detection limit, the mean and standard deviation

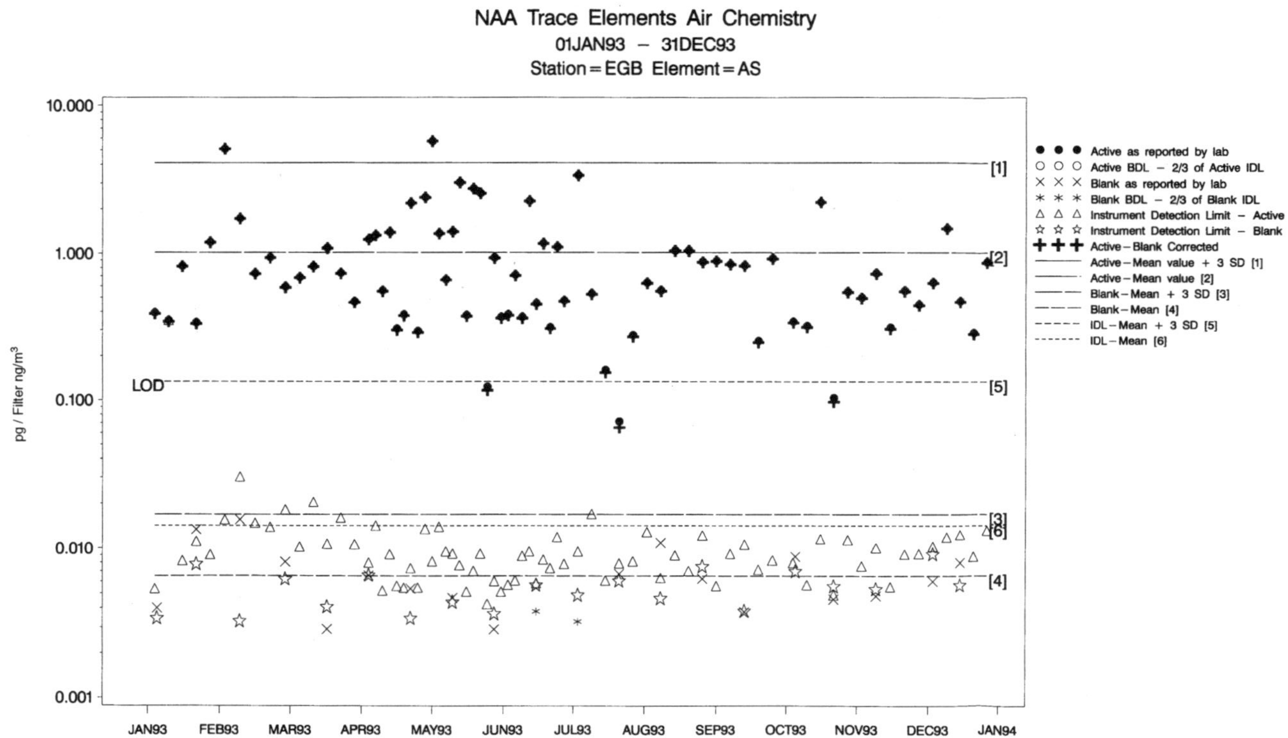

Figure 10. Determined amount of element on a filter compared to the blank corrected, mean, mean plus 3 standard deviations, mean blank, instrumentation detection limit and instrumentation detection limit plus three standard deviations values.

of the elemental concentrations are calculated based on the analytical detection limit of those blank filters. The critical value for testing sample concentration is:

$$A = 3B \tag{10}$$

where:

A — critical value of the sample being analyzed,
B — mean of the elemental concentration of the blank filter.

The blank corrected value is:

$$D = (2/3)B \tag{11}$$

where D — is the blank corrected value.

The method of calculation of the critical value and the blank corrected value comes from classical treatment of detection limits (Currie, 1968). Finally, the aerosol concentrations are obtained from following:

$$\begin{aligned} &\text{If: } X_{\text{sample}} > A \\ &\text{Then: } C = 8\ (X_{\text{sample}} - D)/V \\ &\text{Else: } C < 8D/V \end{aligned} \tag{12}$$

A random number generator can also be used instead of 2/3 of the blank value. Usually how the blank value is treated when it is less than the detection limit, does not greatly affect the environmental interpretation of the results.

A way of clarifying the significance of a set of results from NAA or another method, is shown in Figure 10. In this illustration the amount of material for a particular element determined in an air filter is compared to a series of instrumentation and blank detection limits (Sukloff, 1995). This gives the researcher a visual interpretation of how reliable the results are as compared to various detection limits and statistical criteria. It can be seen in the figure that the results for arsenic expressed in ng/m^3 are very reliable.

QUALITY CONTROL (QC)

An ongoing protocol to ensure reliability of analytical data must be established in each analytical laboratory. For NAA these procedures may include but are not limited to the following:

1. Analysis of Certified Reference Materials

Numerous reference materials are available from international organizations including: National Institute for Standards and Technology, International Atomic Energy Agency, European Community of Reference, and other national laboratories. A number of them are well suited to the kind of application considered here, since they represent coal, fly ash, bulk particulate matter, or artificially prepared filters. Faltejsek and Obrusnik (1994) have written an elaborate PC-based computer program for the quality assessment of NAA of airborne particles using certified reference materials. In the recent past there have been several attempts to produce simulated air filters either by pipetting known elemental concentrations on filters (National Institute for Standards and Technology — NIST SRM 3087 Metals on Filter Media) or re-suspending fly ash on filters (Community Bureau of Reference BCR No. 128 Fly Ash on Artificial Filter). Watjen *et al.* (1993) prepared an ambient aerosol filter using conventional sampling procedures with cellulose nitrate and Nuclepore membrane filters for PIXE analysis. Landsberger *et al.* (1997b) prepared simulated air filters by resuspending contaminated soil on Nuclepore filters.

2. Re-irradiation and Recounting

It is useful to re-irradiate and recount selected samples in order to determine short-lived radionuclides to ascertain analytical reproducibility of the methods used. Equally important is to recount selected samples for medium- and long-lived radionuclides.

3. Spectrum Evaluation

A careful evaluation of each individual spectrum is helpful in determining if any potential nuclear or spectral interferences exist such as uranium fission or unexpected high concentrations of certain elements.

4. Analysis by Different Isotopes or Different Gamma-Rays

In NAA elemental determination can be accomplished using various isotopes as well as various gamma-rays from the same isotope and therefore achieving an internal QC. Several good possibilities for different isotopes for airborne particles include: ^{80m}Br and ^{82}Br; ^{64}Cu and ^{66}Cu; ^{77m}Se and ^{74}Se; ^{122m}Sb, ^{122}Sb, and ^{124}Sb, ^{110m}Ag and ^{110}Ag; ^{69m}Zn and ^{65}Zn; ^{87m}Sr and ^{85}Sr; ^{239}U and ^{239}Np; ^{197m}Hg and ^{203}Hg; ^{125m}Sn and ^{113}Sn; and ^{59}Fe and ^{54}Mn. The radionuclides ^{82}Br, ^{42}K, and ^{24}Na can sometimes be determined both in short- and medium-lived NAA. Finally many isotopes have several gamma-rays from the same decaying isotope that can be employed as a further measure for QC. This procedure is particularly important when one of the gamma-rays used exhibits a spectral or nuclear interference. Some of these gamma-rays can be seen in Tables 2–4.

5. Detection Limits

During any long-term sampling program, instrumentation and analytical procedures may vary, resulting in improvement or regression of detection limits. A plot of detection limits over a long period of time may provide a clue in any deterioration of methodology. The NAA detection limits for arsenic from three different Great Lakes sampling stations over a period of six years using Whatman air filters is illustrated in Figure 11. There are four distinct results. During 1989 samples were counted without any lead shielding and during 1990 shielding was added. In 1991 a different detector was used with the original detector being employed again in 1992.

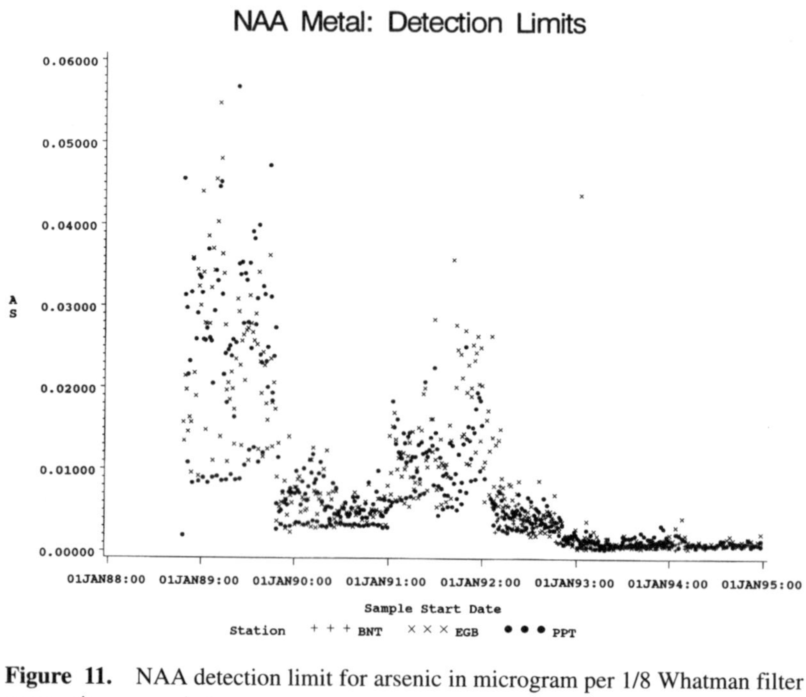

Figure 11. NAA detection limit for arsenic in microgram per 1/8 Whatman filter over a six year period.

Finally, during 1993 and 1994 the implementation of epithermal NAA along with Compton suppression drastically reduced the detection limit.

6. Interlaboratory and Multi-method Comparison Studies

The area of QC employed more frequently in past years is the use of other independent chemical determinations on the same filters. Potocek *et al.* (1991) compared PIXE and XRF results for air samples collected in mountain areas. Similarly, Winchester *et al.* (1990) compared PIXE and XRF for samples collected in Russia. Wang *et al.* (1995) compared results between XRF and NAA, and between XRF and inductively coupled plasma-mass spectrometry (ICP-MS). Larsen *et al.* (1992) compared airborne dust samples by ICP-MS, NAA, and PIXE. In all these studies there have been varying results, often depending on the sensitivity or the lower limit of detection of a particular element related to a specific technique. While NAA, XRF, and PIXE are non-destructive methods, chemical analyses such as ICP, ICP-MS, atomic emission spectroscopy or ion chromatography need dissolution techniques. These procedures may add contamination or result in filters not being completely dissolved. Landsberger *et al.* (1997b) have reported on a detailed intercomparison for XRF, NAA, and PIXE for resuspended airborne particles from contaminated soil. Results from the intercomparison for aluminum is seen in Figure 12. A comparison of fine and coarse particle chlorine, and ion chromatography (IC) and NAA is also seen in Figures 13 and 14. The correlation and regression for the fine particles ($<2.5\,\mu m$) is in excellent agreement between the two methods. However for coarse particles, the IC results are some 10% lower than those for NAA. This may be due to incomplete dissolution of these larger particles ($>2.5\,\mu m$). These are just a few of the studies appearing in the literature that emphasize the multi-method approach for the analysis of air filters, both for completeness and QC.

FUTURE ANALYTICAL DEVELOPMENTS

Competition for NAA of airborne particles comes primarily from ICP and ICP-MS, as well as from PIXE and XRF. Of all the above mentioned methods, NAA has the slowest turn around time due to the long decay and counting times required. To make NAA more competitive and attractive, the incorporation of more efficient detectors including well-type germanium counters must be incorporated in any large sampling program. Through the judicious use of short- and medium-lived thermal and epithermal NAA, in conjunction with Compton suppression methods almost all

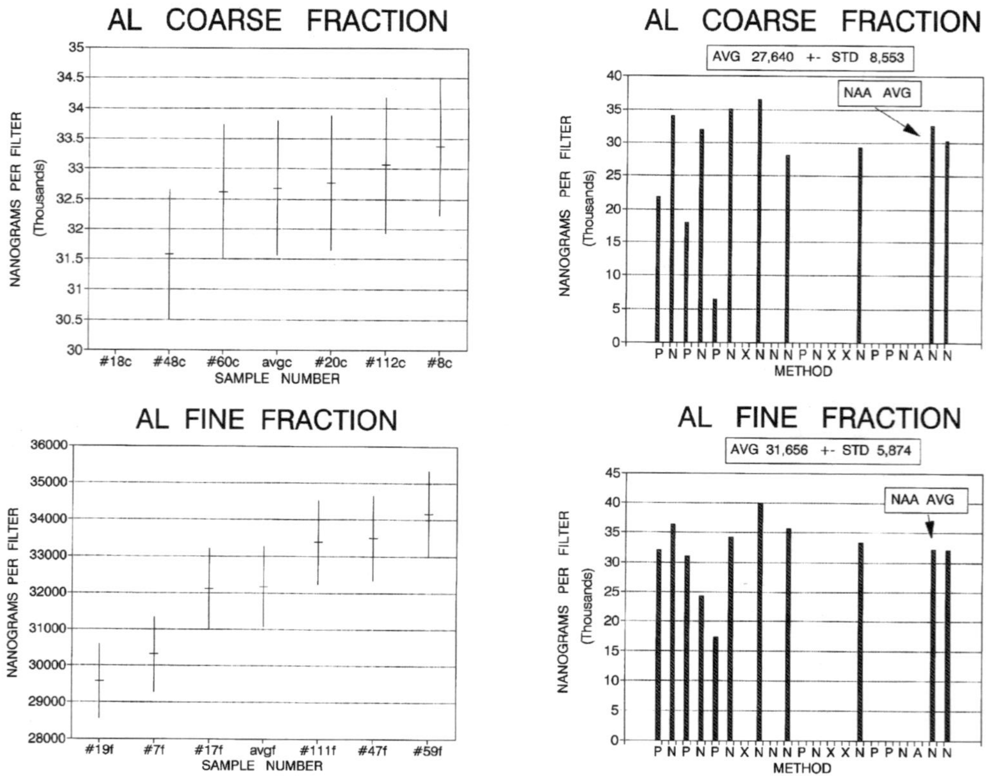

Figure 12. Comparison of different methods (P-PIXE; X-XRF, and N-NAA) for the determination of aluminum on simulated air filters. Graphs on the left side are results for reproducibility for six individual results using NAA.

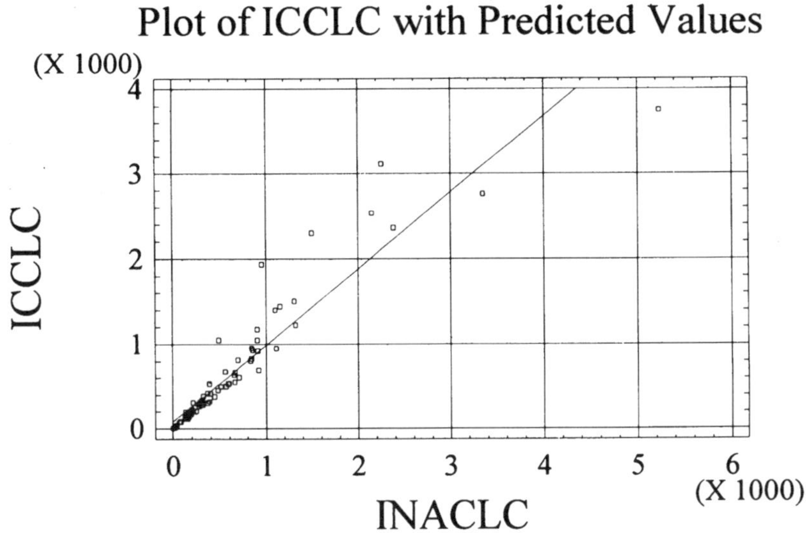

Figure 13. Ion chromatography vs NAA results for Arctic coarse chlorine particles collected on Whatman air filters. IC coarse Cl = 85.06 + 0.90 * NAA coarse Cl; $r^2 = 0.89$.

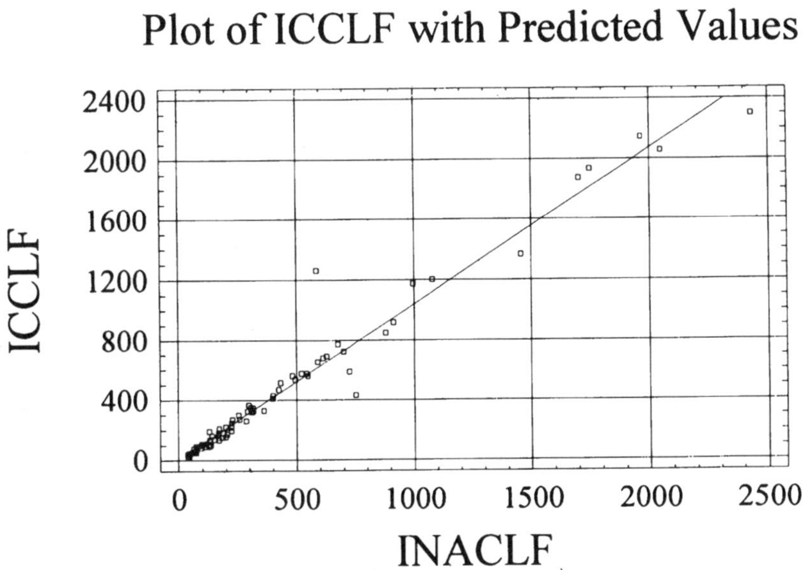

Figure 14. Ion chromatography vs NAA results for Arctic fine chlorine particles collected on Whatman air filters. IC fine Cl = 6.18 + 1.03 * NAA fine Cl; $r^2 = 0.97$.

Table 5. Analytical techniques for sample analysis of Arctic air filters.

Method	Element	Irradiation facility	Ti	Td	Tc
Rapid short	Se	Slowpoke	1 min	10 s	30 s
Epithermal short-lived	I, In, Si	CLPS	5 min	7 min	15 min
Thermal short-lived	Al, Mn, V,	PS	1 min	10 min	10 min
Epithermal medium-lived	As, La, Sb, Sm, Zn,	CLNAT	1 h	2–3 days	1 h
	Cd, Fe, Ni		1 h	4–6 days	5 h

PS — Pneumatic System.
CLPS — Cadmium-Lined Pneumatic System.
CLNAT — Cadmium-Lined Neutron Activation Tube.
Slowpoke — Slowpoke Reactor, University of Dalhousie, Canada.

the elements of interest can be determined without the use of long decays and counting periods. Landsberger *et al.* (1997a) have shown a combination of the above mentioned techniques used in an Arctic study (Table 5). The other techniques of prompt-gamma NAA and "cold neutron" activation analysis should also be explored. Even though only a few such facilities exist, they can be considered to be more research oriented and of possible use for future applications in routine anlaysis. NAA is still a highly effective method for the analysis of airborne particles, especially for very low mass samples, like those encountered in rural and remote regions of the world.

REFERENCES

Aarnio, P.A., M.T. Nikkinen and J.T. Routti, *J. Radioanal. Nucl. Chem.,* **160** (1992) 289.

Alfassi, Z.B., *J. Radioanal. Nuc. Chem.,* **90** (1985) 151.

Alfassi, Z.B., *Activation Analysis*, Vol. 1 and 2, CRC Press, Boca Raton, Florida (1989).

Alfassi, Z.B., *Chemical Analysis by Nuclear Methods,* John Wiley (1994).

Alian, A. and B. Sansoni, *J. Radioanal. Nucl. Chem.*, **89** (1985) 191.

Alian, A. and B. Sansoni, *Activation Analysis of Air Particulate Matter, Central Department for Chemical Analysis*, Report Jul–Spez-475, ISSN 0343-7639, Kernforschungsanlage Julich GmbH, D-5170 Julich, Federal Republic of Germany (1988).

Alian, A. and B. Sansoni, "Activation Analysis of Air Particulate Matter", in *Activation Analysis,* ed. Z.B. Alfassi, CRC Press, Boca Raton, Florida (1989) pp. 503–565.

Becker, D.A., *J. Radioanal. Nucl. Chem.,* **113** (1986) 5.

Berg, T., Royset and E. Steinnes, *Atmos. Environ.,* **27A** (1993) 2435.

Biegalski, S. and S. Landsberger, *J. Radioanal. Nucl. Chem.*, **183** (1994) 313.

Biegalski, S. and S. Landsberger, *J. Radioanal. Nucl. Chem.*, **192** (1995) 195.
Biegalski, S.R., S. Landsberger and K. Heydorn, *J. Trace Micro. Tech.*, **14** (1996) 87.
Blaauw, M. *J. Radioanal. Nucl. Chem.*, **191** (1995) 387.
Bode, P., *Instrumental and Organizational Aspects of a Neutron Activation Analysis Program,* PhD Thesis, Department of Radiochemistry of the Interfacultair Instituut, Delft University of Technology, Mekelweg 15, 2629 JB Delft, Netherlands (1996).
Brune, D., *Anal. Chim. Acta*, **46** (1969) 17.
Chilesa, F., D. Gawlik and P. Bratter, *J. Radioanal. Nucl. Chem.*, **112** (1987) 293.
Christensen, L.H. and K. Heydorn, *J. Radioanal. Nucl. Chem.*, **113** (1987) 19.
Csikai, J., *Handbook of Fast Neutron Generators*, Vols. 1 and 2, CRC Press, Boca Raton, Florida (1987).
Currie, L.A., *Anal. Chem.*, **40** (1968) 586.
Dams, R., J. Billiet and J. Hoste, *Inter. J. Environ. Anal. Chem.*, **4** (1975)141.
Dams, R., *J. Radioanal. Chem.*, **61** (1981) 13.
Dams, R., *Pure and Appl. Chem.*, **64** (1992) 991.
Dams, R., *Application of Multi-Elemental Neutron Activation Analysis in Environmental Research,* in Instrumentelle Multielementanalyse, ed. B. Sansoni, VCH, Weinheim (1985).
Das, H.A. and H.A. Van der Sloot, *Radioanalysis in Geochemistry* in Developments in Geochemistry, Vol. 5, Elsevier, Amsterdam (1989).
Das, H.A., A. Faanhof and H.A. Van de Sloot, *Studies in Environmental Science* in Environmental Radioanalysis, Vol. 22, Elsevier, Amsterdam (1983).
Debertin, K. and R.G. Helmer, *Gamma- and X-ray Spectrometry with Semiconductor Detectors*, North Holland (1988).
DeCorte, F., A. Simonits, A. DeWispelaere and J. Hoste, *J. Radioanal. Nucl. Chem.*, **113** (1987) 145.
DeCorte, F., A. Simonits, F. Bellemans, M.C. Freitas, S. Jovanovic, B. Smodis, G. Erdtmann, H. Petri and A. DeWispelaere, *J. Radioanal. Nucl. Chem.*, **169** (1993) 125.
DeSilva, K.N. and A. Chatt, *J. Trace and Micro. Tec.*, **1** (1982) 307.
Ehmann, W.D. and D.E. Vance, *CRC Critical Reviews in Anal. Chem.*, **20** (1989) 405.
Ehmann, W.D. and D.E. Vance, "Radiochemistry and Nuclear Methods of Analysis, Chemical Analysis" in *A Series of Monographs on Analytical Chemistry and Its Applications*, Vol. 116, eds. J.D. Winefordner and I.M. Kolthoff, John Wiley (1990).
Ehmann, W.D. and S.W. Yates, *Anal. Chem.*, **58** (1986) 49R.
Ehmann, W.D. and S.W. Yates, *Anal. Chem.*, **60** (1988) 42R.
Ehmann, W.D., J. Bruckner and D.M. McKown, *J. Radioanal. Nucl. Chem.*, **57** (1980) 491.
Ehmann, W.D., J.D. Robertson and S.W. Yates, *Anal. Chem.*, **62** (1990) 50R.
Ehmann, W.D., J.D. Robertson and S.W. Yates, *Anal. Chem.*, **64** (1992) 1R.
Erdtmann, G. and H. Petri, "Nuclear Activation Analysis: Fundamentals and Techniques" in *Treatise on Analytical Chemistry*, Part 1, Vol. 14, Section K, Nuclear Activation and Radioisotopic Methods, John Wiley, New York (1986).
Faltejsek, J. and I. Obrusnik, *Biol. Trace Elem. Res.* Vol. 43–45 (1994) 605.
Fazekas, B., T. Belgya, L. Dabolczi, G. Molinar and A. Simonits, *J. Trace Micro. Tech.*, **14** (1996) 167.
First International Workshop on Activation Analysis with Short-Lived Nuclides, *J. Radioanal. Chem.*, **61** (1981).
Gilbert., R.O. and R.R. Kinnison, *Health Phys.*, **40** (1981) 377.
Gilmore, G. and J. Hemmigway, *Practical Gamma-Ray Spectrometry*, John Wiley (1995).
Glascock, M.D., W.Z. Tian and W.D. Ehmann, *J. Radioanal. Nucl. Chem.*, **92** (1985) 379.
Greenberg, R.R., *J. Radioanal. Nucl. Chem.*, **179** (1994) 131.
Helsel, D.R., *Environ. Sci. Technol.*, **24** (1990)1766.
Heydorn, K., *Aspects of Precision and Accuracy in Neutron Activation Analysis*, RISO National Laboratory, Roskilde, Denmark, Report R-419 (1980).
Heydorn, K., *Neutron Activation Analysis for Clinical Trace Element Research*, Vols. 1 and 2, CRC Press, Boca Raton, Florida (1984).
Heydorn, K., *Isotopenpraxis*, **24** (1988) 45.
Heydorn, K. and E. Damsgaard, *J. Radioanal. Nucl. Chem.*, **179** (1994) 87.

IAEA, *Comparison of Nuclear Analytical Methods with Competitive Methods*, IAEA-TECDOC-435, International Atomic Energy Agency, Vienna (1987a).
IAEA, *Applications of Short-Lived Activation Products in Neutron Activation Analysis of Bio-Environmental Samples*, IAEA/RL/141, International Atomic Energy Agency, Vienna (1987b).
IAEA, *Handbook on Nuclear Activation Data* Technical Reports Series No.273, International Atomic Agency, Vienna (1987c).
IAEA, *Practical Aspects of Operating a Neutron Activation Analysis Laboratory*, IAEA-TEDOC-564 (1990).
Janssens, M., B. Desmet, R. Dams and J. Hoste, *J. Radioanal. Chem.*, **26** (1975) 305.
Knoll, G.F., *Radiation Detection and Measurement*, 2nd edition John Wiley (1989).
Kolthoff, I.M., P.J. Elving and V. Krivan (eds), "Nuclear Activation and Radioisotopic Methods of Analysis", Part 1, Vol. 14, in *Treatise on Analytical Chemistry*, John Wiley, New York (1986).
Korun, M. and R. Martinic, *J. Trace Micro. Tech.*, **14** (1996) 77.
Krivan, V., M. Franek, H. Baumann and J. Pavel, *Fres. J. Anal. Chem.*, **338** (1990) 583.
Landsberger, S., *Chem. Geol.*, **57** (1986) 415.
Landsberger, S., *Anal. Chem.*, **60** (1988) 1842.
Landsberger, S., *J. Trace and Micro. Tech.*, **10** (1992) 1.
Landsberger, S., "Delayed Instrumental Neutron Activation Analysis", in *Chemical Analysis by Nuclear Methods*, ed. Z.B. Alfassi, John Wiley (1994a) pp. 141–142.
Landsberger, S., *J. Radioanal. Nucl. Chem.*, **179** (1994b) 67.
Landsberger, S. and S. Biegalski, "Analysis of Inorganic Particulate Pollutants by Nuclear Methods", in *Handbook of Environmental Chemistry, Airborne Particulate Matter*, eds. Th. Kouimtzis and C. Samara, Springer, Volume 4, Part D (1995) pp. 175–200.
Landsberger, S. and S. Peshev, *J. Radioanal. Nucl. Chem.*, **202** (1996) 203.
Landsberger, S. and A. Simsons, *Chem. Geol.*, **62** (1987) 223.
Landsberger, S. and D. Wu, *J. Radioanal. Nucl. Chem.*, **167** (1993) 219.
Landsberger, S. and D. Wu, *Sci. Tot. Environ.*, **173/174** (1995) 323.
Landsberger, S., P.K. Hopke and M.D. Cheng, *Nucl. Sci. Eng.*, **110** (1991) 79.
Landsberger, S., S. Larson and D. Wu., *Anal. Chem.*, **65** (1993) 65.
Landsberger, S., W.D. Cizek and R.H. Campbell, *J. Radioanal. Nucl. Chem.*, **180** (1994) 155.
Landsberger, S., P. Zhang, D. Wu and A. Chatt, *J. Radioanal. Nucl. Chem.*, **217** (1997a) 153.
Landsberger, S., D. Wu, S.J. Vermette and W. Cizek, *J. Radioanal. Nucl. Chem.*, **217** (1997b) 117.
Larsen, E., A.B. Hansen, L.V. Kristensen, P. Solgaard, E. Damsgaard, K. Heydorn and K. Kemp, *J. Trace. Micro. Tech.*, **10** (1992) 43.
Lavi, N., F. Lux and Z.B. Alfassi, *J. Radioanal. Nucl. Chem.*, **129** (1989) 93.
Lindstrom, R.M., M.J.J. Ammerlaan and S.S. Then, *J. Trace Micro. Tech.*, **14** (1996) 67.
Lodge, J.P., *Methods of Air Sampling*, Lewis Publishers (1989) pp. 143–150.
Long, G.L. and J.D. Winefordner, *Anal. Chem.*, **55** (1983) 713A.
Maenhaut, W., "Analytical techniques for Atmospheric Trace Elements", in *Control and Fate of Atmospheric Trace Metals*, (eds.) J.M. Pacyna and B. Ottar, Kluwer Academic Publishers (1989).
Maenhaut, W., *Int. J. PIXE,* **2** (1992) 605.
Moody, J.R., R.R. Greenberg, K.W. Pratt and T.C. Rains, *Anal. Chem.*, **11** (1988) 1203A.
Mughabghab, S.F., M. Divadeenam and N.E. Holden, *Neutron Cross Sections, Volume 1: Neutron Resonance Parameters and Thermal Cross Section, Part A: Z=1–60,* and *Part B: Z=60–100*, Academic Press (1984).
Nakanishi, T. and B. Sansoni, *J. Radioanal. Chem.*, **37** (1977) 945.
Nelson, G.W., *J. Radioanal. Nucl. Chem.*, **98** (1987) 231.
Obrusnik, I. *Chemia Analitycana*, **35** (1990) 221.
Olmez, I, *Turkish J. Nucl. Sci.*, **8** (1981) 15.
Parry S., *J. Radioanal. Chem.*, **72** (1982) 195.
Parry, S.J., "Activation Spectrometry in Chemical Analysis", in *Chemical Analysis: A Series of Monographs on Analytical Chemistry and its Application*, Vol. 119, eds. J.D. Winefordner and I.M. Klothoff, John Wiley (1990).

Perry, R. and R.J. Young, *Handbook of Air Pollution*, Chapman and Hall (1977) pp. 166–170.
Potocek, V., R. Brenner, F. Hodik and J. Voltr, *J. Radioanal. Nucl. Chem.*, **149** (1991) 205.
Rao, S.T., J.-Y. Ku and K.S. Rao, *J. Air and Waste Manag. Assoc.*, **41** (1991) 442.
Rowe, J.J. and E. Steinnes, *J. Radioanal. Chem.*, **37** (1977) 849.
Slymen, D.J., A. de Peyster and R.R. Donohoe, *Environ. Sci. Technol.*, **28** (1994) 898.
Smitt, B.F. and C. Segebade, "Instrumental Analysis of Airdust by Activation with Photon and Photoneutrons", in *Instrumentelle Multielementanalyse*, ed. B. Sansoni, VCH, Weinheim (1985).
Sukloff, W. (private communication, Atmospheric Environment Service, Canada) 1995.
Spyrou, N.M. and S.A. Kerr, *J. Radioanal. Chem.*, **48** (1979) 169.
Tian, W.Z., *J. Radioanal. Nucl. Chem.*, **103** (1986) 225.
Tian, W.Z. and W.D. Ehmann, *J. Radioanal. Nucl. Chem.*, **84** (1984) 89.
Tobler, L., V. Furrer and A. Wyttenbach, *Biol. Trace Elem. Res.*, **26–27** (1990) 623.
Tolgyessy, J. and E.H. Klehr, *Nuclear Environmental Chemical Anlaysis*, John Wiley (1987).
Tolgyessy, J. and M. Kyrs, *Radioanalytical Chemistry* Vols. 1 and 2, Ellis Horwood Ltd., Chichester, United Kingdom (1989).
Wang, C.F., E.E. Chang, P.C. Chiang and N.K. Aras, *Analyst*, **120** (1995) 2521.
Watjen, U., M. Kriews and W. Dannecker, *Nucl. Instrum. Meth. Phys. Res.*, **B75** (1993) 257.
Westpahl, G.P., *J. Trace Micro. Tech.*, **14** (1996) 57.
Winchester, J.W., V.A. Ivanov, M.A. Prokofyev, D.A. Zhukovski, A. Shlikhta, V.F. Zhalev, A.N. Zhukovski, D.M. Strogfanov, J.W. Nelson, S.E. Bauman and L.N. Nakhgaltsev, *Nucl. Instrum. Meth. Phys. Res.*, **B49** (1990) 351.
Zhang, P., *Relative Composition of Crustal and Anthropogenic Trace Element Constituents in Canadian Arctic Aerosol From 1982–1992 as Determined by Neutron Activation Analysis,* MS Thesis, Department of Engineering, University of Illinois, Urbana, Illinois, USA (1996).

6 QUALITY ASSURANCE, QUALITY CONTROL, AND DATA VALIDATION IN ENVIRONMENTAL ANALYSIS OF AIRBORNE PARTICLES

P.K. HOPKE

Department of Chemistry, Clarkson University, Potsdam, New York, USA 13699-5810

INTRODUCTION

One of the most important developments in the past decade with respect to air quality management has been the use of quality assurance/quality control (QA/QC) and data validation as an integral part of the environmental sampling and analytical processes. In the past this effort was seen as unnecessary paperwork or a loss of resources spent for these purposes. It is valuable to perform a formal QA/QC and validation program to consider how samples are taken and analyzed, and to determine the quality of the results. This process also provides documentation that demonstrates the quality of the data being used for decision making purposes.

In general, scientists engage in some level of QA/QC (e.g. instruments are calibrated, and some duplicate analyses are run or samples taken). However, it is not generally performed in a planned systematic fashion with careful documentation of the results. Typically 10–15% of the project effort should be devoted to QA/QC, and these procedures should be built into the ambient sampling and analysis programs.

The US Environmental Protection Agency (EPA) now mandates a formal QA/QC plan be submitted by any of its contractors and approved by the Agency before any samples can be obtained or analyses performed. Detailed guidance on these programs are found in guidelines published by the American National Standards Institute (1987). An excellent reference specifically for the analytical laboratory portion of the QA/QC program is found in Garfield (1984).

QUALITY ASSURANCE PROGRAM PLAN (QAPP)

The first step in developing a QAPP is to identify the program category into which the project falls. The level of detail, documentation, and effort in QA are dependent on the nature of the project measurements and objectives for taking and analyzing samples. The US EPA has defined four categories in which sampling and analyses are performed. They are:

Project Category

1. Category 1 projects have a high level of priority. Therefore to be certain that program objectives are met, they have the most stringent requirements for QA. The results must stand up to legal scrutiny in formal hearings or in resulting litigation.
2. Category 2 measurements are projects whose results complement other studies with sufficient scope and depth that can be used to build a case for rulemaking, regulation promulgation, or policy decisions. These are studies that have a high visibility with the public even though they do not have a direct effect on regulatory decisions.
3. Category 3 projects are those whose results will be used for evaluating and choosing basic options, performing feasibility studies, or initial assessments of areas in which there is no existing data.
4. Category 4 projects are basically research oriented without significant immediate regulatory implications.

A QAPP contains most of the same elements irrespective of the category into which the project has been placed. However, the rigor in which the calculation of uncertainties and bias are obtained and the level of detail regarding representativeness and comparability of the data vary widely depending on the category. The elements of a QAPP are listed in Table 1.

Project Description

The project description provides enough details of the study so that a technical person unfamiliar with it can understand the sampling/measurement program to be implemented. In many cases an external review of the

Table 1. Project elements needed for given program categories (AEERL, 1988).

QA plan element	Category applicability
1. Project description	1,2,3,4
2. Project and QA/QC organization	1,2*
3. Data quality objectives for critical measurements	1,2,3,4
4. Sampling procedures	1,2,3
5. Sample custody	1,2**
6. Calibration procedures and frequency	1,2**
7. Analytical procedures	1,2,3
8. Data reduction, validation, and reporting	1,2,3
9. Internal QC checks	1,2**
10. Calculation of data quality indicators	1,2**,3**
11. Preventive maintenance	1
12. Plans for performance and systems audits	1,2,3
13. Corrective action	1,2,3
14. QC reports to management	1

* Organization chart only.
** These elements may *not* be applicable for all projects in the specific category. The applicability or importance of these elements depends on the project descriptions and the data quality indicators.

QAPP is performed and the reviewer must be provided with sufficient detail of the project so that its validity can be determined. Other details of the project include: general flow of samples between sampling sites and laboratories, when, where, and for how long the project will take place, and the various analytical requirements.

Project Organization and Responsibilities

For most projects, there should be a clear delineation of individual responsibilities. The key individuals in the program are identified and are presented in a table or organization chart so that each participant knows the flow of management responsibilities throughout the various aspects of the project.

Data Quality Indicator Goals for Critical Measurements

There are five data quality indicators that are basic to all studies:

1. precision
2. bias

3. completeness
4. representativeness
5. comparability

In addition, other indicators such as limits of detection may be added as needed. Expected or desired values for accuracy and precision can be established for each measured variable. These values are based on prior knowledge of the capabilities of a measurement system, method validation studies using replicates, spikes, reference standards, calibrations, recovery studies, etc., and the requirements of the specific project. It is critical to be sure that the data quality at least meets the project needs. Only then can one consider whether or not to exceed those goals, and what the costs will be. In order to determine that these precision and accuracy goals have been met, quality control limits are set and a system of QC checks established.

1. Precision

For analytical measurements the precision of the process is determined by the standard deviation of the individual measurements. For problems such as the measurement of elemental analysis of airborne particles, this usually means collection and analysis of collocated samples. These are defined as two or more samples collected at the same location and time in that they can be considered identical. These results will provide quantitative measures of the random errors in sampling, handling, storage, and analysis. Simply splitting a sample such as cutting strips from an 20.3 cm × 25.4 cm (8 inch × 10 inch) filter does not fully duplicate all of the sources of error in the total measurement process. The QAPP should consider objectives for both the total measurement and the analytical precision.

Another aspect of precision that needs to be considered and addressed in the QAPP is the repeatability of the analysis, i.e. the ability of a single laboratory to repeat an analysis on identical samples. It can be both short-term repeatability (e.g., a single day) or the long-term precision that might occur over a period of days to months.

For studies containing a large number of samples, typically a fixed frequency of repeated measurements is recommended. Typical frequencies would be in the order of one in seven to one in ten. When dealing with smaller batches, more frequent duplicate analyses are performed to be certain that there is a sufficient sample to permit the precision to be determined. If there are a number of samples in which the analyte of interest is below detection limits, then samples with measurable concentrations are selected as those to be used for the precision assessment. When that is not possible, then duplicate samples where standard additions are made to yield measurable values can be used. In this case, both recovery and precision are calculated for each sample.

2. *Bias*

It is not possible to directly measure the bias in the sampling process. Part of it in the handling, storage, transportation, aging, etc can be assessed by spiking the sample in the field (usually from precipitation samples); but this will not identify bias in the sampler itself. Assuming that the flow rates in the samplers have been accurately assessed, sampling biases have to be determined using test aerosols, typically in a wind tunnel. However, a systematic difference (one flow rate always lower than the other) between nominally duplicate samplers would indicate the potential for a bias in one sampler or the other, but would not indicate which one is correct. Standard additions in the laboratory can be used to assess the bias of the analytical method. When possible, additions should be made so that the analyst is not aware of the spiking of a sample or the amount added. Alternatively, bias can be assessed by the analysis of standard reference materials sufficiently similar in matrix characteristics to the actual samples, and therefore represent a fair test of the method.

3. *Completeness*

Data completeness is the degree in which data has been obtained during the sampling and analysis program relative to the original identified needs. For example, if annual average values are to be estimated with a given level of confidence and samples are taken throughout the year at fixed intervals (i.e., every 6th day), then an adequate number of samples are needed, both in total over the year and within each calendar quarter. Again the program goals and the planned use of the results will determine how complete the process must be. In some cases essentially 100% completeness may be necessary. If the cost of sampling is significantly less than that of analysis, then it may be prudent to take more samples than that needed for analysis. In this way, if samples are lost or damaged in the analytical stage, others can be used to provide sufficient completeness to meet the program needs. Part of the QAPP is to identify the likely causes of data loss so that they can be minimized or eliminated.

4. *Representativeness*

Usually the objective of an ambient monitoring program is to determine an average or worst case concentration over a region for a given time interval. The idea is to randomly collect samples from the possible target areas that exist in the location during that time. This objective can be achieved with a probability based sampling plan. To obtain such a design may require the collaboration of a statistician. There may not be sufficient information at the beginning to permit the development of a detailed design, and a smaller program will be needed to provide the basis for the larger design. There are often pressures to collect and analyze samples for the sake of showing the public that a problem is being addressed. However, without a proper design, the data may not be representative of the actual situation. This leads to improper actions that either do not solve the problem, or may even make the situation worse (for example, by eliminating jobs when there is no improvement in environmental quality). In such circumstances, the credibility of the environmental quality managers is severely diminished and thus, their future effectiveness reduced. Although it may be very difficult, an effort must be made to obtain truly representative data so that the best possible decisions can be made about future courses of action.

5. Comparability

Comparability is the confidence in which one data set can be compared to another. It is often necessary to compare environmental quality before and after action is taken that is expected to improve the situation. Again it is important that the results of such a comparison be credible with all of the interested parties. Historical data may have problems that are discovered later as analytical methods improve. For example, a bias may be known to have been present in the sampling and/or analysis process relative to current practice. It is acceptable to correct for such bias, but the basis for the correction and exactly what is actually done must be incorporated in the program reports. Therefore the QAPP includes an indication as to the comparability of the data to be obtained from the planned project with other data sets.

Sampling Procedures

The first step in sampling is site selection. A set of written criteria for choosing sites based on the objectives of the project is necessary, and sampling sites that have been identified, should be carefully described. It if often useful to include photographs of the sampling site and the location of the sampler. Include detailed written instructions as to how the sampling is to be done. If standard procedures are being used, then a reference to the appropriate set of standard methods must be provided. For non-standard procedures, describe the process in detail. Although a reference to standard methods can be made in the QAPP, the technician who is actually performing the sampling needs a copy of the complete detailed procedure. Any special conditions for the preparation of the sampling equipment (e.g. removal of filters form sampler) are carefully noted so that sample contamination or similar difficulties are minimized. Prepare forms or notebooks with space to record pertinent sampling information: start and stop times, flow rates, manually measured temperature, etc. Also include comments about any unusual circumstances related to that sample.

Sample Custody

For samples, whose data may have legal implications, it is necessary to establish a chain of custody from the initial sampling materials (blank filter or empty container) to final analytical results. This is accomplished with written confirmation of the exchange of the sample or material from one step of the process to the next. Various legal jurisdictions have different requirements, but usually dates and times of the transfers need to be noted and signatures obtained by both parties in the exchange.

Calibration Procedures and Frequency

Written procedures describing the process in which the instruments are calibrated are necessary. Calibration includes both the sampling (flow rates in the sampler) and the analysis processes, and the frequency that they are preformed. Some initial knowledge of the behavior of the various components of the measurement process is required in order to make judgements about calibration frequencies. Calibration is required after any modification of the system or maintenance operation, however routine. The documentation describes the standards used and their traceability to primary standards or basic physical principles.

Analytical Procedures

Routine analyses should have carefully written standard operating procedures (SOPs). For exploratory research projects (Category 4 and sometimes Category 3), there should be evolving written procedures so that as methods are developed they are incorporated into the SOPs for consistent future use. It is particularly important that information be obtained in writing from the key individuals who are developing and testing the methods so that if they leave the laboratory their knowledge can easily be transferred to those who take their place. This is critical in university settings where students and post-doctoral personnel are not permanent staff. Even when detailed written procedures are not required for research publications, they need to be completed.

Internal QC Checks

As part of the integrated measurement process, there are a number of pieces of information regarding available data quality: collocated samples, replicates, spiked samples, split samples, blanks, detection limit checks, calibration checks, and use of control charts. It is important to examine these data while the program is still on going so that problems can be identified as early as possible, and corrected. Control limits for these checks are established based on the desired levels of precision and accuracy. The limits should be large enough so that they are rarely exceeded when the system is working within the desired boundaries for precision and accuracy. Conversely they need to be stringent enough so that there is a high probability of being exceeded when the measurement system is operating outside of acceptable limits. Often these considerations are quite complicated and require a combination of both scientific and statistical expertise.

To illustrate the idea of control charts, a hypothetical example of examining the mass of a "check weight" used to test the accuracy and precision of a balance is shown in Figure 1. The measured values of the standard weight can be plotted as a function of the session number. Before each session in which filters are weighed (either unexposed or exposed), the balance is checked by weighing a standard class-S weight of between three and five grams. The measured and actual weights are recorded. Normally one sets a control limit of ± 0.001 g (1 mg). If the balance reads outside of this control limit, it has a bias. If the balance is functioning properly, there should be an equal number of values above and below the nominal value, and ideally no values outside the 1.0 mg limit. In this instance there are three values outside the limit when it is assumed that errors in the weighing process are normally distributed. If the limit of 1 mg is exceeded in $<1\%$ of the cases, it corresponds to a value greater than three time the standard deviation. In this case, this control level has not been achieved, particularly since two outliers occur in the first 10 samples. In general, if a value shows a systematic trend or an unacceptable number of uncontrolled variables, then appropriate remedial action is required before additional samples are taken and analyzed.

Calculation of Data Quality Indicators

Report the precision and accuracy of all measurements, and include them in the QAPP. All errors arising from the different techniques used should be propagated throughout the calculation of air concentrations.

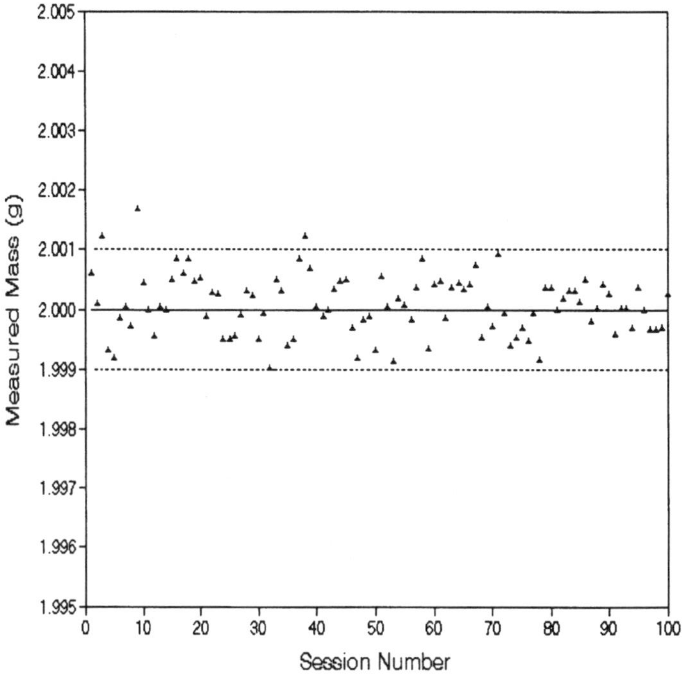

Figure 1. Hypothetical data for the weight of a 2 g check weight measured at the start of 100 separate weighing sessions.

Preventive Maintenance

Many pieces of equipment require routine maintenance. For example the brushes on high volume sampling pumps need replacement after 400–500 hours of operation. A schedule of preventive maintenance activities should be prepared with estimates of the number of replacement parts that will be needed over the course of the entire project. As a result of maintenance requirements, it may be necessary to purchase extra parts in advance (e.g. spare pump motors that can be quickly changed in the field without interruption of the sampling protocol). The preventive maintenance schedule helps to identify the personnel needs of the project so that a complete cost estimate can be determined.

Performance and Systems Audits

Another important element for a large scale measurement program is to have an external auditor examine the compliance of the data quality management program. Someone who is not directly involved in the measurements should periodically review the results and more importantly the QAPP processes. The objective is to ensure that what was planned to be done regarding data quality, is in fact being completed in a timely and appropriate manner, and is properly documented.

Corrective Action

The key to using a QA/QC program is to quickly identify problems that may interfere with the data quality goals and correct them. Individuals should be responsible for the different aspects of the program, (e.g. sampling, analytical system, data basing, etc.) The QAPP with its written procedures provides personnel with the tools for rectifying a problem if the program manager is unavailable. Ultimately this results in better response planning.

QC Reports to Management

As part of any a study, a report of data quality indicators and the degree in which the project is meeting its objectives is necessary. A final report allows the reader to evaluate the results, and provides confidence in its validity. Interim reporting of the data quality indicators is necessary in some cases. If there are changes in program needs, the project can be modified to meet the new objectives. If there are major changes, then a new planning process should be performed. This kind of disruption occurs in a limited number of cases since it requires an additional expenditure of time and effort.

Data Validation, Reduction, and Reporting

One of the problems of modern instrumentation and the computing systems built into them is that the process in which raw data is turned into analytical results becomes a "black box". It is therefore possible to develop the GIGO syndrome (Garbage In ⇒ Garbage Out). Therefore an examination of the analytical results as they are being produced is necessary. The experienced analyst can often identify data that do not seem right. Although such intuitive approaches to data validation do not replace more rigorous examination, they are often a good starting point for identifying a problem before meaningless numbers are generated. Frequency distributions on data processing can be helpful in identifying potential outliers. At the same time, there should be an effort to minimize the amount of manual transfer of data. Digits may be transposed or decimal points misplaced. There is a need to develop software programs that read the files produced by the analytical instrument to eliminate this potential source of errors.

Data validation are those activities performed after the data have been collected, and are different from QA/QC techniques. QA/QC methods are employed to minimize the amount of bad data collected, while data validation tries to prevent bad data from being processed. It is the process by which data are filtered and either accepted or flagged for further investigation following a predetermined set of criteria, based on the study objectives and type of data being examined. It should be performed as soon as possible after analysis so that questionable results can be checked by recalling unusual events and/or specific meteorological conditions that may have affected the sampled environment. Validation can be manual or computerized, and is performed to isolate spurious values that should not be automatically rejected. Although records of "invalid" data do not appear in the final data set, they need to be archived. Most of the statistical tests described in this chapter are found in Barnett and Lewis (1978). Data validation procedures can be conveniently divided into four sections:

- Use routine check and review procedures to some extent in every data validation process.
- Test for internal consistency of the data.
- Test for consistency with previous data, (historical or temporal).
- Test for consistency with other data sets, collected at the same time or under similar conditions.

Routine Validation Checks

There are four types of routine validation checks:

1. data identification
2. unusual event review
3. deterministic relationship
4. data processing procedures

1. Data identification checks

Data have to be properly identified as to the location and time that the sampling occurred, and labelling is important. This includes making certain that the units ($\mu g\,m^{-3}$, or $ng\,m^{-3}$, etc) are properly used.

2. Unusual event review

A log should be maintained to record extrinsic events such as construction activity in the sampling area, dust storms, traffic jams, and other activities that might result in high or unusual sample composition. For example, a plot of fine particle K ($ng\,m^{-3}$) measured during a sampling campaign in St. Louis, Missouri in the middle 1970s in shown in Figure 2. The particular series of data points come from July and August 1976 (Alpert and Hopke, 1981). It can be seen that there are two points that are very high compared to all of the other data values. In this case, the sampling site was at a sports stadium at a university in the city. On July 4 1976, in recognition of the United States' National Independence Day, there was a major fireworks display. Potassium nitrate is a major ingredient in gun-powder and more than 11 $\mu g\,m^{-3}$ of K was observed during the sample obtained from 12 noon to 12 midnight on that date. Knowing the activities that were conducted near the sampler made it easy to identify the source of the anomalous results. Since this was an unusual event, not expected to be repeated on a regular basis, it was reasonable to exclude these samples from the analysis of the data. The difference in the mean values calculated over all of the data set and with the two unusual event data points are shown in Table 3 in chapter 8 of this book.

Another way to plot data is as a frequency distribution as shown in Figure 3, a conventional frequency distribution plot, and Figure 4, a box and whisker plot. It can be seen in Figure 3 that the distribution is strongly positively skewed and does not fit a normal or even a symmetric

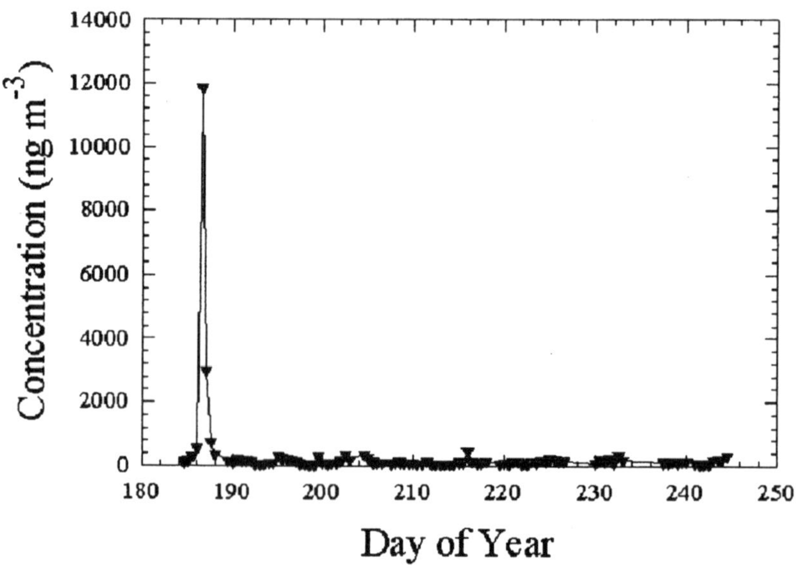

Figure 2. Plot of K concentration ($ng\,m^{-3}$) as a function of the day of the year for July and August 1976 at Regional Air Pollution Study site 112.

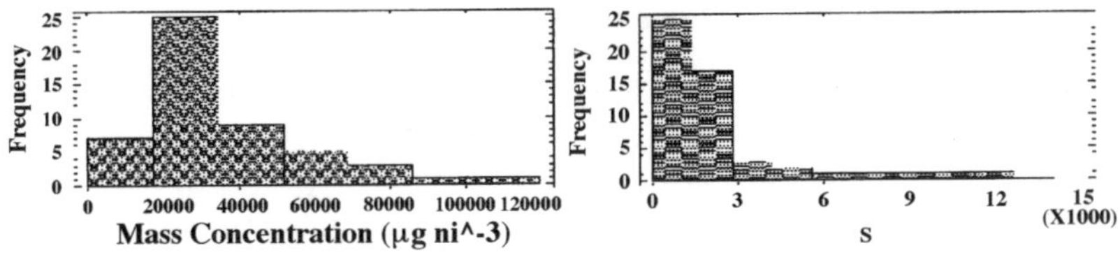

Figure 3. Frequency distributions for fine particle mass (left) and sulfur (right) concentrations.

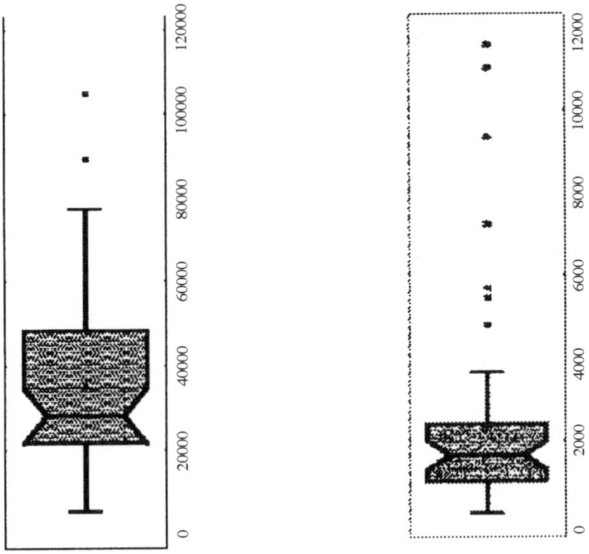

Figure 4. Box and whisker plots for fine particle mass (left) and fine sulfur (right) concentrations.

distribution function. The box and whisker plot emphasizes the same result by showing the median value as a notch in the box. This notch corresponds to the width of the confidence interval for the median. The central box covers the middle 50% of the data values, between the lower and upper quartiles. The whiskers extend to the extreme (maximum and minimum values), only to those points that are within 1.5 times the interquartile range. Values lying outside of those bounds are plotted as individual points. The mean value is denoted by the point within the box, and typically lies above the median for environmental data given the commonly observed positive skewness of the distributions.

3. *Deterministic relationship checks*

Routinely check data sets that have two or more physically or chemically related variables to ensure that the measured values for an individual variable does not exceed the corresponding measured value of the total sample. For example, no individual elemental concentration or even the sum of the measured elements should exceed the total measured sample mass. This check is one of the reasons why it is useful to determine the total airborne mass concentration as well as the concentrations of the individual elements. A plot of fine particle mass as well as black carbon and sulfur as measured in a major city is shown in Figure 5. It can be seen that the

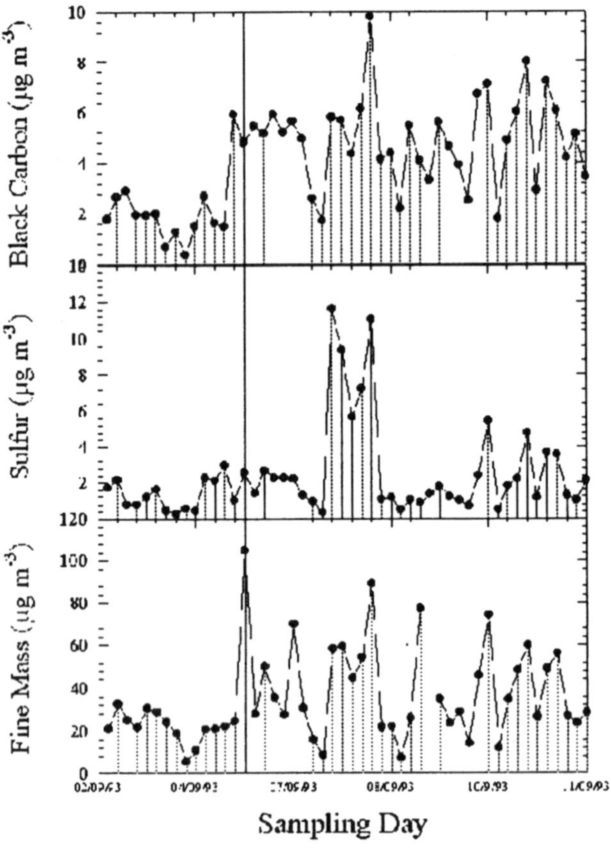

Figure 5. Fine particle, black carbon, sulfur, and total mass for a series of samples.

individual species concentrations are always less than the total measured mass concentration.

4. Data processing procedures

Sixty-seven procedures in use for detecting and, when possible, correcting errors as they occur in the computer systems are listed in *Performance Assurance and Data Integrity Practices* (US Department of Commerce, 1978).

TESTS FOR INTERNAL CONSISTENCY OF THE DATA

These tests check values that are atypical when compared to the whole data set. Common anomalies of this type include unusually high or low values or large differences in adjacent values. The following tests for internal consistency are listed in order of increasing statistical sophistication:

1. data plots
2. Dixon ratio test
3. Grubbs test
4. gap test
5. Johnson *p* test
6. multivariate test

1. Data Plots

Data plotting is one of the most effective means of identifying possible anomalies. Figure 2 is a good example of being able to identify an unusual event in the data. An anomalous data point is also demonstrated in Figure 5. There is a sample that shows over $100\,\mu g\,m^{-3}$ of total fine particle mass. However, the corresponding S value is approximately $2\,\mu g\,m^{-3}$, while the corresponding black carbon concentration is about $4.8\,\mu g\,m^{-3}$. Even if all of the S is present as $(NH_4)_2SO_4$, there is still a large fraction of the fine particle mass missing when one examines all of the measured elemental species. This value appears to be anomalous and requires additional examination to be certain that the mass concentration values have been correctly reported. There are many software packages available that permit rapid and easy plotting of large quantities of data. It is recommended to make both a set of time series plots (value against day during which the sample was taken) and bivariate plots (one element versus another) for each possible pair of elements.

2. Dixon Ratio Test

The Dixon ratio test is the simplest of the statistical test recommended for evaluating the internal consistency of data. The test for the largest value requires only the identification of the lowest (x_1) and the two highest values (x_{n-1} and x_n) in the data set. The ratio

$$R=\frac{x_n-x_{n-1}}{x_n-x_1} \tag{1}$$

is calculated and then compared with tabulated values in the appropriate table (Barnett and Lewis, 1978). Consistency is indicated by a ratio near zero and a possible data anomaly is indicated by a ratio near unity. The critical values of the ratio are derived on the assumption of a normal distribution and thus, a logarithmic transformation is usually needed for most environmental data. Examining the $\log_e$ of the airborne fine mass and fine S concentrations plotted from Figure 5 yields values of the Dixon test of 0.05 and 0.01, respectively. The value corresponding to the number of samples in the data set is obtained from tables available in Barnett and Lewis (1978). The calculated values are compared to these tabulated ones, and in this case are not anomalous.

3. Grubbs Test

This test like the Dixon ratio assumes normally distributed values. However, it only requires computation of the mean, $\bar{x}$, and the standard deviation (s) of the data. The test statistic is

$$T=\frac{x_n-\bar{x}}{s}, \tag{2}$$

where x_n is the largest value in the data set..The calculated T is compared to a tabulated value at an appropriate level of risk. Again for these data sets, the results suggest that they are not anomalous.

4. Gap Test

This test identifies possible data anomalies by examining the length of the gap or distance between the two largest values (x_n and x_{n-1}), the second and third largest values (x_{n-1} and x_{n-2}), and similarly for other gaps. The two parameter exponential distribution is fitted to the upper tail of the distribution of the sample data, and the probabilities of the observed gap sizes determined. If the probability is very small, the larger value is considered as a possible data anomaly.

5. Johnson *p* Test

This test developed at the US EPA (Nelson *et al.*, 1980) fits a distribution function (e.g. lognormal or Weibull distributions) to the upper tail of the sample data distribution, and then compares the consistency of the largest value with that predicted by the fitted distribution.

6. Multivariate Test Procedures

The procedures given above can be used for testing data sets with more than one variable by applying them independently to each other. However, this approach may be inefficient, particularly when the variables are statistically correlated. In some cases, a multivariate test will show that a value of one variable that appears to be an outlier using a single variable test, is consistent with the data set when one or more other variables are considered. Conversely, there may be values of a variable consistent with the other data in the set when considering only one variable, but which is definitely a possible outlier when considering two or more variables. Principal components analysis can often be helpful in identifying these kinds of problems (Roscoe *et al.*, 1982).

TESTS FOR CONSISTENCY WITH PREVIOUS DATA (HISTORICAL OR TEMPORAL)

These tests check the consistency of the data set with respect to similar data recorded in the past. In particular these procedures will detect changes where each result is increased (decreased) by a constant or by a multiplicative factor. The tests for historical consistency include:

1. gross limit
2. pattern and successive difference tests
3. parameter relationship tests

1. Gross Limit

Gross limit checks are useful in detecting data values that are either highly unlikely or generally considered impossible. Upper and lower limits are developed by examining historical data for a site or others in the area/region. Whenever possible the limits should be specific for each monitoring site, and consider both the variable and instrument/method characteristics.

2. Pattern Tests

These tests check the data for pollutant behavior which has never or very rarely occurred in the past. Like gross limit checks, they require that a set of limits be determined empirically from prescreened historical data. Values that lie outside of these predetermined limits are flagged for further investigation. It is recommended that the following patterns be examined:

1. individual concentration values
2. difference in adjacent concentration values
3. difference or percentage difference between a value and both of its adjacent values
4. average of four or more consecutive values

3. Parameter Relationship Tests

Parameter relationship tests can be divided into two types: a deterministic one, involving the theoretical relationships between variables (e.g., fine S and fine particle mass), or empirical test that determines whether or not a variable is behaving normally in relationship to observed behavior of one or more others. For example, coarse crustal elements including Al, Si, Ti, and Fe should normally covary. The relationship between Pb and Br in areas where leaded gasoline is still used in motor vehicles is shown in Figure 6.

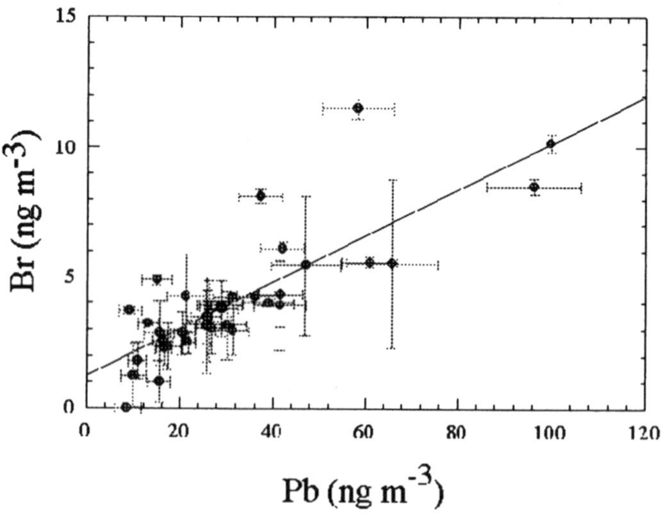

Figure 6. Br concentration versus Pb for fine particles in a city where leaded gasoline is used in cars.

TESTS FOR CONSISTENCY WITH OTHER DATA SETS, COLLECTED AT THE SAME TIME OR UNDER SIMILAR CONDITIONS (CONSISTENCY OF PARALLEL DATA SETS)

Tests for internal consistency implicitly assume that most of the values in the data set are correct. Consequently, if all of the values in a data set incorporate a small positive bias, tests such as the Dixon ratio test would not indicate that the data set was inconsistent. A method of identifying systematic bias is to compare the data set with others that presumably have been derived from the same environment (i.e., same air mass at the same time) and to check for differences in the average value or overall distribution of values. A plot of a comparison of values for fine particle mass between the Gent sampler and an cyclone impactor used in the ASP program in New South Wales, Australia (Liu *et al.*, 1995; 1996) is shown in Figure 7. In these cases, measurements at nearby sites in an air monitoring

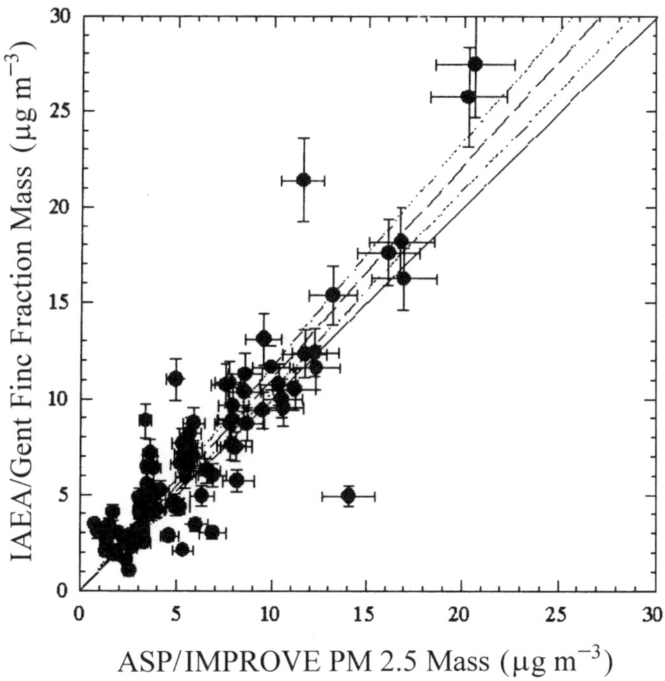

Figure 7. PM-2.5 measurement done with the ASP/IMPROVE cyclone and the IAEA/Gent sampler. The solid line is the 1:1 correspondence. The dashed line is the regression line and the dash-dot-dash lot represents the 95% confidence interval.

network or side-by-side intercomparisons in conjunction with other independent investigators can be used to look for systematic differences. It cannot be determined *a priori* which values are correct, but if differences exist, it indicates that careful checks of procedures are needed.

SUMMARY

It is important to have clearly written QA/QC procedures in any environmental program. It is also possible to learn a considerable amount about the quality of a data set by examining it. Data validation methods can be used to identify potential problem areas that require further examination. These problems can arise because of unusual events that produce very high or low values. It may be that these samples should be excluded from further analysis because they do not represent the normally observed situation. They also may be present because of errors in the sampling and analysis process. If the data permit, new values can be calculated after the process has been rectified. Alternatively it may only be possible to ensure that future data do not include such errors. The key point is that the data requires examination for its quality before it is used for other purposes such as source-receptor modeling.

REFERENCES

AEERL, *Quality Assurance Procedures Manual for Contractors and Financial Recepients*, Final Report Contract, 68-02-4291, US Environmental Protection Agency, Research Triangle Park, North Calorina, USA 27711 (1988).

American National Standards Institute, ANSI/ASQC-Q94 (1987).

Alpert, D.J. and P.K. Hopke, *Atmos. Environ.*, **15** (1981) 675.

Barnett, V. and T. Lewis, *Outliers in Statistical Data*, John Wiley & Sons, Inc., New York, NY (1978).

Garfield, F.M., *Quality Assurance Principles for Analytical Laboratories*, Association of Official Analytical Chemists (1984).

Liu X., P.K. Hopke, D. Cohen and G. Bailey, *Sci. Total Environ.*, **175** (1995) 65.

Liu, X., N. Gao, P.K. Hopke and D. Cohen, *Atmos. Environ.*, **30** (1996) 9.

Nelson, A.C., D.W. Armentrout and T.R. Johnson, *Validation of Air Monitoring Data*, Report No. EPA-450/2-80-030, US Environmental Protection Agency, Washington, DC (1980).

Roscoe, B.A., P.K. Hopke, S.L. Dattner and J.M. Jenks, *J. Air Pollut. Control Assoc.*, **32** (1982) 637.

US Department of Commerce, *Performance Assurance and Data Integrity Practices*, Computer Science and Technology, National Bureau of Standards, Washington, DC (1978).

7 QUALITY ASSURANCE AND QUALITY CONTROL IN THE ELEMENTAL ANALYSIS OF AIRBORNE PARTICLES

S.R. BIEGALSKI

Department of Nuclear Engineering, University of Illinois,
214 Nuclear Engineering Laboratory, 103 South Goodwin Avenue,
Urbana, Illinois, 61801, USA

INTRODUCTION

The objectives of quality assurance (QA) and quality control (QC) programs are to understand the sources of error in analytical measurements, ensure that analytical results are within statistical control, and analytical variations are within acceptable limits. Quality assurance is used to guarantee that the analytical system is functioning within satisfactory bounds; while quality control is an established mechanism to control errors in an analytical method (Taylor, 1981).

The analysis of airborne particles often involves the use of sophisticated analytical techniques such as neutron activation analysis (NAA), energy dispersive X-ray fluorescence (EDXRF), proton induced X-ray emission (PIXE), ion-chromatography (IC), atomic absorption spectroscopy (AAS), inductively coupled plasma — mass spectroscopy (ICP-MS), inductively coupled plasma — atomic emission spectroscopy (ICP-AES), and others. These techniques have demonstrated their potential to yield both accurate and precise measurements of airborne particles. It is the role of the scientist to make sure that the results are free of bias and within statistical control. Therefore, it is important that every laboratory and scientist have set guidelines to ensure that their results are within the control that they desire. Some laboratories have even developed computer software to aid in the control of data quality assessment (Faltejsek and Obrusnik, 1994).

A number of papers have been written that have detailed guidelines for and descriptions of quality assurance programs (Juran, 1974; EPA, 1976, 1977; Faltejsek and Obrusnik 1994; Parr and Zeisler, 1994). While many of these guides give specifics on this subject, they often lack a demonstration of the application of real data to QA/QC procedures. The objective of this chapter is to demonstrate this application for the elemental analysis of airborne particles. While there are many QA/QC measures used during the collection of airborne particles (e.g. sampling procedures), the focus here will be on steps that may be taken in an analytical laboratory.

Achieving laboratory results that are within the desired statistical control and free of bias is not always an easy task. Analytical procedures and laboratory conditions strongly influence experimental outcome. Simple tasks like instrumental calibration and the prevention of sample contamination should not be taken lightly. It is often the case where a simple step in an analytical procedure will affect the outcome of the entire system. It is important to have consistent guidelines that cover every step of the analytical procedure from the time the laboratory receives a sample, to when the results are sent off. Every step must be evaluated and measures must be taken to reduce the chance of experimental error. Heydorn (1993) has shown that by evaluating the statistical uncertainty of each analytical step, statistics may be used to determine errors in micro and trace analysis. It is vital that experiments be run in controlled environments and that their results are reproducible.

Many measures can be taken to prevent errors during analytical procedures. This chapter will not emphasize methods to prevent experimental error; instead methods of detecting experimental error will be discussed. While this will serve as a guide on how to look for experimental errors, the task of determining the source of errors must be approached on an individual basis.

A key to any QA program is making sure that analytical results are reproducible. Reproducibility is the ability to obtain, within statistical uncertainty, the same analytical result by independent means. This can be investigated in a number of ways. With respect to aerosol sampling, multiple samples can be collected and analyzed from the same sampling period, an aerosol filter can be divided into more than one sample and each one can be analyzed separately, and consistency may be checked between different size cuts of size-segregated samples. From the analytical technique point of view, reproducibility can be checked by analyzing a set of samples more than once with the same method, determining the same element by different methods within the same technique, and utilizing different analytical techniques to analyze the same sample. Reproducibility may also be investigated with reference materials.

COLLECTION OF SIMULTANEOUS AEROSOL SAMPLES

If coordination between an aerosol sampling network and the analysis laboratory is possible, an excellent QA procedure is to collect two aerosol samples at the same sampling site during the same sampling period. Analysis of these samples should yield results within acceptable uncertainty of one another. This QA procedure will serve to check the aerosol collection technique as well as the method of analysis.

The particulate air concentrations obtained from the collection of six Gent stacked filter unit (SFU) samples (Maenhaut, 1992) are shown in Table 1. This sample set represents three sampling periods where two air samplers were operated at the same site for the same sampling period. The SFU sampler collects both a fine ($d_p < 2\,\mu m$) and a coarse ($2\,\mu m < d_p < 10\,\mu m$) sample. The results compare very well indicating that the two samplers are operating in a way that is reproducible. The particulate concentrations were determined by gravimetric analysis. These results also serve as QA for the gravimetric analytical procedure. For the gravimetric analysis, filters were weighed in a humidity controlled chamber at least four times before and after sample collection. Aerosol mass uncertainty was calculated by taking the standard deviation of the multiple mass measurements. A two percent uncertainty was assumed in the volumetric flow measurement. These uncertainties were propagated to calculate the particle concentration uncertainty.

Table 1. PM-10 particulate concentrations with Gent stacked filter unit (SFU) sampler.

Sampler	Sampling Period	Fine Stage ($\mu g\, m^{-3}$)	Coarse Stage ($\mu g\, m^{-3}$)	Total PM-10 ($\mu g\, m^{-3}$)
SFU-1	1	11.900 ± 0.323	6.623 ± 0.176	18.523 ± 0.499
SFU-2	1	11.323 ± 0.317	6.745 ± 0.195	18.069 ± 0.513
SFU-1	2	8.413 ± 0.205	4.508 ± 0.166	12.922 ± 0.365
SFU-2	2	8.755 ± 0.179	4.050 ± 0.090	12.806 ± 0.268
SFU-1	3	6.206 ± 0.381	4.057 ± 0.293	10.263 ± 0.674
SFU-2	3	6.150 ± 0.132	3.778 ± 0.218	9.928 ± 0.320

For additional QA, the same samples were analyzed by neutron activation analysis (NAA). In addition to the collected aerosol filters, blank filters had to be analyzed so that a blank correction could be performed. The results for analysis of sodium are shown in Table 2. All the samples have statistically equal Na concentrations except the coarse filters for sampling period one. This difference is difficult to explain since Table 1 indicates that similar aerosol masses were deposited on both coarse filters for sampling period one and the remainder of the Na concentrations compare well. While this discrepancy could be the result of a statistical deviation greater than one standard deviation, it could also be due sample contamination or aerosol loss from a filter. This difference highlights an area where additional QA measures should be taken.

The following types of QA tests are excellent in the investigation of an entire aerosol sampling system.

- Similar tests could be performed by collecting samples with two different types of aerosol samplers, however, adjustments would have to be made for variations between the samplers (like aerosol cut-points).
- An aerosol sample could also be divided in half and analyzed separately.
- The analysis of two fractions of an aerosol filter would serve to verify the homogeneity of the aerosol sample, as well as to check how reproducible the analytical method is.
- Additional comparisons can be made between the different size fractionated stages from the same sample.

Quality control measures such as the ones mentioned above are also very informative when setting up an aerosol sampling operation. The feasibility of the experiment is verified by making sure that the aerosol samplers are operating correctly, the desired components of the aerosol sample are above the analytical detection limits, and the sample blank concentrations

Table 2. Sodium concentrations from Gent stacked filter (SFU) sampler.

Sampler	Sampling Period	Fine Na (ng m^{-3})	Coarse Na (ng m^{-3})	Total Na (ng m^{-3})
SFU-1	1	41.80 ± 5.49	103.42 ± 4.82	145.22 ± 10.30
SFU-2	1	37.84 ± 1.96	116.98 ± 4.80	154.82 ± 6.77
SFU-1	2	16.41 ± 4.04	51.17 ± 2.67	67.59 ± 6.71
SFU-2	2	19.85 ± 1.19	51.14 ± 2.34	70.99 ± 3.54
SFU-1	3	19.48 ± 4.06	20.06 ± 1.42	40.11 ± 5.48
SFU-2	3	24.64 ± 1.26	19.74 ± 1.31	44.38 ± 2.58

are not unacceptably high. Filter blank problems will be discussed later in this chapter.

REPRODUCTION OF ANALYTICAL RESULTS

An important aspect of quality assurance is the verification of analytical results by independent means i.e. the investigation of the reproducibility of the analytical result. This can be checked by methods such as analyzing a set of samples more than once with the same method, determining the same quantity by different methods within the same technique, and utilizing different analytical techniques to analyze the same sample. A review of the different isotopic reactions in NAA which can be utilized for QA purposes is given by Byrne (1994).

When comparing a few samples, agreement may be determined as shown in Tables 1 and 2. Statistical techniques may be used for comparisons of larger data sets. In many cases a comparison is made between analytical results whose precision is of similar magnitude. For these cases, the statistic β is very useful (Heydorn, 1991). This is a test for the absence of bias in the determination of concentrations utilizing two separate analytical procedures and was used by Heydorn (1980) to ascertain the absence of a blank value in a method for the radiochemical determination of arsenic. For most comparisons, the slope of the correlation line is close to one, so an approximation for β is used to compare two sample sets:

$$\beta=\sum_{i=1}^{n}\frac{(y_{i1}-\rho y_{i2})^2}{\sigma_{i1}^2+\sigma_{i2}^2}$$

where:

y_{i1} — the individual results for sample set one
y_{i2} — the individual results for sample set two
σ_{i1} — the estimated standard deviation for the sample set one result
σ_{i2} — the estimated standard deviation for the sample set two result
ρ — the ratio between the average of the two sample sets $\overline{y}_1/\overline{y}_2$

Note that if the relative standard error of the estimated precision is sufficiently small, then the values of β have a χ^2 distribution with N degrees of freedom. The value, ρ is approximately the slope of a correlation line between the two sample sets and can adjust for minor calibration differences. The use of the β statistic will be demonstrated below.

One of the most basic tests that demonstrates the QC of an analytical procedure is to analyze a set of samples more than once by the same analytical procedure. This test for 12 samples that were analyzed by NAA for the element Br is demonstrated in Figure 1. The first sample analysis is represented by the values on the x-axis and the second sample on the y-axis. Adequate time was allowed for sample decay between sample irradiations. Since there were 12 samples, the data set contained 11° of freedom and the statistic β was calculated to be 14.6. The one-tailed probability of β having a χ^2 distribution with 11° of freedom was determined to be 0.201. This was calculated with a Microsoft Excel spreadsheet using the "chidist" function. If the statistic β is too high, the one-tailed probability would be close to one and represent a lack of statistical control in the analytical system. If the statistic β is too small, the one-tailed probability would be close to zero and indicate an over estimate of the statistical variations within the analytical system. As a general rule, the one-tailed probability of β having a χ^2 distribution with N degrees of freedom should be between 0.05 and 0.95. Data will normally either pass or fail this test by a significant margin.

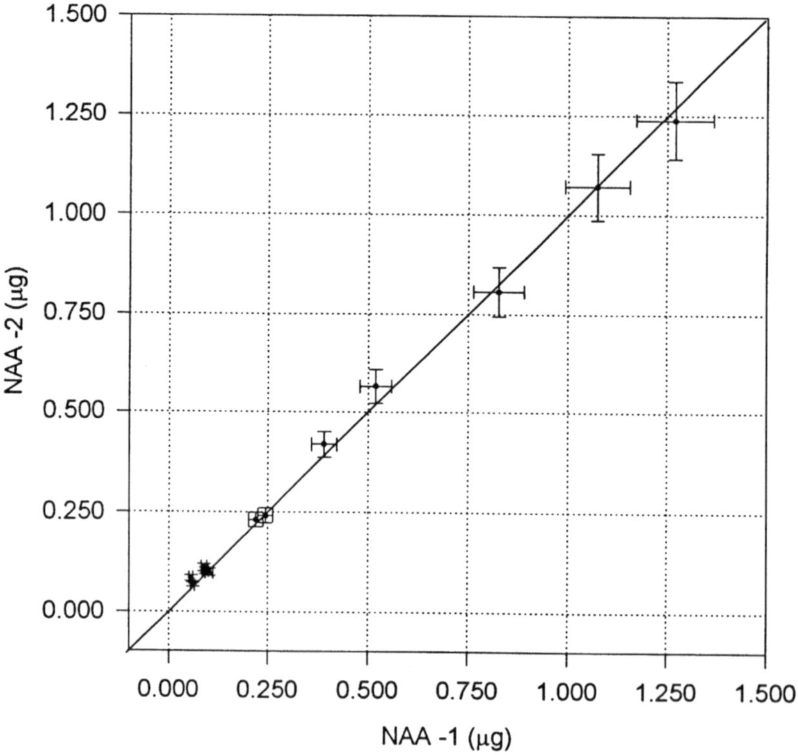

Figure 1. Two NAA analyses at the same laboratory for Br (Degrees of Freedom 11, $\beta = 14.6, P(\chi^2 > \beta) = 0.201$).

Concentrations can be determined by different signals with many analysis techniques. In NAA, various nuclear reactions may be used as well as the analysis of different photo-peaks. Using X-ray techniques, such as XRF and PIXE, different X-ray peaks can be extracted for analysis. A comparison between the analysis of aerosol filters for Br via two separate nuclear reactions is displayed in Figure 2. The $^{79}Br(n, \gamma)^{80}Br$ reaction produces a short-lived radioisotope and is determined through a short irradiation methodology. Since ^{80m}Br is simultaneously produced along with ^{80}Br a correction term was utilized to compensate for ^{80m}Br decay to ^{80}Br (Biegalski and Landsberger, 1994). The $^{81}Br(n, \gamma)^{82}Br$ reaction produces a long-lived radioisotope that must be determined through a longer irradiation, decay, and counting methodology. Thus, Br concentrations were determined for the same samples utilizing two separate irradiations, gamma-ray spectrum collections, and nuclear reactions. There were

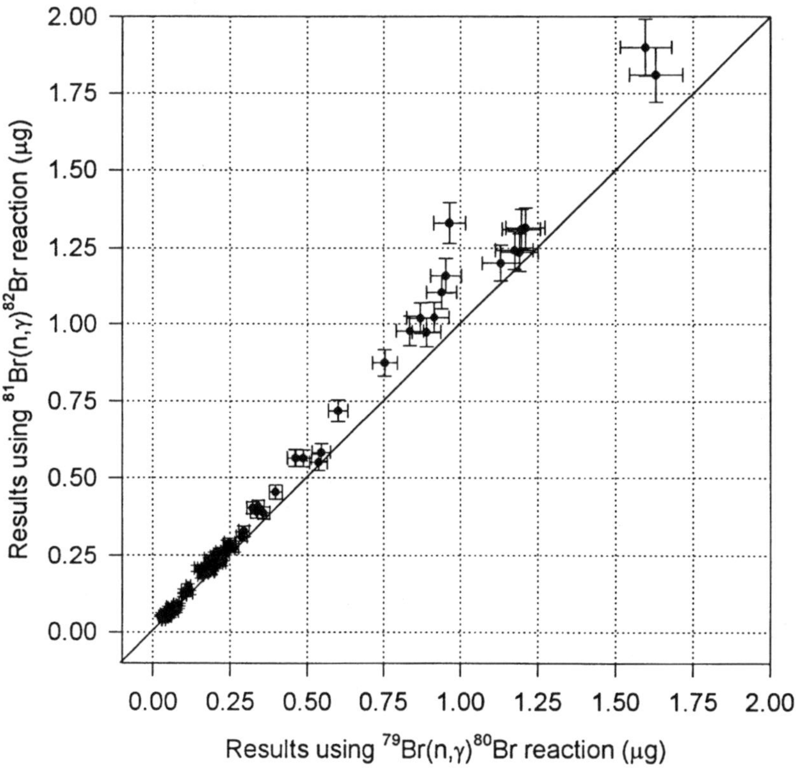

Figure 2. Comparison between two NAA analyses for Br utilizing two independent reactions. (Degrees of Freedom = 117, $\beta = 105$, $P(\chi^2 > \beta) = 0.779$).

118 samples in the data set which equates to 117 degrees of freedom. The statistic β was calculated to be 105 and the one-tailed probability of β having a χ^2 distribution with 117 degrees of freedom was calculated to be 0.779. It is important to note that the one-tailed probability of β having a χ^2 distribution is a function of both β and the number of degrees of freedom. Therefore, larger data sets will have larger β values.

Inter-laboratory comparisons are also useful for QA/QC purposes (Landsberger *et al.*, 1996; Wang *et al.*, 1995). It is helpful to compare results between laboratories that utilize the same analytical method, as well as ones that use different analytical techniques to determine the same aerosol constituent. A unique comparison depicting the results of Br analyzed at two different laboratories for the same samples is shown in Figure 3. A comparison between NAA and PIXE for the elements Zn and Si, is found in Figures 4 and 5 respectively. For Zn, the comparison included 61 samples

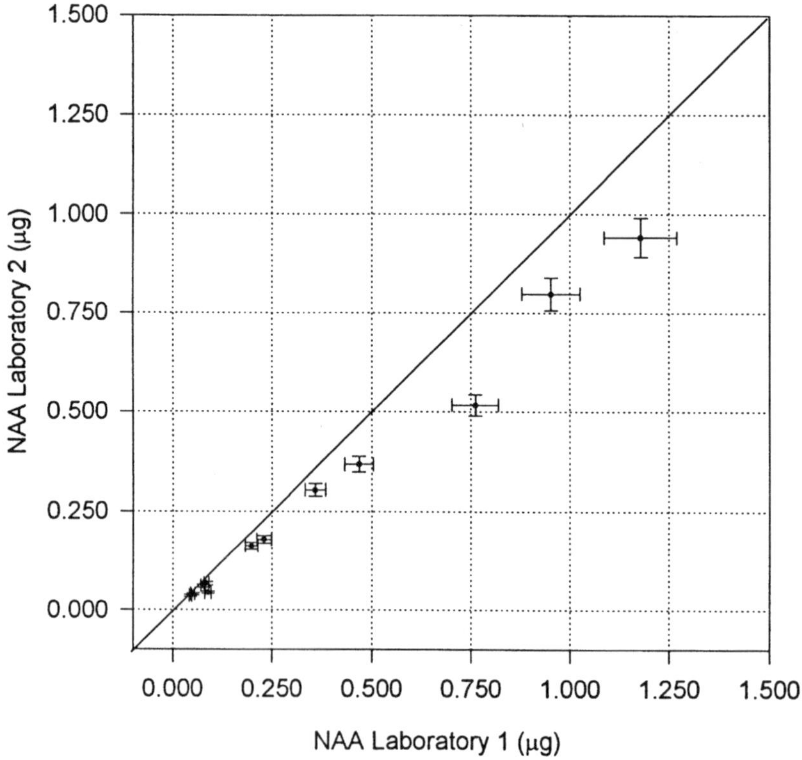

Figure 3. Comparison between two NAA laboratories for Br (Degrees of Freedom = 11, β) = 13.6, $P(\chi^2 > \beta) = 0.255$).

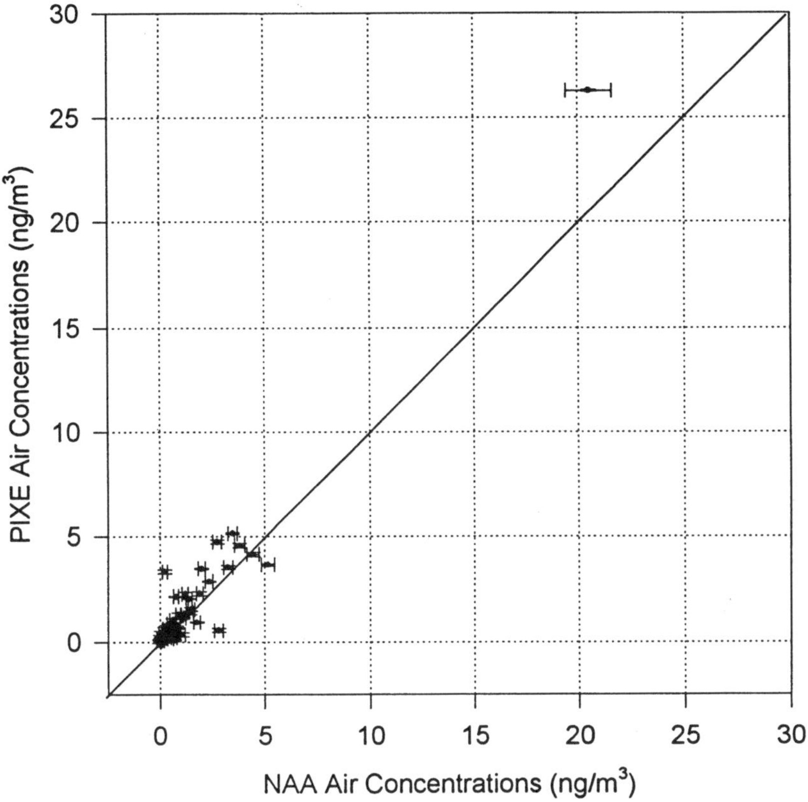

Figure 4. NAA vs. PIXE comparison for Zn (Degrees of Freedom = 60, $\beta = 645$, $P(\chi^2 > \beta) = 6.7$ E-99). Air concentrations are blank corrected.

and the one-tailed probability of β having a χ^2 distribution is 6.7×10^{-99} For Si, the comparison included 12 samples and the one-tailed probability of β having a χ^2 distribution is 0.229. These results from NAA and PIXE for the elements Zn and Si will be used to demonstrate the usefulness of the β statistic test. All too often, these types of comparisons are made and a linear regression is calculated to quantify the correlation. A linear regression fit is valid if there is more than an order of magnitude difference between the precision of the two analytical methods, but for most comparisons this is not the case. The β statistic test accounts for the uncertainty of each value while linear regressions normally require one variable to be error free. Linear regression fits are also heavily weighted by the high values. Linear regression fits to the results shown in Figures 4 and 5 yield

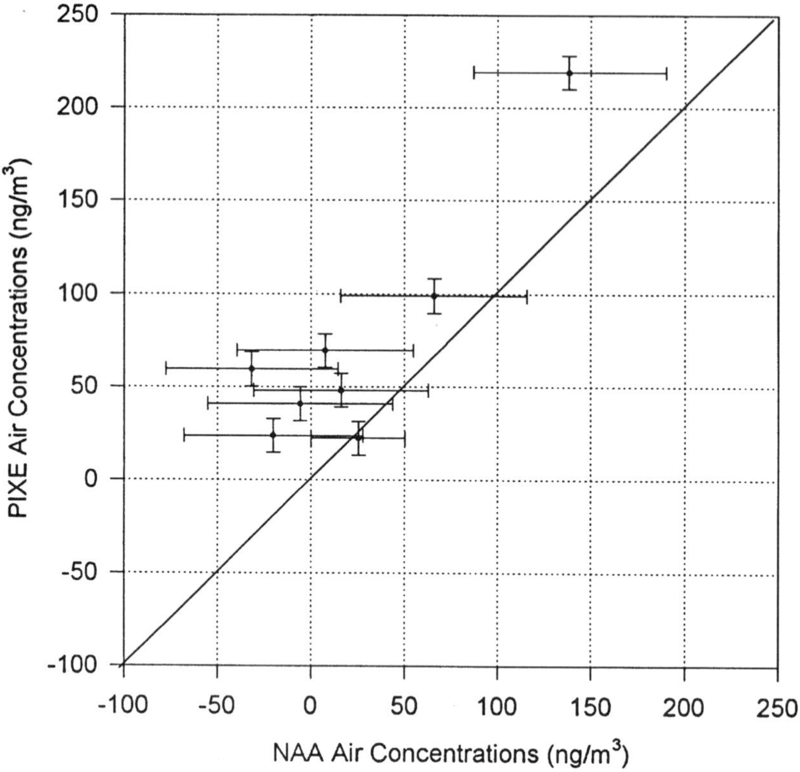

Figure 5. NAA vs. PIXE comparison for Si (Degrees of Freedom $=7, \beta=9.34$, $P(\chi^2>\beta)=0.229$). Air concentrations are blank corrected.

R^2 values of 0.946 and 0.275, respectively. The linear regression fit to the Zn values denotes a good correlation, however we know from the β statistic test that the Zn values are not within statistical control. This is opposite for the Si values: the linear regression fit indicates a bad correlation, but the β statistic test indicates that the values are within statistical control. Therefore, we see that the β statistic test should be used in these cases and why a linear regression fit has little meaning.

REFERENCE MATERIALS

A reference material is *a material or substance where one or more properties of which are sufficiently well established to be used for the calibration of an apparatus, the assessment of a measurement method, or for assigning values to materials* (International Standards Organization, 1981). Reference materials can be classified as certified reference materials (CRM) and standard reference materials (SRM). A CRM is a reference material which *is accompanied by, or traceable to, a certificate stating the property values concerned issued by an organization, which is generally accepted as technically competent* (International Standards Organization, 1981). A SRM is a CRM when it is issued by a reputable organization, such as the U.S. National Institute of Standards and Technology. Reference materials play an fundamental role in the QA of chemical measurements (Taylor, 1993). They serve to validate the accuracy of analytical results (Heydorn *et al.*, 1995) and in some cases are used for comparison (Garg *et al.*, 1995, Glascock and Anderson, 1993). A good reference material is chosen to closely match the major and important minor elemental concentrations of the material analyzed. For nuclear related techniques a good match is desirable to simulate sample self-shielding, rate-related effects, and spectral interferences (International Atomic Energy Agency, 1990).

A reference material is certified by using definitive, mostly absolute, or reference analytical methods with proven accuracy at a given laboratory, or on data obtained in a qualified network of experienced laboratories (Kucera, 1995). Detailed descriptions of how reference materials are certified may be found in a number of references (Becker *et al.*, 1994; Wilson *et al.*, 1994; Becker, 1995; Greenberg *et al.*, 1995). Reference materials are normally certified for amounts not less than 100 mg (Mao and Chai, 1995) and may be purchased through reputable organizations. Many sell suitable standard reference materials for the analysis of atmospheric aerosols for use in NAA, but those for XRF, PIXE, and other techniques are less common. In fact there are no certified reference materials for real air filter samples.

The preparation of reference materials is a difficult task. The material must be stable, available in sufficient amounts, homogeneous, have matrix and analytic levels that match the intended purpose, and have sufficient precision and accuracy (Kucera, 1995). Greenberg (1994) discusses problems in multielemental standard preparation such as purity, non-stoichiometric composition, moisture, stability, evaporative losses, volumetric errors, contamination, volatility losses, and errors. Micro-analytical techniques pose a special problem in the preparation of reference materials since the mass measured by these techniques is so small. The sample mass that is conventionally employed in ion beam analyses is nominally 100 times smaller than the suggested one determined by the certifying body that insures homogeneity (Valkovic, 1995). One method used to prepare a homogeneous samples for micro-analysis utilizes vacuum evaporation in order to create a uniform film of a reference material (Hanru and Qing, 1995).

For QA/QC purposes, CRM's are analyzed in conjunction with other samples whose composition is to be determined. The CRM's concentrations are determined as if they are not known and then the results are compared with the certified values. This QA/QC procedure should be regularly employed to test the accuracy and precision of analytical results.

FILTER BLANKS

The most common method in aerosol measurement uses filter media to collect the aerosol particles. This technique is popular due to its flexibility, simplicity, and economics (Lee and Ramamurthi, 1993). Particles are collected by a filter through the mechanisms of impaction, interception, and diffusion onto the filter media. The choice of an aerosol filter type is an important aspect of QC. Significant aspects of filter media selection include filtration efficiency, pressure drop across the filter, cost, and compatibility with the experiment. The most consequential of these aspects is the compatibility of the filter media with the aerosol experiment. It will be shown below how chemical concentrations in the filter blank have the potential to severely hinder the detection of certain aerosol components. Lippmann (1989) gives a detailed outline of filter types, characteristics, and manufactures.

Filter types can be classified as fibrous, porous-membrane, straight-through pore, and granulated bed (Lee and Ramamurthi, 1993). Fibrous filters are popular and made of mat or woven fibers. They have good mechanical strength, low pressure drops, and are composed of materials such as paper, glass, quartz, and cellulose. Porous-membrane filters have pores that follow a circuitous path through the filter media, high collection efficiencies, and large pressure drops. Straight-through pore filters are made of thin films with uniform sized pores bored straight through the filter. Their collection efficiency and pressure drop is similar to the porous-membrane filter type. Granular-bed filters are used for special applications in order to recover the aerosols when the particle-laden filter is washed after sampling.

Blank data for a rural sampling station is illustrated in Table 3. The values represent a data set with nearly 100 blank filters and over 300 aerosol samples. The following concentrations are displayed: average blank, standard deviation of the blank, average blank plus three standard deviations of the blank, percent of aerosol samples above NAA detection limit, percent of aerosol samples above the average blank, and percent of aerosol samples above the average blank plus three standard deviations of the blank concentration. This data shows how crucial the trace metal concentrations

Table 3. Blank values at rural sampling site Whatman 41 filters.

Element	Blank Average (ng/filter/8)	Blank Std. (ng/filter/8)	Blank Average + 3σ (ng/filter/8)	Sample % Above NAA Detection Limit	Sample % above Average Blank	Sample % above Average Blank + 3σ
Ag	0.0002	0.0010	0.0032	39.8	39.8	39.8
Al	2.2498	1.4369	6.5605	99.7	99.4	93.3
As	0.0023	0.0020	0.0084	98.9	98.9	98.6
Au	0.0360	0.0409	0.1588	80.1	48.8	11.4
Br	0.3590	0.2636	1.1497	99.7	93.6	49.0
Ca	7.6351	3.0718	16.8506	99.7	98.6	93.0
Ce	0.0101	0.0060	0.0283	80.2	79.6	64.8
Cl	36.5658	7.1108	57.8983	99.7	97.8	73.1
Co	0.0088	0.0054	0.0249	99.7	95.5	64.0
Cr	0.1564	0.1116	0.4912	96.9	79.6	29.1
Cs	0.0066	0.0044	0.0198	92.5	65.9	13.1
Cu	0.8355	0.6769	2.8663	87.7	77.0	26.6
Dy	0.0042	0.0030	0.0133	39.8	39.3	23.0
Eu	0.0011	0.0003	0.0018	87.0	83.0	53.0
Fe	3.0966	2.6010	10.8996	99.7	99.2	85.5
Hf	0.0003	0.0006	0.0022	60.9	60.9	49.7
I	0.0133	0.0057	0.0303	100.0	99.4	96.4
In	0.0004	0.0005	0.0018	74.1	49.4	8.4
K	0.0000	0.0000	0.0000	32.5	32.5	32.5
La	0.0020	0.0022	0.0085	96.9	96.9	90.1
Lu	0.0047	0.0051	0.0201	6.0	4.5	0.0
Mn	0.0564	0.0401	0.1767	99.7	99.7	98.6
Na	11.5756	5.5417	28.2006	99.7	95.0	59.9
Nd	0.1687	0.1115	0.5033	15.6	13.5	5.7
Ni	0.1199	0.0792	0.3575	58.7	58.1	38.8
Rb	0.0085	0.0193	0.0663	49.7	49.7	42.7
Sb	0.0038	0.0050	0.0189	99.7	99.4	87.4
Sc	0.0008	0.0009	0.0034	99.7	98.0	72.5
Se	0.0158	0.0088	0.0421	99.4	98.6	92.5
Si	15.0312	10.4887	46.4971	89.8	89.2	69.9
Sm	0.0022	0.0059	0.0200	99.2	75.7	13.1
Sn	0.0801	0.0415	0.2045	79.5	64.5	19.9
Ta	0.0007	0.0003	0.0017	39.7	39.7	16.5

Table 3. (Continued).

Element	Blank Average (ng/filter/8)	Blank Std. (ng/filter/8)	Blank Average + 3σ (ng/filter/8)	Sample % Above NAA Detection Limit	Sample % above Average Blank	Sample % above Average Blank + 3σ
Tb	0.0004	0.0002	0.0010	31.8	31.8	22.4
Th	0.0003	0.0006	0.0022	87.7	87.7	79.3
Ti	0.5643	0.4622	1.9511	86.0	83.8	61.3
U	0.0036	0.0030	0.0126	100.0	78.9	12.7
V	0.0064	0.0051	0.0218	99.7	99.4	96.9
W	0.0074	0.0059	0.0252	92.8	62.0	16.3
Yb	0.0036	0.0023	0.0103	13.5	12.5	5.2
Zn	0.1445	0.0509	0.2974	99.7	99.7	99.2

in the filter blank are for the determination of rural aerosols. The filters represented in Table 3 are Whatman 41 fibrous filters and were analyzed for trace elements by NAA. The NAA detection limit is as defined by Currie (1968).

There is a need for careful blank evaluation in aerosol sampling. Trace metal concentrations in blank filters are displayed in Table 3. Since this is unavoidable, efforts are made to both reduce and quantify the magnitude of the contamination. This effect on the blank filter is worse for sampling sites with low aerosol concentrations. For this particular rural site, the aerosol sample results for the elements Au, Cs, In, Lu, Nd, Sm, U, and Yb are indistinguishable from the blank values.

The element Au is a good example that demonstrates how certain air concentrations may not be determined with Whatman 41 air filters at this site. The average Au blank concentration is 0.0360 ± 0.0409 ng per eighth of a filter (only one eighth of the filter was analyzed by NAA). The large standard deviation of the Au blank concentration indicates that the variability of Au in Whatman 41 air filters is high. Out of all the aerosol samples analyzed for Au, 80.1% had Au concentrations above the NAA detection limit, 48.8% above the average blank value, and only 11.4% were above the average blank plus three standard deviations of the blank. The limit of detection (LOD) is defined to be 3σ above the mean of the background distribution (Baron and Heitbrink, 1993). However, data below this LOD value can still contain useful information (Helsel, 1990). These results indicate that the majority of the variation observed for

Au concentrations in the aerosol samples is dominated by the variations in the blank concentration. The Au concentration in the Whatman 41 filter needs to be significantly reduced before air concentrations can be measured with these filters at this sampling location. The blank value is not a limiting factor for the detection of all trace elements. For elements such as Al, I and Zn, the blank concentration is minimal when compared to the aerosol deposit. This allows for a blank correction to be performed on the aerosol data with minimal introduction of additional uncertainty.

The ideal blank concentration is zero, however, those above zero can be tolerated for some elements. For other elements, high blank concentrations make sampling difficult. One method to decrease the blank problem is to sample for longer periods, but, this is associated with a loss of time resolution.

DETECTION LIMIT TREATMENT

During analysis of aerosol samples, values below the LOD are often encountered. The treatment of these values is always of concern. The difference from the method LOD and the total LOD must be distinguished. The method LOD is in reference to the analytical detection limit, while the total LOD accounts for the method LOD as well as the background introduced by the filter blank.

One could easily ignore all the values below the LOD, but that would bias the data on the high side. It would be just as easy to set all the below LOD values to zero. This method is also not acceptable because it would bias the data low. One technique frequently used in the past is to set all the below LOD values to half the LOD (Cohen and Ryan, 1989). This assumes a uniform distribution of possible data points below the LOD. There is no reason to believe that the values below the LOD would be uniformly distributed if that is not the case for those above. For tighter distributions with the geometric standard deviation about 1.5, Hornung and Reed (1990) suggest filling in the below LOD values by the LOD divided by the square-root of two. This assumes that the data below the LOD values would be triangularly shaped like the left tail of a log-normally distributed data set. Much more rigorous methods exist where the points measured above the LOD are used to extrapolate the concentrations below. These methods range from Monte Carlo statistical predictions (Trivikrama *et al.*, 1991), to randomly drawing data points from an assumed distribution (Helsel, 1990). The standard deviation of the blank correction is determined as explained by Taylor (1984).

CONCLUSIONS

Data quality and control are fundamental for any analytical laboratory. For this to exist, detailed QA/QC plans are essential. However, these plans are ineffective if there is a lack of commitment to them or if they are poorly implemented. Proper resources, training and facilities are all necessary for a laboratory to perform quality work.

Verification of the precision and accuracy of analytical results is of great importance. Reference materials as well as other comparative techniques are necessary to ensure data quality. The techniques and tests mentioned in this chapter demonstrate the capability of QA/QC procedures. Quality assurance and quality control programs should be individually tailored to the needs of each analytical method.

REFERENCES

Baron, P.A. and W.A. Heitbrink, Factors affecting aerosol measurement quality, in *Aerosol Measurements: Principles, Techniques and Applications*, eds. W. Willeke and P.A. Baron, Van Nostrand Reinhold, New York (1993).

Becker, D.A., D.L. Anderson, R.M. Lindstrom, R.R. Greenberg, K.M. Garrity and E.A. Mackey, *J. Radioanal. Nuc. Chem.*, **179** (1994) 149.

Becker, D.A., *J. Radioanal. Nuc. Chem.*, **193** (1995) 25.

Biegalski, S. and S. Landsberger, *J. Radioanal. Nuc. Chem.*, **183** (1994) 313.

Byrne, A.R., *Biol. Trace Elem. Res.*, **43** (1994) 529.

Cohen, M.A. and P.B. Ryan, *JAPCA Note-Book, J. Air and Waste Manag. Assoc.*, **39** (1989) 328.

Currie, L.A., *Anal. Chem.*, **40** (1968) 586.

E.P.A., *Quality Assurance Handbook for Air Pollution Measurements Systems: Principles*, Vol. 1, Pub. No. 600/9-76-005 (1976).

E.P.A., *Quality Assurance Handbook for Air Pollution Measurements Systems: Principles*, Vol. 2, Pub. No. 600/4-77-027a (1977).

Faltejsek, J. and I. Obrusnik, *Biol. Trace Elem. Res.*, **43** (1994) 605.

Garg, A.N., N.L. Chutke, M.N. Ambulkar and A.L. Aggarwal, *J. Radioanal. Nuc. Chem.*, **192** (1995) 307.

Glascock, M.D.and M.P. Anderson, *J. Radioanal. Nuc. Chem.*, **174** (1993) 229.

Greenberg, R.R., *J. Radioanal. Nuc. Chem.*, **179** (1994) 131.

Greenberg, R.R., E.A. Mackey and D.A. Becker, *J. Radioanal. Nuc. Chem.*, **193** (1995) 7.

Hanru, S. and X. Qing, *Nuc. Instrum. Meth. in Phys. Res. B*, **104** (1995) 201.

Helsel, D.R., *Environ. Sci. Technol.*, **24** (1990) 1767.

Heydorn, K., *Aspects of Precision and Accuracy in Neutron Activation Analysis*, p. 180, Risø National Laboratory, Roskilde, Denmark (1980).

Heydorn, K., *Mikrochim. Acta*, **3** (1991) 1.

Heydorn, K., *Anal. Chim. Acta*, **283** (1993) 494.

Heydorn, K., Z. Alfassi, E. Damsgaard, B. Rietz and Psolgaard, *J. Radioanal. Nuc. Chem.*, **192** (1995) 321.

Hornung, R.W. and L.D. Reed, *Appl. Occup. Environ. Hyg.*, **5** (1990) 45.

International Atomic Energy Agency, *Practical Aspects of Operating a Neutron Activation Analysis Laboratory*, IAEA-TECDOC-564, Vienna (1990).

International Organization for Standardization, *Terms and Definitions Used in Conjunction with Reference Materials*, ISO Guide 30, Geneva (1981).

Juarn, J.M., *Quality Control Handbook*, 3rd edn., McGraw Hill, NY (1974).

Kucera, J.M., *Chem. Anal.*, **40** (1995) 405.

Landsberger, S., D. Wu, S.J. Vermette and W. Cizek, *J. Radioanal. Nuc. Chem.*, **215** (1996) 117.

Lee, K.W. and M. Ramamurthi, *Filter Collection* in Aerosol Measurement Principles, Techniques, and Applications, eds. K. Willeke and P. A. Baron, Van Nostrand Reinhold, NY (1993).

Lippmann, M., *Sampling Aerosols by Filtration*, in Air Sampling Instruments for Evaluation of Atmospheric Contaminants, 7th edn., Cincinnati, OH: American Conference of Governmental Industrial Hygienists, (1989).

Mao, X. and C. Chai, *Fresenius J. Anal. Chem.*, **352** (1995) 174.

Maenhaut, W., *The "GENT" Stacked Filter Unit (SFU) Sampler for the Collection of Atmospheric Aerosols in Two Size Fractionations: Description and Instructions for Installation and Use*, IAEA Co-ordinated Research Program: CRP E4.10.08, (1992).

Parr, R.M. and R. Zeisler, *Biol. and Trace Elem. Res.*, **43** (1994) 651.

Taylor, J.K., *Anal. Chem.*, **53** (1981) 1588A.

Taylor, J.K., *J. Test. and Eval.*, **12** (1984) 54.

Taylor, J.K., *Standard Reference Materials: Handbook for SRM Users*, NIST Spec. Pub. 260–100, US Gov. Printing Office, Washington, D.C. (1993).

Trivikrama Rao, S., J.Y. Ku and K. Shankar Rao, *J. Air and Waste Manage. Assoc.*, **41** (1991) 442.

Valkovic, O., M. Jaklovic, S. Fazinic, V. Valkovic, G. Moschini and E. Menapace, *Nuc. Instrum. Meth. in Phys. Res. B*, **99** (1995) 372.

Wang, C.F., E.E. Chang, P.C. Chaing and N.K. Aras, *Analyst*, **120** (1995) 2521.

Wilson, S.A., P.H. Briggs, J.S. Mee and D.F. Siems, *Geostandards Newsletter*, **18** (1994) 85.

8 AN INTRODUCTION TO SOURCE RECEPTOR MODELING

P.K. HOPKE

Department of Chemistry, Clarkson University, Potsdam, New York 13699-5810, USA

INTRODUCTION

The management of air quality is a difficult but important problem. In general it involves: identification of the sources of materials emitted into the air, quantitative estimation of emission rates of the pollutants, understanding of the transport of substances from sources to downwind locations, and knowledge of the physical and chemical transformation processes that can occur during that transport. All of these factors can be put together into a mathematical model and used to estimate the changes in observable airborne concentrations that might occur if various actions are taken. Such actions could include the initiation of new sources as new industries are built and begin to function, and imposing of emission controls on existing facilities in order to reduce the pollutant concentrations.

The atmosphere is a very complex system and it is necessary to greatly simplify the descriptions of physical and chemical properties in order to produce a mathematical model capable of being calculated on even the largest and fastest computers. Significant improvements have been made over the past twenty years in mathematical modeling of pollutant dispersion in the atmosphere. There are still many instances when these models are insufficient in providing the information required to fully development effective air quality management strategies. It is necessary to have alternative methods available to assist in the identification of sources, and apportionment of the observed pollutant concentrations to those sources. These methods are called receptor-oriented or receptor models since they focus on the behavior of the ambient environment at the point of impact. This is opposed to source-oriented dispersion models that focus on transport, dilution, and transformations occurring at the source and follow the pollutants to the sampling or receptor site. A comprehensive view of the field can be found in Hopke (1991).

PRINCIPLE OF MASS CONSERVATION

All of the currently used receptor models are based on the assumption of mass conservation and use of a mass balance analysis. For example, assume that the total airborne particulate zinc concentration (ng/m^3) measured at a site can be considered the sum of contributions from independent source types such as: motor vehicles, incinerators, non-ferrous metal smelters, etc.

$$\mathrm{Zn_T} = \mathrm{Zn}_{\mathrm{smelter}} + \mathrm{Zn}_{\mathrm{auto}} + \mathrm{Zn}_{\mathrm{incin}} + \cdots \quad (1)$$

A smelter emits particles containing materials other than zinc. Therefore, the atmospheric concentration of zinc from smelters in ng/m^3, $\mathrm{Zn}_{\mathrm{smelter}}$, can be considered the product of two cofactors; the gravimetric concentration (ng/mg) of zinc in smelter particulate emissions, $a_{\mathrm{Zn,\,smelter}}$, and mass concentration (mg/m^3) of smelter particles in the atmosphere, f_{smelter}:

$$\mathrm{Zn}_{\mathrm{smelter}} = a_{\mathrm{Zn,smelter}} f_{\mathrm{smelter}} \quad (2)$$

The normal approach for obtaining a data set for receptor modeling is to determine a large number of chemical constituents such as elemental concentrations in a number of samples. The mass balance equation can then be extended to account for all m elements in the n samples as contributions from p independent sources

$$x_{ij} = \sum_{k=1}^{p} a_{ik} f_{kj} \quad i = 1, m, \; j = 1, n \quad (3)$$

where:

x_{ij} — is the ith elemental concentration measured in the jth sample,
a_{ik} — is the gravimetric concentration of the ith element in material from the kth source,

f_{kj} — is the airborne mass concentration of material from the kth source contributing to the jth sample.

Equation (3) can be solved in a variety of ways depending on the amount of information that is initially known about the system. In all cases, the ambient concentrations of the species of interest have to be measured for a series of airborne particle samples. Although there are some methods that can resolve the sources for a single sample, one sample does not characterize the air quality in the area, nor does it provide the basis for making air quality management decisions. Multiple samples need to be obtained and analyzed.

If the number of sources, p, and composition profile of each source, a_{ik}, are known, then the only unknowns are the contributions of each source to each sample. Those values can be estimated using the least-squares method. This calculation is commonly called a Chemical Mass Balance (CMB) analysis. Alternatively, the receptor modeling problem can be recast as a multivariate calibration problem and various methods can be employed to solve it including: Partial Least Squares (PLS), Artificial Neural Networks (ANN), Simulated Annealing, and the Genetic Algorithm (GA). Finally, if only the ambient concentration values are known, then one of several forms of factor analysis can be used to estimate the number of sources, their concentration profiles, and mass contributions.

NATURAL PHYSICAL CONSTRAINTS

There is a set of natural physical constraints on the system that must be considered in developing any model for identifying and apportioning the sources of airborne particle mass (Henry, 1991). The fundamental, natural physical constraints that must be obeyed are:

1. The original data must be reproduced by the model — the model must explain the observations.
2. The predicted source compositions must be non-negative — a source cannot have a negative percentage of an element.
3. The predicted source contributions to the aerosol must all be non-negative — a source cannot emit negative mass.
4. The sum of the predicted elemental mass contributions for each source must be less than or equal to the total measured mass for each element — the whole is greater than or equal to the sum of its parts.

These constraints are similar to those imposed on the spectrochemical mixture resolution problem. Both the absorbance at any wavelength, and contribution of each component to the absorbance of the total sample are non-negative. There are some important differences between the spectrochemical mixture resolution problem and receptor modeling for source apportionment. These are described in the next section.

MIXTURE RESOLUTION PROBLEM FOR AIRBORNE PARTICLES

In many chemical studies, the measured properties of the system can be considered to be the linear sum of terms representing the fundamental effects in that system times appropriate weighing factors. For example, the absorbance at a particular wavelength of a mixture of compounds for a fixed path length, z, is considered the sum of the absorbencies of the individual components

$$\frac{\alpha(\lambda)}{z} = \varepsilon(\lambda)_1 C_1 + \varepsilon(\lambda)_2 C_2 + \cdots + \varepsilon(\lambda)_p C_p \tag{4}$$

where ε_i is the molar extinction coefficient for the ith compound at wavelength and C_i is the corresponding concentration. If the absorbencies of a mixture of several absorbing species are measured at m various wavelengths, a series of equations can be obtained:

$$\frac{\alpha(\lambda_j)}{z} = \sum_{k=1}^{p} \varepsilon(\lambda_j)_k C_k \tag{5}$$

If we know which components are present and what the molar extinction coefficients are for each compound at each wavelength, the concentrations of each compound can be determined using a multiple linear regression fit to these data.

In many cases neither the number of compounds nor their absorbance spectra may be known. For example, several compounds may elute from an HPLC column at about the same retention time so that a broad elution peak or several poorly resolved peaks containing these compounds may be observed. Thus, at any point in the elution curve, there would be a mixture of the same components but in differing proportions. If the absorbance spectrum of each of these different mixtures could be measured for example by using a diode array system, then the resulting data set would consist of a number of absorption spectra for a series of n different mixtures of the same compounds:

$$\frac{\alpha_i(\lambda_j)}{z} = \sum_{k=1}^{p} \varepsilon(\lambda_j)_k C_{ki}, \quad j=1,m, \; i=1,n \tag{6}$$

For such a data set, factor analysis can be employed to identify the number of components in the mixture, absorption spectra of each component, and concentration of each compound for each of the mixtures. Similar problems are found in analytical and environmental chemistry where there are mixtures of unknown numbers of components and the properties of each pure component are not known *a priori*.

There are several important differences between receptor modeling for airborne particle source apportionment and the mixture resolution problem. These discrepancies have an impact on the ability of receptor models to accurately identify the sources and apportion the measured mass to those sources. Some of the important differences are:

1. Variable Source Composition Profiles

In the spectrochemical mixture, the properties of each component are fixed or varying in some reproducible manner. If the UV-visible absorption spectrum of a compound is measured properly, it is the same whether measured in the United States, Germany, or anywhere else. The absorbance as a function of wavelength does not change from day to day. However, source composition profiles are not constant. A given particle source like a coal-fired power plant may change the source of its coal. Changing from one mine to another could substantially change the elemental composition pattern produced by burning that coal. Even across a single coal bed, there are discrepancies in the relative amounts of mineral matter, and they contain very different compositions. For example, the frequency distribution of SiO_2 for 72 samples of coal from a single mine are plotted in Figure 1 (Roscoe *et al.*, 1984). There can be day-to-day variations in the composition of the emitted particles. Also potentially major differences in composition from different locations can occur due to variations in input materials from these sites with their different geochemical histories. Unlike spectrochemical mixtures, there is variability in the source profile values, a_{ik} as in Equation (6). This variability will cause difficulties for any of the mathematical methods used to solve Equation (6) and must be considered as an integral part of the processes that give rise to the observed atmospheric concentrations.

2. Modification of the Source Composition in the Atmosphere

Mixtures of components that are being measured spectroscopically have the spectral properties of the components that are fixed or vary consistently with their environment. For example, components may alter their absorbance spectrum with changes in pH; but the determination of these

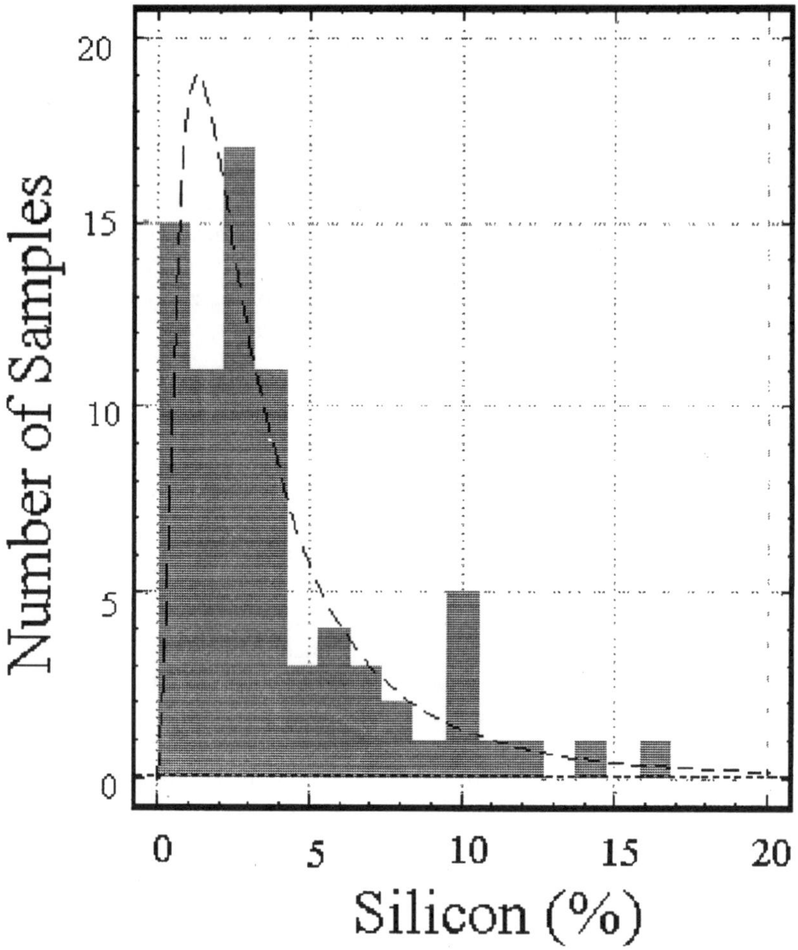

Figure 1. Frequency distribution for SiO_2 concentration in coal samples from a single coal face.

changes can be reproduced. In the case of airborne particles, the processes that give rise to the sample collected on a filter are much more complex. These processes are represented schematically in Figure 2. Particles and its supporting gas are emitted into the atmosphere. Some of the particulate material may evaporate and be lost. Evaporative loss is a particularly important problem for small particles (Leong *et al.*, 1983), and deposition an principal process for larger particles. If particle composition is dependent on size, then the loss of a particular size fraction by any mechanism

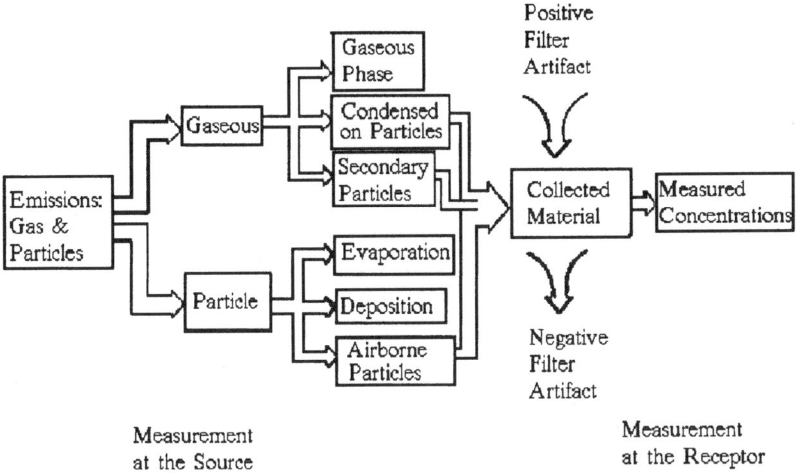

Figure 2. Schematic diagram of the emission, transport, and transformation of airborne particulate matter leading to samples used for receptor modeling.

will change the composition profile from that which was emitted to something different collected at the receptor site. In addition there can be chemical reactions that result in losses from particles. For example, gaseous acids like HNO_3 and H_2SO_4 can diffuse to the particle surface and cause reactions that change the particle composition. Motor vehicles burning leaded gasoline emit lead as particles containing bromine and chlorine. The adsorbed acid could cause displacement of the halide as the volatile HBr or HCl leaving the particle with a reduced Br/Pb ratio at the receptor site, as compared to tailpipe emission measurements (Dzubay *et al.*, 1979; Harrison and Sturges, 1983).

Particle compositions can also be modified by the adsorption of materials onto particle surfaces. Many times samples are taken to the stack of a source when the stack temperature is still elevated. Volatile components may still be in the gas phase at this elevated temperature, but they condense onto the particles once they begin to mix with the ambient air in the stack plume. Various sampling techniques have been developed to either collect the gaseous materials or mimic this dilution and cooling process to provide more realistic source samples (Houck, 1991). Care must be exercised in the source sampling so that the emission composition profiles measured reflect what is likely to be the composition once the particles reach the receptor site.

Particles can also form from initial gaseous emissions through chemical reactions in the atmosphere. Emitted SO_2 can be converted to $(NH_4)_2SO_4$

or NH_4HSO_4 particles, and NO is transformed into NH_4NO_3. Because the gaseous material dispersed differently than the co-emitted particles and takes time for them to form, it is usually difficult to identify the sources of these secondary particles. They are generally not attributed to specific sources, unless other information besides particle composition can be incorporated into the data analysis. Therefore, unlike the spectrochemical mixture resolution problem, properties of the component profiles corresponding to the spectra are not constant and are a function of the atmospheric conditions under which the sample is obtained.

3. Signal-to-Noise Ratio

A final major difference between the spectrochemical measurements used to resolve mixtures into their components and measurements used in receptor modeling is the level of noise in the data. With modern spectrophotometers, absorbance can typically be measured to levels of a few tenths of a percent except in the case where the absorbance is quite close to zero. In ambient airborne particle sample data, there are several sources of error. Some of them include positive and negative filter artifacts. A glass fiber filter is highly retentive for particles, but has an alkaline surface that can catalyze the conversion of SO_2 to SO_4^{2-}. More sulphate is measured on the filter than was in the particles present in the air sampled. This overestimation represents a positive filter artifact. Conversely, NH_4NO_3 after being collected on a filter can dissociate to NH_3 and HNO_3 and be lost from the sample so that a negative artifact is observed. It is possible to overcome this with carefully designed sampling systems (Dzubay and Stevens, 1991; Wolff *et al.*, 1991). There can be difficulties in the data analysis when such sampling problems were not carefully considered in the design of the sampling program.

Even with reliable samples, the complexities of analyzing for multiple species with the associated interferences and limits on detectability of low concentration leads to analytical uncertainties. These are generally at least 5% for species well above detection limits and 20% or more as the concentrations approach the limit of detectability. In general, receptor modeling has to deal with at least an order of magnitude greater uncertainty in the analytical results that provide the input data to the mixture resolution algorithm. The modeling output then typically reflects this higher level of uncertainty. To date most receptor modeling has focussed on particulate matter, because the composition of particles was thought to be more conserved than for gaseous species. There have been some efforts to examine gaseous materials, but those studies has been limited to less reactive organic compounds. Examples will be presented later.

There are several approaches to receptor model analysis that have been successfully applied. These include:

- Multiple linear regression called chemical mass balance (CMB).
- Factor analysis methods including principal components analysis (PCA).
- Target transformation factor analysis (TTFA).
- Factor analysis with imposed external constraints (Source Apportionment by Factors with Explicit Constraints, SAFER, and Positive Matrix Factorization, PMF).

It is also possible to use multivariate calibration methods including partial least-squares, genetic algorithm, and neural networks. These models can be applied to both particulate and gaseous species. The basis for each of these methods will be presented in subsequent sections with examples of their application to the identification and quantitative apportionment of pollution sources in the atmosphere.

CHEMICAL MASS BALANCE

Introduction

CMB sometimes called the chemical element balance solves Equation (3) directly for each sample by assuming that the number of sources and their compositions at the receptor site are known. This approach was first independently suggested by Winchester and Nifong (1971) and by Miller *et al.* (1972). The composition of an ambient sample is then used in a multiple linear regression against source compositions to derive the mass contribution of each source to that particular sample. Miller *et al.* (1972) modified Equation (3) to include changes in the composition of the source material while in transit to the receptor

$$x_{ij} = \sum_{k=1}^{p} \alpha_{ik} a'_{ik} f_{kj}$$

where α_{ik} is the coefficient of fractionation so that if a'_{ik} were the composition of the particles as emitted by the source, a_{ik} is the composition of the particles at the receptor site ($a_{ik} = \alpha_{ik} a'_{ik}$). In practice, it is generally impossible to determine the α_{ik} values and they are assumed to be unity ($a_{ik} = a'_{ik}$).

Ordinary Least Squares

Initially the CMB problem was solved using ordinary least squares (Friedlander, 1973). Kowalczyk *et al.* (1978) used a weighted least squares regression analysis to fit six sources with eight elements for ten ambient samples. In these analyses, the ambient elemental concentrations are weighted by the inverse of the square of the analytical uncertainty in that measurement.

Subsequently, Kowalczyk *et al.* (1982) examined 130 samples using seven sources with 28 elements included in the fitting process. Although 28 elements were used in the fitting process, the fit did not change appreciably with varying numbers of elements included with the exception of some of the key tracer elements such as Na, Pb, and V. Cheng and Hopke (1989) have re-examined these data using a variety of regression diagnostics. They found that these marker elements can be clearly identified, and that their influence on the quality of the fit to the ambient data and the source mass contributions can be quantitatively estimated.

Effective Variance Least Squares

In 1979, Watson (1979) and Dunker (1979) independently suggested a mathematical formula called effective variance weighting that included the uncertainty in the measurement of the source composition profiles, as well as those in the ambient concentrations. As part of this analysis, a method was also developed to permit the calculation of the uncertainties in the mass contributions. The approach has been described by Cooper *et al.* (1984) and Watson *et al.* (1991). Effective-variance least squares has been incorporated into the standard personal computer software developed by the US EPA for receptor modeling. However, Cheng *et al.* (1988) found that for most situations there was no significant improvement in the results of the CMB analysis between the Ordinary Weighted Least Squares approach and Effective Variance Least Squares.

Illustrative Example

To illustrate the use of the CMB method, an example will be taken from the Glover *et al.* (1991) study of the sources of airborne particulate matter in Granite City, Illinois. With the promulgation of the 1987 National Ambient Air Quality Standards for Particulate Matter-10 µm (PM-10), it has been necessary to review the State Implementation Plan (SIP) in each state for those areas most likely not to be in compliance with the new standard. In Illinois, one such area is Granite City, an industrial city northeast of St. Louis, Missouri, with a history of total suspended particulate and airborne lead non-attainment.

The locations of the major industries in Granite City and the ambient airborne particulate sampler are shown in Figure 3. The local industries

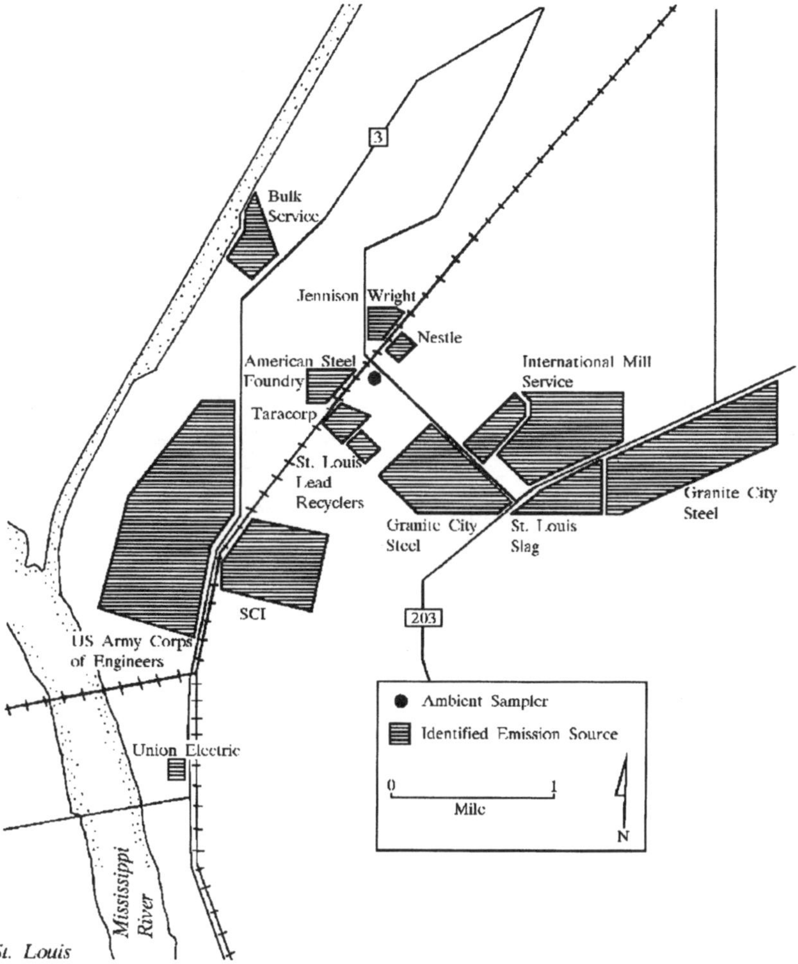

Figure 3. Granite City, Illinois local point sources.

include steel mills (American Steel and Granite City Steel), a former secondary lead smelter (Terracorp), aluminum smelter (SCI), and chemical plant (Jennison Wright). There is also a US Army Corps of Engineers storage facility located at the edge of town. The location of the major industries in the greater St. Louis metropolitan area in relation to the ambient airborne particulate sampler is illustrated in Figure 4.

As a part of the studies necessary to prepare an effective and efficient SIP, receptor modeling was applied to elemental composition data for 24 h

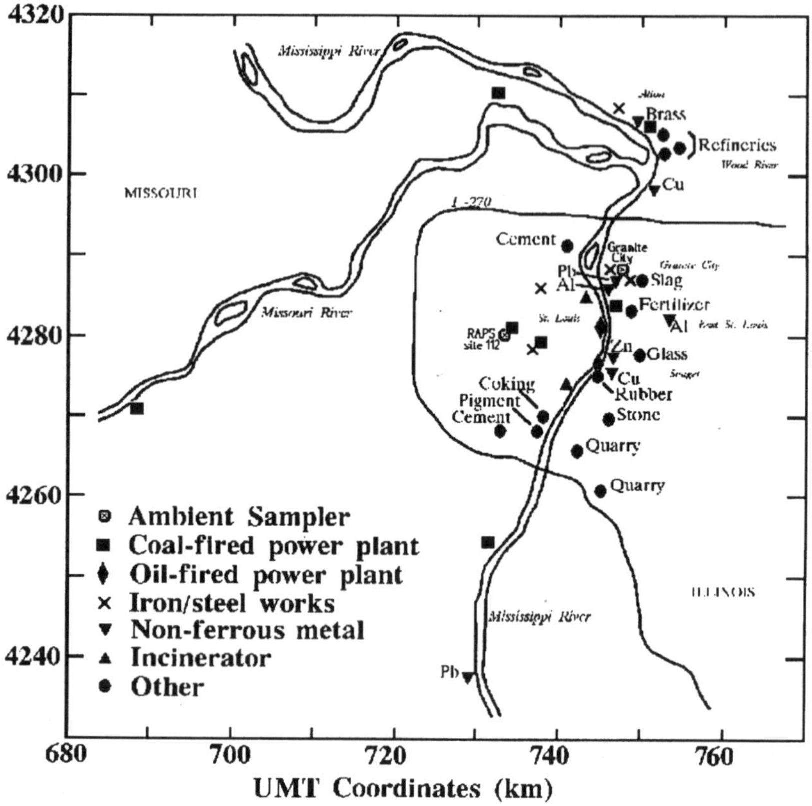

Figure 4. Point source locations in the Greater St. Louis, Missouri area.

airborne particle samples taken in Granite City by the Illinois State Water Survey using an automated dichotomous sampler. This sampler collects particles with an inlet that excludes large particles by having a 50% transmission efficiency for 10 μm particles. The particles that penetrate into the sampler are separated into two aerodynamic size fractions; <2.5 μm (fine) and 2.5–10 μm (coarse). They are collected on Teflon® filters that are then available for chemical analysis.

The particle samples were subjected to both X-ray fluorescence (XRF) and instrumental neutron activation analysis (NAA) in order to provide the input data for receptor modeling. Forty-eight sample pairs (fine and coarse) were thus analyzed for 33 elements. Each of these samples were then subjected to two CMB analyses. For the first analysis, the source profiles were taken from libraries available in the literature. To supplement these source profiles, 12 dust samples were collected in and around Granite City. These were aerosolized, sampled, and analyzed by XRF and NAA to provide site specific source profiles for the second CMB analysis.

In an attempt to account for more of the mass on each ambient filter, total carbon was measured seven times during the ambient sampling period. A Sierra PM-10 sampler equipped with quartz fiber filters was co-located with the dichotomous sampler for this purpose. Each quartz filter was analyzed for total PM-10 and carbon mass. After the PM-10 mass of each filter was determined, the filter was treated with HCl to remove any carbonate. Each filter was then oxidized at 800°C, converting the elemental and organic carbon to CO_2. The amount of CO_2 released was measured with a Dohrmann carbon analyzer. A linear regression was used to relate the mass of total carbon to the total PM-10 mass of each quartz filter. This regression is represented by:

$$\mathrm{TC} = 0.074 \cdot \mathrm{PM\text{-}10} + 3.129 \qquad (8)$$

where TC (total carbon) and PM-10 are both measured in μg/m^3.

Computer controlled scanning electron microscopy (CCSEM) (Hopke and Casuccio, 1991) was used to partition the total carbon measurements between the fine and coarse fractions. The first and last quartz filters collected were analyzed by CCSEM. The number, physical mass, and aerodynamic mass distribution of the particles on each filter were determined along with an elemental analysis of the particles. The CCSEM measurements determined that the total carbon was apportioned between the fine and coarse fractions by:

$$\mathrm{TC}_{\mathrm{fine}} = 0.919 \cdot \mathrm{TC} \qquad (9)$$

$$\mathrm{TC}_{\mathrm{coarse}} = 0.082 \cdot \mathrm{TC} \qquad (10)$$

The PM-10 mass on each of the quartz filters was scaled to the PM-10 mass collected on the Teflon® filters. The mass of each pair of fine and coarse Teflon® filters were added to find the total PM-10 mass on the Teflon® filters. $\mathrm{TC}_{\mathrm{fine}}$ and $\mathrm{TC}_{\mathrm{coarse}}$ for the Teflon® filters were found by multiplying the scaling factor for each sample with Equations (5) and (6), respectively.

An initial CMB analysis was performed using source profiles derived from analysis of data from the 1975–1977 Regional Air Pollution Study (RAPS) of St. Louis (Alpert and Hopke, 1981; Liu *et al.*, 1982; Severin *et al.*, 1983). These results showed that the composition of air pollution in the St. Louis area had changed over the 10 years between the two sampling periods. Only one-fifth of the fine and one-fourth of the coarse profiles used in the first CMB analysis were taken from the profiles derived from the 1975 to 1977 RAPS results. These profiles accounted for 11% and 20% of all of the identified fine and coarse mass, respectively.

By including the local dust samples among the potential source profiles in a second CMB analysis, a marked improvement in the quality of the predicted results was achieved. The re-analysis did not change the types of sources identified by the CMB analysis. However, the apportionment between sources varied enough to cause the relative importance of sources to change. The improvement in the results could be seen in the adjusted squared correlation coefficient for the fits, R^2, which increased significantly for both fractions. This increase was especially apparent for the coarse fraction where the average negative change was less than one-quarter of 1%, while the average positive change was above 7%. The predicted mass of the fine fraction became closer to the observed mass with only a slight increase in error as seen in Figure 5. (The error in the initial predicted results was influenced by the use of an artificial sulfur component, a source containing only sulfur, which caused the initial error to be fairly low.) The predicted fine fraction fitting elements is shown in Figure 6. The addition of the locally derived source profiles changed the results from an

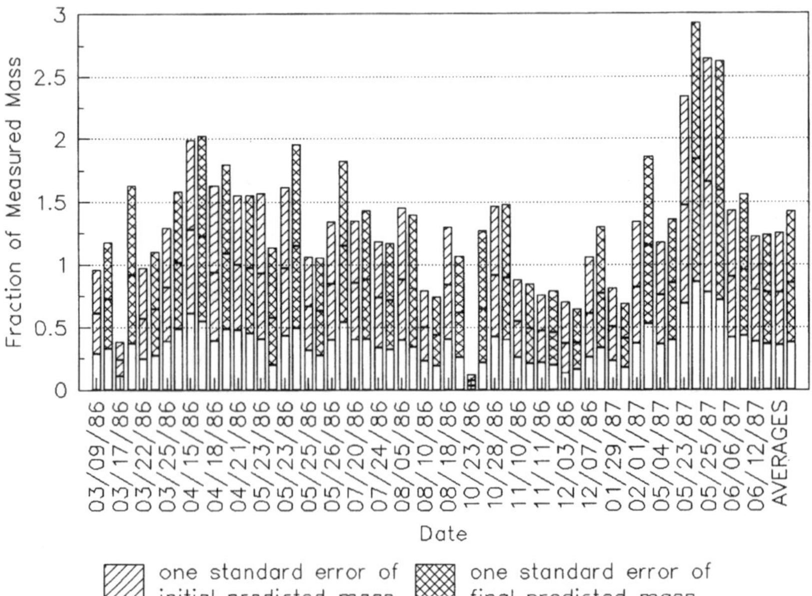

Figure 5. Predicted mass fraction of selected fine fraction CMB results, March 1986–June 1987, Granite City, Illinois.

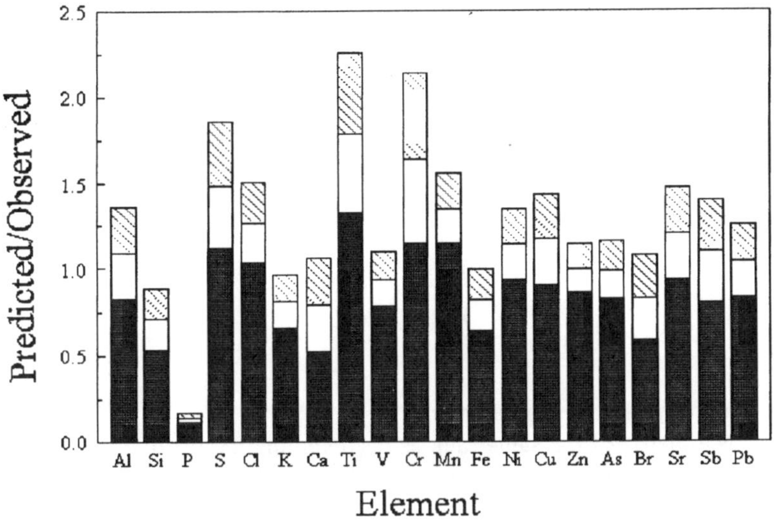

Figure 6. The predicted to observed ratio for the fine fraction elements for March 25, 1986. The lines separating the two segments at the top of each bar are values. The bar segments represent + or − one standard deviation of the estimated uncertainties.

average over-prediction to an average under-prediction. Similar results were obtained for the coarse fraction samples. Under-prediction is a more desirable error. During the fitting process it is more difficult to explain unobserved mass than not explain all of the observed mass. There are always other unidentified sources that might explain the unaccounted for mass.

MULTIVARIATE CALIBRATION METHODS

Another class of methods, multivariate calibration, has been used in a limited number of research applications. In these methods, it is assumed that there are a series of mixtures for when amounts of each component are known and a series of properties has been measured. The methods include partial least-squares, genetic algorithm, and neural networks. Although there has not yet been extensive results or use of these methods, there appears to be possibilities for further application of these methods.

Partial Least-Squares

Partial least-squares (PLS) has been applied in a number of multivariate calibration problems (e.g., Martens and Naes, 1989). Frank and Kowalski (1985) first suggested the use of PLS for solving the receptor modeling problem. They did this by using PLS in a manner that is analogous to Equation (6) and got good results for the first data set (Currie *et al.*, 1984). However, the model can be extended as (Vong *et al.*, 1988):

$$\begin{aligned} X &= TP^{\mathrm{t}} + \varepsilon_x \\ Y &= UQ^{\mathrm{t}} + \varepsilon_y \end{aligned} \tag{11}$$

where:

X — is the matrix of source profiles,
Y — is a matrix of the same rank as X containing the relative contribution of each source to each sample in the training set.

The matrices, P and Q are the latent variable structures for the X and Y matrices, respectively. The interrelationship between X and Y is examined by regressing each column of U against the corresponding column of T:

$$\hat{b}_j = u_j \cdot t_j \tag{12}$$

where u_j and t_j are the jth columns of U and T, respectively. A predicted Y can now be calculated:

$$\hat{Y} = \begin{bmatrix} b_1 & & \\ & b_j & \\ & & b_m \end{bmatrix} TQ^{\mathrm{t}} + \varepsilon \tag{13}$$

These relationships are developed from the training set and then applied to the new set of mixtures, the unknowns. For the unknown sample data, the $\hat{Y}$ values then provide the contributions of the various sources to the samples.

The key to using PLS as well as most multivariate calibration methods is to have a series of mixtures of known proportions and their corresponding spectral responses. In the receptor modeling problem, this would mean a set of samples where the amounts of the given source contributions and corresponding compositions of each of these mixtures were known. This requirement means that we need a set of ambient samples where the source apportionment is known and such samples usually do not exist. However, Vong *et al.* (1988) solved this problem by using source profiles for the suspected sources in the system with unity and zero values as their fractions in the various mixtures, the identity matrix. Thus, a training set is provided that should enclose the space of all of the ambient samples if the correct set of profiles has been obtained. Wang and Larson (1993) extended the model by adding a weighting scheme dependent of the error values. Subsequent studies on some real data (Vong, 1993) have shown that PLS can be used to test the input sources to determine if all of the ambient samples fall within the convex bounds and find directions to modify source profiles in order to properly incorporate the variations that might have occurred between the emission source and receptor site. The PLS method has considerable appeal because it can provide a clear test for the inclusion of a source profile into the analysis as well as indicate if there are samples that fall outside the boundaries established by the known source profiles (Vong, 1993). Such results indicate the need for the inclusion of additional sources in the analysis. Further testing and application of PLS is clearly needed.

Genetic Algorithm

Genetic algorithms (Goldberg, 1989) were developed by Holland (1975) as part of a study of adaptive processes. It is different from deterministic approaches in that the GA tries to mimic the statistical evolutionary process in nature. The procedure builds a random population of strings, each of which represents one possible solution to the problem being investigated. In general, this string can be a simple bit sequence, vector of sequence numbers, or any other suitable linear representation of the parameters of the problem to be optimized. For each member of the population, a fitness value is calculated by transferring it to the function to be optimized. This fitness value determines the string's relative position within all members of the population regarding their closeness to their optimum value.

A new generation of values is then created from copies of the strings of the old population, considering their relative fitness. Strings with a fitness higher than the average are copied more times than strings with lower values. As an iterative application of this reproduction rule, it would only select the best string from the initial population. Two innovative features are therefore introduced that change the values of strings chosen arbitrarily. A mutation event can randomly change one or more information unit of a string with a specific probability. A second change is made to some randomly chosen members of the population in which two strings are mixed by cutting them, and recombining them crosswise providing that a considerable amount of information of each string remains intact.

This new population then enters the next cycle and is exposed to the selective pressure. Within each cycle, the string showing the highest fitness value is memorized. The process is repeated until a predefined end condition is met. It represents a versatile method for intelligent searching of the complex surface of solutions that may exist in receptor modeling. A decimal coding has been developed for strings that seems well suited to the receptor modeling problem (Lucasius and Kateman, 1989; Cartwright and Mott, 1991).

An initial application of the GA has been performed by Cartwright and Harris (1993) in which they analyze simulated data created from the profiles provided by Currie *et al.* (1984). They analyze error-free and relatively low error (2.5%) data. For these situations they are able to get good agreement between the model results and the known values for both the source profiles and contributions. These results are encouraging, but the GA has yet to be employed on any actual air quality data and therefore, requires further testing before its utility can be fully assessed.

Artificial Neural Networks

A schematic diagram of an artificial neural network is shown in Figure 7. The large circles are artificial neurons and the squares represent weights that describe the importance of the signal being transmitted along a given path. The net input to a given neuron would be given by:

$$\mathrm{Net}_j = \sum_i \mathrm{Inp}_i \cdot w_{ij}, \tag{14}$$

where:

Inp — the total signal being transmitted from one neuron to the next along a single axon,
w_{ij} — the weight function for that connection.

The conceptual basis for artificial neural networks has been present in the literature for a long time (McCulloch and Pitts, 1943; Pitts and McCulloch, 1947; and Hebb, 1949). Hopfield (1982) introduced the novel concept of non-linearity between the total input received by a neuron from other neurons, and the output produced and transferred onward to other neurons. The nonlinear output and feedback coupling of outputs with inputs gave new flexibility to this old architecture and sparked a major new field of interest across a number of disciplines. A detailed description of artificial neural networks in chemistry is provided by Zupan and Gasteiger (1991). A number of applications have been made to a variety of problems including multivariate calibration (Kateman, 1992). There is still considerable uncertainty as to their usefulness outside the bounds of the input training set.

The self-organizing method (Kohonen, 1989) has been presented for local scale problems with single (Wienke and Hopke, 1994a,b) and multiple

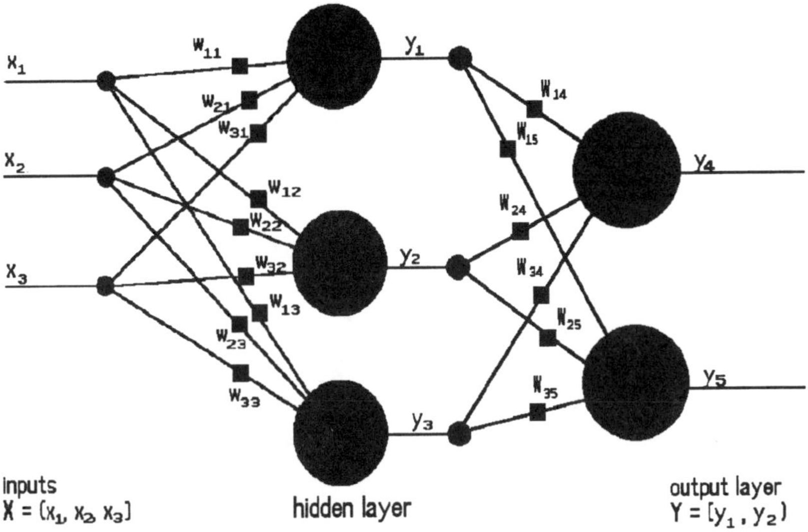

Figure 7. Schematic diagram of a neural network with one hidden layer.

sampling sites (Wienke *et al.*, 1994). This method can analyze a three dimensional data block as a whole and yield both source profiles and geographical information on the identified emission sources.

Figure 8 is a schematic diagram of a multiple site model system and its projection into a rectangular array of neural network nodes (Wienke *et al.*, 1994). The method assumes that there are r receptor sites, x_l with $l = 1, \ldots, r$, distributed over a given geographical area. An unknown number, p of air pollution sources, S_p with unknown geographic locations emit patterns of chemical constituents that are determined as m particulate or gaseous species. Neighboring pollution sources and meteorological factors (wind speed, direction, and duration, rain, etc.) change the pure source profile a_m to a multivariate vector of sample concentrations c_m. This variation of a pure source profile, a_{kj} with $k = 1, \ldots, p$ and $j = 1, \ldots, m$, during its transport in the atmosphere can be considered weighted by the mass contribution of the particular source, b_{ik}, according to:

$$c_{ij} = \sum_{k=1}^{p} b_{ik} \cdot a_{kj} + e_{ij} \tag{15}$$

where e_{ij} represents noise. If written in a matrix notation that is transposed to the classical receptor model, the matrix A_{pm} of p source profiles is weighted by a source contribution matrix B_{np} according to:

$$C_{nm} = B_{np} \cdot A_{pn} + E_{nm} \tag{16}$$

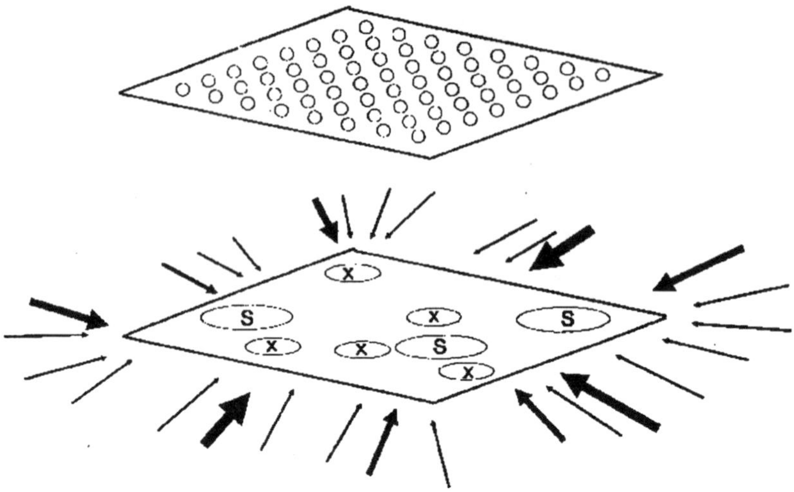

Figure 8. A model arrangement of a sampling network of five sites (x) situated between three emissions sources (S). The arrows symbolize distinct sample vectors, c_m that have been emitted by a source as a source profile, weighted by the meteorological conditions and finally transported through the atmosphere to a sampling site. The geographical area can be superimposed onto the nodes of a Kohonen neural network.

If the n samples were taken simultaneously at r receptor sites, all of the individual r data matrices C_{nm} can be combined to a three dimensional data block (tensor), C_{nmr}. This ideal case requires synchronized measurements between all r sampling sites and this constraint can only be fulfilled by a fully synchronized sampling network. If not, the data cube would be incomplete in several of the $i=1,\ldots,n$ rows. The present model does not require synchronized measurements.

For the present neural network approach, the individual r data matrices, $C_{n_l m}$, with distinct sample numbers n_l, are combined to form a two-dimensional total sample data matrix C_{Nm} with:

$$N=\sum_{l=1}^{r} n_l \tag{17}$$

The total number of samples, N from the r sampling sites can be arranged in this matrix in an arbitrarily chosen order.

For the internal representation of the present neural network model in the computer, the geographical area in Figure 8 (bottom) is thought to be superimposed onto a two dimensional Kohonen Neural Network containing u neurons (Figure 8, top). This superposition creates a two dimensional data projection that can be compared with a real geographical map. In Figure 8, a square of 8×8 neurons has been chosen ($u = 64$). Alternative two dimensional neural arrangements can be, for example, a circle, rectangular, or ellipse. In reality, the u neurons are only an index expression for $t = 1, \ldots, u$ weight vectors, w_m of the same length m as the N sample vectors.

The essence of Kohonen's algorithm is a repeated comparison of the $i = 1, \ldots, N$ sample vectors, c_{im}, with these $t = 1, \ldots, u$ weight vectors, w_m, using a distance metric such as the Euclidean distance or correlation coefficient. During each single comparison, a winner, t among all u weight vectors can be found that has the highest degree of similarity (smallest distance) to c_{im}. After that, the $j = 1, \ldots, m$ elements, $w_{t,j}$(old), of the winning weight vector t are adapted a small step closer to the elements c_{ij} of the actual ith input sample vector using the Kohonen learning rule:

$$w_{tj}(\text{new}) = w_{tj}(\text{old}) + \eta \cdot [c_{ij} - w_{tj}(\text{old})] \tag{18}$$

The learning rate, η is chosen as a positive real number with a value less than 0.1.

Together with the winner, t, further weight vectors are modified around the winner within a limited topological neighborhood of radius, R. Squares, hexagons or circles can be used as topological neighborhoods. To start, R is large but decreases slowly with the training time. After the adaption, the subset of weight vectors within R became slightly more similar to the input vectors in terms of the distance metric employed. A comparison of all N sample vectors with all u neural weight vectors and their modifications is called one epoch. This process is repeated over many epochs yielding a self-organized behavior of the N samples in the low-dimensional neural array. The samples form a visual structure related to the original m-dimensional variables space.

The final result of this training process is the collection of subsets of sample vectors aggregated to some of the weight vectors. The number of these loaded neurons is v, with v less than or equal to u. The trained weight vector for such an aggregated cluster of input vectors comes numerically very close to their mean vector. The remaining $u - v$ weight vectors do not have any associated input vectors. Their corresponding neurons are defined as unloaded neurons.

Only an unsupervised pattern recognition result, similar to those obtainable from hierarchical cluster analysis, non-linear mapping, or principal component scores plots has been obtained at this point. Although the topology of the m-space will be preserved by this special type of projection, the quantitative interpretation of such a Kohonen map remains difficult especially when information about the correct neighborhood disappears during such a projection into a 2-D plane. The interpretability of the results is improved by adding a visualization of the global relationships among the loaded neurons in the map. The Minimal Spanning Tree (Prim, 1958) is added to provide this visualization. Wienke and Hopke (1994a) give a complete description of the combined algorithm. Details of the calculation and choices for the training parameters are found in Wienke and Hopke (1994b).

As described in Wienke and Hopke (1994a,b) and Wienke *et al.* (1994), this methodology has proven quite useful in extracting information from both single and multiple site data for urban scale source/receptor modeling. Wienke and Hopke (1994a) present several illustrative examples that demonstrate the nature of the analysis results. A set of airborne particle compositions for samples obtained at a single sampling site (coarse fraction samples from Granite City, Illinois) were analyzed. After transferring the combined map to a Geographical Unit Circle (GUC), a correct pattern of the main industrial emission sources around the sampling site in Granite City was obtained by decoding a 48-dimensional space of chemical-analytical variables into a visual and geographical 2-D space. For a further quantitative evaluation of such a combined mapping of Minimal Spanning Tree and Kohonen Neural Network, Wienke and Hopke (1994b) examined the fine fraction data and described the concept of a GUC in greater detail. The GUC around the single sampling site in Granite City yielded emission levels, trace element profiles, and geographic directions of several air pollution sources.

Wienke *et al.* (1994) extended the analysis to multiple site data with both simulated data and particle composition data from the 1987 SCAQS study of the South Coast Air Basin of southern California. This projection of a 17-dimensional space of chemical variables provided a two-dimensional, visually evaluated, geographical arrangement of air pollution sources in the South Coast Air Basin. This combination of two exploratory data analysis methods appears to be a powerful tool for extracting considerable information about the location and composition of sources contributing to ambient particle samples. There remain questions as to how many neurons are really needed to provide a good representation of the topology of the system under study. Since the time to process the data goes up rapidly with increasing numbers of neurons, it would be helpful to examine the relationship between the number of samples, variables, and neurons needed to provide reasonable maps. One can explore the possibilities of sources being both inside and outside the space defined by a set of sampling sites. The sources located within the space surrounded by sites provides interpolation for location identification. However, sources located outside the area spanned by the sources requires extrapolation and a major question arises as to how far that extrapolation can be successfully made. Careful simulation studies will be needed to test the resolving power of this approach.

This is not the only application of neural networks. One can also employ the more commonly used back-propagation neural networks to solve the CMB problem with the possibility of incorporating non-linearities into the system. The mass balance problem is analogous to multivariate calibration. Again the problem of having a set of calibration samples where the contributions of the identified sources are *a priori* is a major limitation. The same approach described previously for solving the PLS problem can be used here as well.

FACTOR ANALYSIS METHODS

Introduction

Alternative approaches have been developed for identifying and quantitatively apportioning sources of airborne particles using multivariate statistical analysis. Eigenvector analysis has been the principal method applied to airborne particle composition data. An eigenvector analysis tries to simplify the description of a system by determining the minimum number of new variables necessary to reproduce the measured attributes of the system. The mathematical basis of these methods has been described by Hopke (1985).

Principal components and factor analysis are names given to several of the forms of eigenvector analysis. It was originally developed and used in psychology to provide mathematical models of psychological theories of human ability and behavior (Harman, 1976). Eigenvector analysis has found wide application throughout the physical and life sciences. Unfortunately, a great deal of confusion exists in the literature in regard to the terminology of this analysis. Various changes in the way the method is applied have resulted in it being called factor analysis, principal components analysis, principal components factor analysis, empirical orthogonal function analysis, Karhunen-Loeve transform, etc., depending on the way the data are scaled before analysis or how the resulting vectors are treated after eigenvector analysis is completed. All of the methods have the same basic objective; compression of data into fewer dimensions and identification of the structure of interrelationships that exist between the variables measured or the cases studied.

Principal Components Analysis

The simplest form of factor analysis is principal components analysis. This method attempts to simplify the description of a system by determining a minimum set of basis vectors that span the data space to be interpreted. In other words, a new set of variables are found as linear combinations of the measured variables so that the observed variations in the system can be reproduced by a smaller number of these causal factors. It has been widely used in studies of airborne particulate matter composition data (e.g., Roscoe *et al.*, 1982; Gao *et al.*, 1994).

Since PCA can only be performed on a set of samples in which the various sources contribute different amounts of particles to each sample, the mass balance needs to be expanded to a matrix equation:

$$Z = A \cdot F \tag{19}$$

where:

Z — is a matrix of sample vectors,
A — is the matrix of loading vectors related to the source compositions,
F — is the matrix of scores that are related to the contribution of that source type to the variance of that particular measured variable.

PCA can produce components that have a strong relationship with only one variable. Since it is quite reasonable for environmental systems to show factors that produce such single variable behavior, it is advisable to use PCA; and extend the number of factors to those necessary to reproduce the original data, within the error limits inherent in the data set.

By design, the eigenvector analysis compresses the information content of the data set into as few eigenvectors as possible. In considering the number of factors used to describe the system, it is essential to carefully examine the problems of reconstructing both the variability within the data, and the actual data itself. Following the diagonalization of the correlation or covariance matrix, it is necessary to make the difficult choice of the number of factors, p, to use in the subsequent analysis. This problem occurs in any application of an eigenvector analysis of data containing noise. In the absence of error, the eigenvalues beyond the true number of sources become zero, except for calculational error. The choice becomes more difficult depending on the error in the data. Several approaches have been suggested (Duewer *et al.*, 1976; Hopke *et al.*, 1980).

A large relative decrease in the magnitude of the eigenvalues is one indicator of the correct number of factors. It can often be useful to plot the eigenvalues as a function of factor number and look for sharp breaks in the slope of the line (Cattell, 1966). If the eigenvalue is a measure of the information content of the corresponding eigenvector, then only sufficiently large eigenvalues need to be retained in order to reproduce the variation initially present in the data. One of the most commonly used and abused criteria for selecting the number of factors to keep is retaining only those eigenvalues greater than one (Gutman, 1954). The argument is made that the normalized variables each carry one unit of variance. Thus, if an eigenvalue is less than one, it carries less information than one of the initial variables and is therefore not needed. Kaiser and Hunka (1973) make a strong argument that although an eigenvalue greater than one does set a lower limit on the number of factors to be retained, it does not set a simultaneous upper bound. There must be at least as many factors as there are eigenvalues greater than one; but there can be more than that number that are important to the understanding of the system's behavior.

Hopke (1982) has suggested a useful empirical criterion for choosing the number of retained eigenvectors. In a number of cases of airborne particulate matter identification it was found that choosing the number of factors containing variance greater than one after an orthogonal rotation, provided a stable solution. Since the eigenvector analysis artificially compresses variance into the first few factors, reapportioning the variance using the rotations described in the next section will result in more factors with total variance greater than one, than there are eigenvalues greater than one. In many cases, this number of factors will stay the same even after rotation. Because the matrices produced by the diagonalization process have been calculated in a way to maximize the amount of variance contained in each factor, they may not reflect the pattern of variables representative of a particle source. The factors are rotated generally to achieve what is termed simple structure (Hopke *et al.*, 1976). A varimax rotation is most commonly used in such analyses.

In order to demonstrate the use of Principal Components Analysis (PCA) for the resolution of sources of urban aerosols, it was applied to a composition data set obtained from aerosol samples collected during the RAPS program in St. Louis, Missouri. In the RAPS program, automated dichotomous samplers were operated over a two year period at 10 sites in the St. Louis metropolitan area. The location of RAPS sampling station 112 is shown in Figure 4. Ambient aerosol samples were collected in fine, $<2.4\,\mu m$ and coarse, 2.4–$20\,\mu m$ fractions, and were analyzed at the Lawrence Berkeley Laboratory for total mass by beta-gauge measurements and 27 elements by X-ray fluorescence. The data for the samples collected during July and August 1976 from station 112 were selected for analysis.

Station 112 is located near Francis Field, the football stadium on the campus of Washington University, west of downtown St. Louis. During the 62 days of July and August, filters were changed at twelve hour intervals, producing a total of 124 samples in each the fine and coarse fractions. Data were missing for 24 pairs leaving a total of 100 pairs of coarse and fine fraction samples. Of the 27 elements determined for each sample, a majority of the determinations had 10 elements below the detection limits. Since a complete and accurate data set is required to perform a factor analysis, these 10 elements were eliminated from the analysis. For example, arsenic was excluded because almost all of the values were below the detection limits. Arsenic determinations by X-ray fluorescence are often unreliable because of an interference between the arsenic K X-ray and the lead L X-ray. NAA of these samples would produce better arsenic determinations. Reliable data for arsenic may be important to the differentiation of coal fly ash and crustal material; two materials with very similar source profiles. The low percentage of measured elements can lead to distortions in the scaling factors produced by the multiple regression analysis. The remaining mass consists primarily of hydrogen, oxygen, nitrogen, and carbon. Although no measurements of carbon are included in the RAPS data, that portion of the sample mass must still be accounted for by the resolved sources. In order to produce the best possible source resolutions, it is vital to have accurate measurements of the mass of total suspended particles (TSP), as well as determinations for as many elements as possible. The fine and coarse samples were analyzed separately and only the fine fraction results will be reported here. The eigenvalues for this data set are given in Table 1. There are four factors with eigenvalue greater than 1.0.

Table 1. Eigenvalues for PCA of July–August 1976 fine fraction data from RAPS site 112.

Factor number	Eigenvalue
1	5.472
2	3.265
3	2.064
4	1.143
5	0.892
6	0.633
7	0.607
8	0.377
9	0.228
10	0.116
11	0.080
12	0.059
13	0.041
14	0.014
15	0.009

However, there are values of 0.8, 0.6, and 0.6. There are at least four factors, but it appears likely that there will be seven factors with total variance greater than one after rotation. The total variance is calculated as the sum of the squared loadings, a_{ij}^2, for a given factor. The varimax rotated loadings are presented in Table 2. The loadings are considered a correlation coefficient between the original variables and new factor variables that represent a linear combination. Loading values over 0.50 are generally considered significant although smaller ones may still provide insights into the nature of the sources. It is now necessary to interpret the pattern of elements for each factor.

The first factor is unusual in that it has high values for Al, K, Ca, and Sr. It is not obvious what might be the source of such an unusual combination of elements. In this case, the examination of the corresponding scores can prove helpful. A time series plot of the scores is given in Figure 9. It can be seen that the score values are low with the exception of two samples. It should be remembered that these scores are in standardized space. A score of zero means that this is an average contribution of the given source to the given sample. The large score suggests a specific source that was only operating during these two sampling periods. These samples were taken during the period of noon to midnight on July 4 and midnight

Table 2. Rotated factor loading matrix for site 112 fine fraction data.

Element	Factor number						
	1	2	3	4	5	6	7
Al	**0.619**	**0.766**	−0.040	0.049	−0.085	−0.054	−0.019
Si	0.067	**0.981**	0.000	−0.034	−0.060	−0.097	−0.052
S	−0.040	−0.041	0.183	0.118	**0.957**	0.104	0.049
Cl	0.248	−0.016	0.187	0.473	**0.735**	0.265	0.137
K	**0.982**	−0.002	0.034	0.075	0.055	0.078	−0.021
Ca	**0.827**	0.379	0.062	0.226	0.001	0.042	0.178
Ti	0.062	0.115	0.118	0.169	0.096	−0.003	**0.955**
Cr	0.143	0.012	**0.778**	0.271	0.177	0.220	0.201
Mn	−0.022	0.137	**0.885**	0.143	0.144	0.240	0.004
Fe	−0.048	**0.830**	0.243	0.124	0.087	0.125	0.419
Cu	0.173	0.016	0.159	0.239	0.130	**0.859**	−0.070
Zn	−0.053	−0.086	0.439	0.087	0.159	**0.762**	0.112
Br	0.295	0.061	0.179	**0.872**	0.154	0.182	0.186
Sr	**0.984**	−0.018	0.023	0.068	0.055	0.051	−0.004
Pb	0.010	0.044	0.446	**0.747**	0.367	0.238	0.097

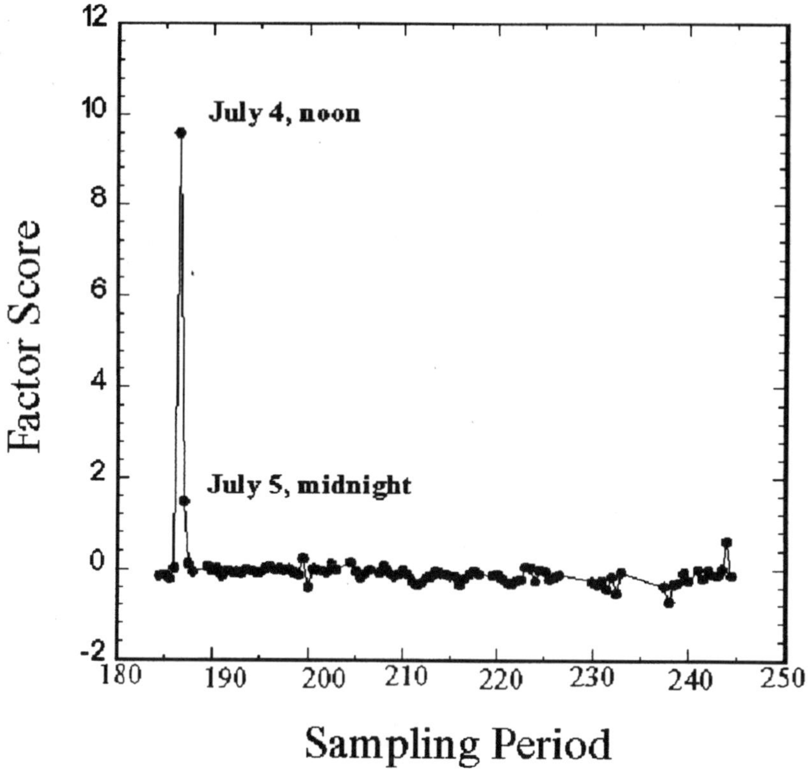

Figure 9. Time series factor scores for factor 1, site 112 fine fraction data.

to noon on July 5. It was the celebration of the bicentennial of the United States and there was a large fireworks show at this stadium. Therefore this factor represents fireworks. One can compare the average values for the measured variables in the complete data sets and in a subset in which the samples for July 4 and 5 have been deleted. This comparison is provided in Table 3.

It is clear that the presence of these two samples within a set of 99 has created a single factor. Such a potent source could significantly distort the whole analysis. Since this source is unique to this particular occasion and is the average properties of the system that are of interest, another analysis with these samples deleted is performed. The eigenvalue results for this analysis is presented in Table 4 and the loadings are presented in Table 5. Now there are only three eigenvalues greater than one, but there still appears to be up to seven factors. To evaluate the best model, each of the

Table 3. Comparison of data with and without samples from July 4th and 5th (ng/m^3) RAPS station 112, July and August 1976, fine fraction.

Element	With July 4th & 5th Mean	Without July 4th & 5th Mean[a]
Al	220 ± 30	220 ± 30
Si	440 ± 60	450 ± 60
S	4370 ± 310	4360 ± 320
Cl	90 ± 10	80 ± 9
K	320 ± 130	150 ± 9
Ca	110 ± 10	110 ± 10
Ti	63 ± 13	64 ± 13
Mn	17 ± 3	17 ± 3
Fe	220 ± 20	220 ± 20
Ni	2.3 ± 0.2	2.3 ± 0.2
Cu	16 ± 3	15 ± 3
Zn	78 ± 8	75 ± 8
Se	2.7 ± 0.2	2.7 ± 0.2
Br	140 ± 9	130 ± 8
Sr	5 ± 4	1.1 ± 0.1
Ba	19 ± 5	15 ± 4
Pb	730 ± 50	720 ± 50

Table 4. Eigenvalues for PCA of July–August 1976 fine fraction data from RAPS site 112 without fireworks samples.

Factor number	Eigenvalue
1	5.495
2	3.746
3	1.446
4	0.950
5	0.865
6	0.747
7	0.581
8	0.326
9	0.296
10	0.238
11	0.125
12	0.080
13	0.057
14	0.036
15	0.012

Table 5. Rotated factor loading matrix for site 112 fine fraction data without fireworks samples.

Element	1	2	3	4	5	6	7
Al	**0.970**	−0.060	−0.110	−0.024	0.118	−0.053	−0.040
Si	**0.976**	0.004	−0.030	−0.075	0.095	−0.055	−0.080
S	−0.072	0.184	**0.956**	0.097	0.008	0.065	0.046
Cl	−0.034	0.198	**0.786**	0.462	0.121	0.210	0.135
K	0.284	0.169	0.081	0.079	**0.889**	0.141	−0.090
Ca	**0.706**	0.037	−0.065	0.360	0.239	−0.013	0.295
Ti	0.141	0.126	0.111	0.171	0.019	−0.022	**0.948**
Cr	0.035	**0.801**	0.183	0.250	−0.004	0.210	0.213
Mn	−0.094	**0.887**	0.156	0.149	0.200	0.120	−0.000
Fe	**−0.800**	0.275	0.131	0.105	0.161	0.104	0.375
Cu	−0.006	0.228	0.147	0.192	0.082	**0.918**	−0.030
Zn	−0.208	0.467	0.179	0.131	**0.524**	**0.526**	0.082
Br	0.082	0.195	0.188	**0.913**	0.091	0.145	0.172
Sr	**0.554**	−0.007	−0.042	0.046	**0.708**	−0.041	0.234
Pb	0.032	0.474	0.404	**0.716**	0.038	0.197	0.072

possible solutions including five, six, and seven factors need to be examined to see which makes the most physical and chemical sense, and which provides a reasonable overall fit to the data. The eight factor solution is clearly one with too many factors since the eighth one does not have any loading with a significant value. After examining these solutions, the seven factor solution is presented in Table 5. The key is to be able to interpret them in terms of likely emission sources.

The first factor is crustal elements with the exception of iron. One cannot distinguish between soil or fly ash based on the elements that are available so it could be either/or a combination of both source types. The Fe loading has a high negative value meaning that when there is significant soil/fly ash input to the samples, there is a decreased iron concentration. There is a large iron/steel complex to the northeast of the site in Granite City, discussed earlier, as well as several iron/steel works on the Missouri side of the Mississippi River which would provide high Fe values. Major sources of fly ash or soil generally lie in other directions, and thus the samples with high Fe from the ferrous metal works are those with low values of the other crustal species and vice-versa.

The second factor has high loadings for Cr and Mn. These elements would be expected to be related to iron/steel works. It would have been more logical if Fe had also had a high positive loading on this factor, but it appears that the negative correlation with the other crustal elements dominates over any correlation between Fe and these other ferrous metals. The third factor is regional sulfate, but again there is a curious additional loading from Cl. It would be expected that S is poorly correlated with the primary particulate emissions since local emissions of S would be in the form of SO_2. It takes time to convert the SO_2 into SO_4^{2-} and the dry deposition rate of SO_2 is much more rapid than that of particulate sulfate.

The fourth factor is clearly motor vehicles. These samples were taken in 1976 when leaded gasoline was still in widespread use in the United States. The lead variance is also partially seen in the second factor indicating that the steel complex being observed is probably the one in Granite City where a secondary lead smelter was still in operation at the site that is now Terracorp (Figure 3). In the fifth factor, Sr, Zn, and K are featured. Although there is no correlation with Pb, this factor is probably an indication of refuse incinerators. So much of the lead variance is associated with motor vehicles and the Granite City source area, that it does not appear here. Potassium and strontium are strongly related to the combustion of paper, particularly when it would be expected that there is no wood combustion in the area.

The sixth factor is related to Cu and Zn. There are two significant sources of non-ferrous metals in the Sauget, Illinois area to the east of site 105. It has been found (Rheingrover and Gordon, 1988) that the Cu plant is a source of fine particles, while the Zn smelter is a source of larger particles. Therefore, we assign this factor to being the Cu products plant. The final factor has a high loading only for Ti. There was a large paint pigment plant that produces TiO_2 and is a major source of fine particle titanium in the area.

The factor analysis can identify major source types and examine some of the relationships among the measured compositions. However, it does not provide a quantitative apportionment in the same way that a CMB analysis can. The problem is that the data are centered and scaled before they are analyzed. It is possible to unscale and uncenter them to produce an Absolute Principal Components Analysis (APCA) using the procedure outlined by Thurston and Spengler (1985). This approach has been successfully applied in a number of cases (Maenhaut and Cafmeyer, 1987; Andrade *et al.*, 1993).

One approach that avoids centering and then uncentering is Target Transformation Factor Analysis (TTFA) discussed in Hopke (1995). Another method providing quantitative source resolutions is Source Apportionment by Factors with Explicit Constraints (SAFER). In this method, the space in which the factor solution lies is truncated by imposing one or more of the natural physical constraints discussed above.

SAFER has been described by Henry and Kim (1989) and by Henry (1991). In addition to the natural constraints, it is also possible to incorporate other ones based on additional knowledge of the system. If the range in which the concentration of an element must lie can be estimated, then these limits can be added to the analysis. SAFER has been recently applied to the interpretation of airborne organic compound data from Atlanta, Georgia (Henry *et al.*, 1994). In this study, they extracted the vehicle-related hydrocarbon source compositions from the ambient sample analytical results.

Positive Matrix Factorization

Several interesting approaches to the problems of undetermined data analysis such as factor analysis have been developed. One method of factor analysis is Positive Matrix Factorization (PMF). It is based on alternating least-squares (Paatero and Tapper, 1993). The factor analysis model has been described in Anttila *et al.* (1995) and Polissar *et al.* (1996). Since the data matrix is scaled element-by-element, Paatero and Tapper (1993) have shown that a meaningful principal components analysis cannot be performed. This form of factor analysis is quite different and is only now reaching the literature to a greater extent. Initially the problem was solved iteratively using alternating least squares. In alternating least-squares, one of the matrices, A or F, is taken as known and the chi-squared is minimized with respect to the other matrix as a weighted linear-least-squares problem. Then the roles of A and F are reversed, so that the matrix that has just been calculated is fixed and the other is calculated by minimizing Q. The process then continues until convergence.

A global optimization scheme has been developed in which the joint solution is directly determined. The method has been applied to data sets of major ion compositions of daily precipitation samples collected over a number of sites in Finland (Anttila *et al.*, 1995). Considerable information on the sources of these ions through the use of PMF is obtainable. It appears promising, but one needs to look carefully at the question of optimum scaling and potential solution methods. It has been applied to Arctic data from the NPS sites and the results are quite encouraging (Polissar *et al.*, 1996) as a method to more quantitatively resolve the major source contributions. The factor analysis problem is written as follows:

$$X = G \cdot F \tag{20}$$

where X is the matrix of observed data. The "Residual Matrix," E is defined as:

$$E_{ij} = X_{ij} - \hat{X} = X_{ij} - \sum_{h=1}^{p} G_{ih} \cdot F_{hj} \tag{21}$$

where:

$i = 1, \ldots, m$ elements,
$j = 1, \ldots, n$ samples.

Define the object function, Q, to be minimized as a function of G and F as:

$$Q(E) = \sum_{i=1}^{m} \sum_{j=1}^{n} \left(\frac{E_{ij}}{\sigma_{ij}} \right)^2 \tag{22}$$

One has to minimize $Q(E)$ with respect to G and F with the constraint that all of the elements of G and F are to be non-negative.

It has been shown (Paatero and Tapper, 1993) that this approach cannot be solved using the singular value decomposition, and a meaningful PCA cannot be performed if the data matrix is scaled element-by-element. The non-negativity constraint is imposed through the use of a penalty function in an enhanced function, $\bar{Q}$. Regularization can also be imposed to reduce rotational freedom so that the enhanced object function is defined as:

$$\begin{aligned}\bar{Q}(E, G, F) &= Q(E) + P(G) + P(G) + R(G) + R(F) \\ &= \sum_{i=1}^{m} \sum_{k=1}^{n} \left(\frac{E_{ij}}{\sigma_{ij}} \right)^2 + \alpha \sum_{i=1}^{m} \sum_{k=1}^{n} \log G_{ih} + \beta \sum_{i=1}^{m} \sum_{k=1}^{n} \log F_{hj} \\ &\quad + \gamma \sum_{i=1}^{m} \sum_{k=1}^{n} G_{ih}^2 + \delta \sum_{i=1}^{m} \sum_{k=1}^{n} F_{ih}^2 \end{aligned} \tag{23}$$

where:

$P(G)$ and $P(F)$ — are penalty functions that prevent the factors G and F from becoming negative.

$R(G)$ and $R(F)$ — are regularization functions that remove some of the rotational indeterminacy.

For the unconstrained case, α and β are set to zero. However, the regularization terms are always present in the analysis.

APPLICATION OF PMF TO A SAMPLING SITE IN ALASKA

The locations of the monitoring sites in Alaska are shown in Figure 10. The aerosol sampling in Alaska is carried out in part, by the Interagency Monitoring of PROtected Visual Environments (IMPROVE) network, a consortium of federal land managing communities (Malm *et al.*, 1994). Denali National Park is an IMPROVE site. The National Park Service performed aerosol measurements at six other locations in Alaska. Locations, periods of observation, and total numbers of samples for the National Park

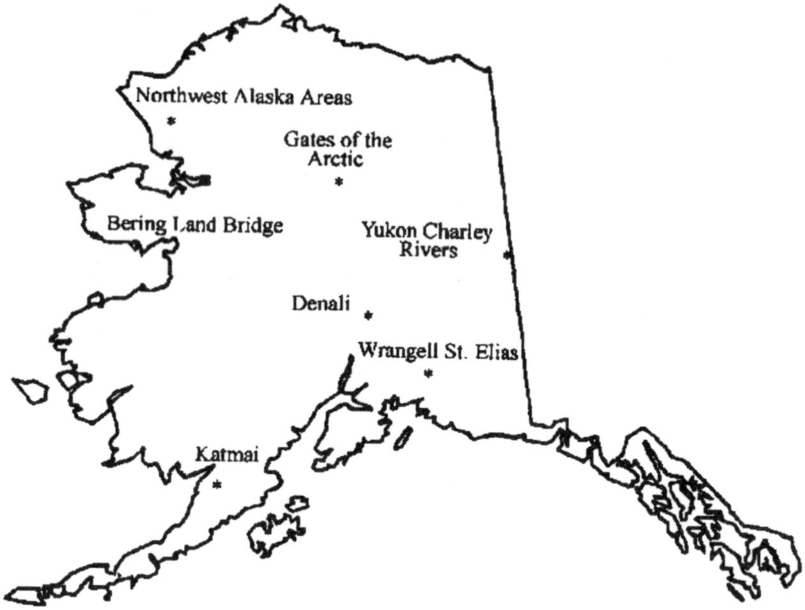

Figure 10. Locations of sites in Alaska National Park Service aerosol sampling network.

Service aerosol sampling network sites in Alaska are shown in Table 6. The aerosol network is managed by the University of California — Davis (Cahill *et al.*, 1986; Malm *et al.*, 1994). All of the samples of atmospheric particles less than 2.5 μm in diameter were collected on Teflon® filters. The duration of aerosol sampling was 24 h with samples collected on Wednesdays and Saturdays. The samples were analyzed by Particle Induced X-Ray Emission (PIXE) to determine aerosol elemental composition, and Laser Integrating Plate Method to measure the aerosol absorption coefficient (Malm *et al.*, 1994). Samples of atmospheric aerosol at the Denali site were also collected on quartz fiber filters and analyzed by Thermal Optical Reflectance (TOR) combustion for organic and elemental carbon measurements (Chow *et al.*, 1993). The peaks in the TOR thermograms from the aerosol samples are defined as low (25–450°C) and high temperature (450–550°C) for organic samples and low (550–700°C) and high (700–800°C) temperature for elemental carbon: OCLT, OCHT, ECLT, and ECHT, respectively.

Table 6. Locations, periods of observation, and total number of samples for the National Park Service aerosol sampling network sites in Alaska.

Site	Abbreviation	Longitude °W	Latitude °N	First sample	Last sample	Total no.
Bering Land Bridge National Preserve	BELA	165.42	64.50	Apr. 1987	Nov. 1992	340
Denali National Park	DENA	149.30	63.45	Mar. 1988	Nov. 1993	583
Gates of the Arctic National Park	GAAR	151.31	66.54	Jun. 1987	Nov. 1992	384
Katmai National Park	KATM	156.38	58.40	Jun. 1987	Nov. 1992	296
Northwest Alaska Areas National Park	NOAA	162.35	67.52	Jun. 1986	Nov. 1992	491
Wrangell St. Elias National Park	WRST	145.70	62.00	Jun. 1987	Nov. 1992	347
Yukon Charley National Preserve	YUCH	141.12	64.47	Jun. 1986	Nov. 1992	514

RESULTS

Sums of one third of average detection limits and average analytical uncertainties for standard deviation matrix σ were used for the PMF calculation (Anttila *et al.*, 1995). Average values with corresponding standard deviations were used for missing ones. Results of the PMF analysis are shown in Figures 11–14. Loadings were not normalized, so that comparison should be made between different factors loadings for each definite element line.

For the original Denali data set, the highest BABS (b-absorption) loading is associated with K, organic and elemental C (factor 4, Figure 11). A lower BABS loading is found in factor 2 along with S, Cu, As, and Br. Factor 4 might represent the wood smoke from forest fires contributions while factor 2 represents the long range transported aerosol. Factor 1 has high loadings for Al, Si, K, Ca, Ti, Mn, and Fe. These elements are generally associated with soil. Factor 5 represents sea salt aerosol (high values for Na, Br and Cl), and factor 6 is probably partially connected with local pollution (high SO_2 and heavy metals values). Finally, factor 3 includes aerosol components with high contribution of organic compounds (high loadings of hydrogen and organic carbon).

The clean data set for the same location is characterized by higher values for BABS and sulfur on the same factor (factor 3, Figure 12). This factor can be attributed to long range transported aerosol. The contribution of BABS to factors 2, 4, and 5 is partially related to the local and regional pollution sources. (Factor 2 has high values for hydrogen and organic carbon, and factor 5 shows high SO_2, heavy metals, and organic carbon loadings). BABS loading in these factors is much lower than the original data set. Factors 6 and 4 represent sea salt and soil aerosol, respectively. It can be seen from these results that the screening of the data set described above by using local meteorological data, reduces local and regional source contribution in the data set to a minimum.

Similar results were obtained for another Alaskan site, Gates of the Arctic National Park (Figures 13–14). The separation procedure of the background data points from other data points, connected with regional pollution using the regression expression obtained for the Denali site, reduced the BABS values in the factors attributed to wood smoke pollution (factors 3 and 5, Figures 13 and 14, respectively). It also increased BABS value in the factor with high sulfur value (factors 4 and 3, Figures 13 and 14, respectively). The last factor probably represents the long range transported aerosol. Both sets of PMF results for the Gates of the Arctic site contained soil factors (factors 1 and 2, respectively) and those connected with heavy metals (factors 5, 6 and 4, 6 respectively).

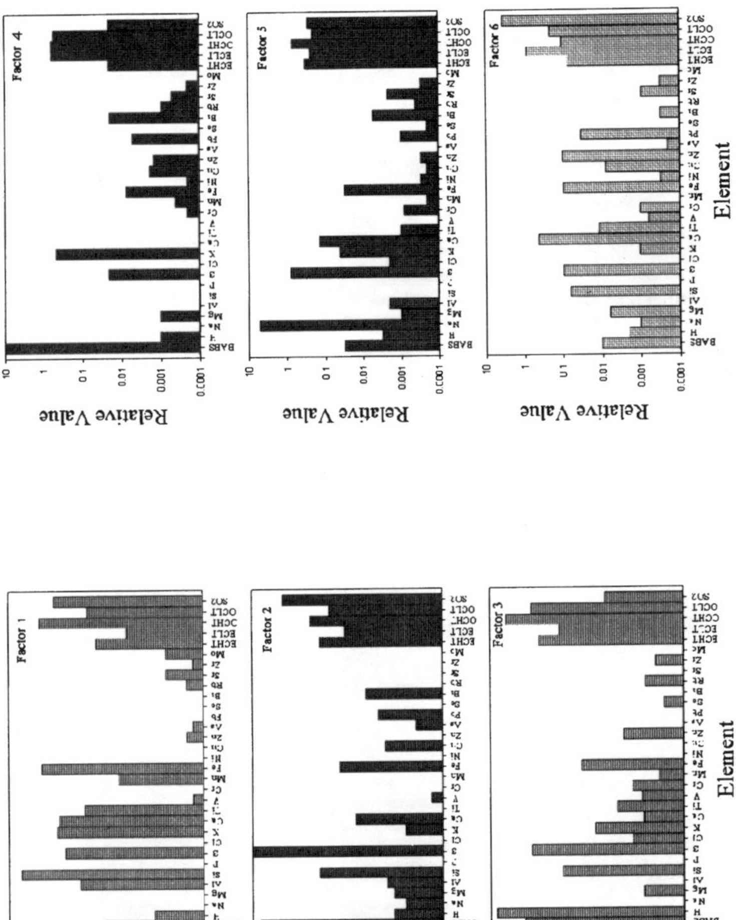

Figure 11. Source profiles as derived from PMF for the complete NPS data set from Dunali.

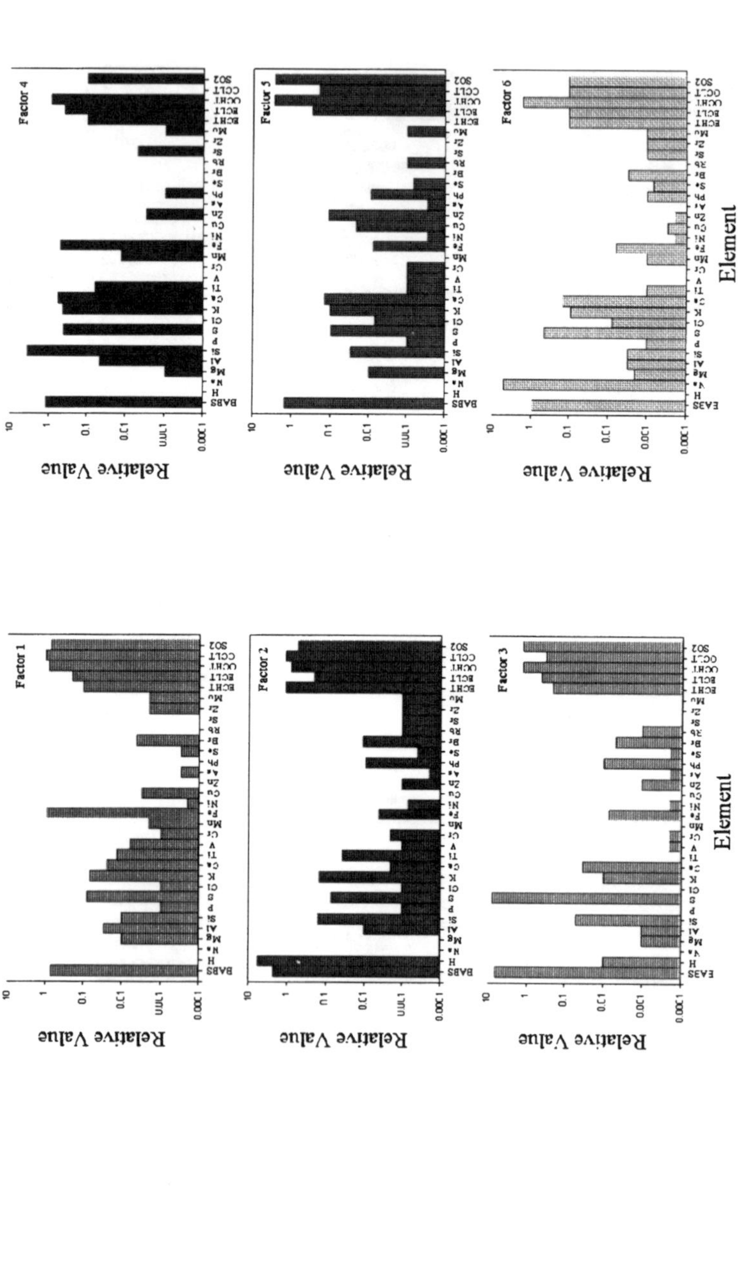

Figure 12. Source profiles as derived from PMF for the background samples in the NPS data set from Dunali.

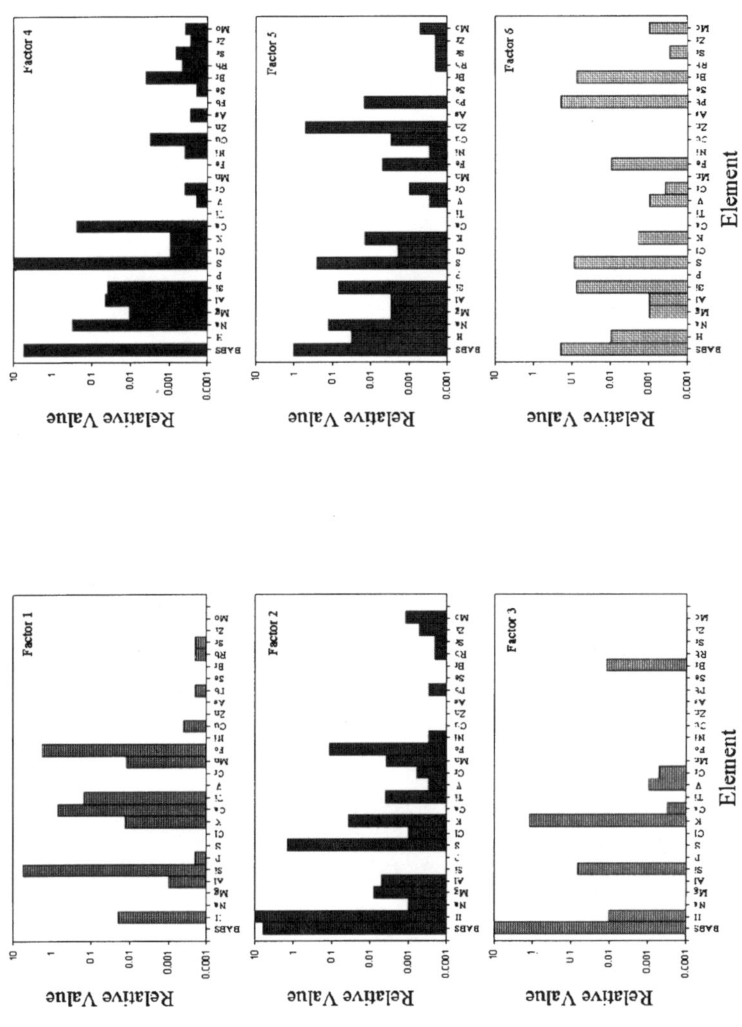

Figure 13. Source profiles as derived from PMF for the complete NPS data set from Gates on the Arctic.

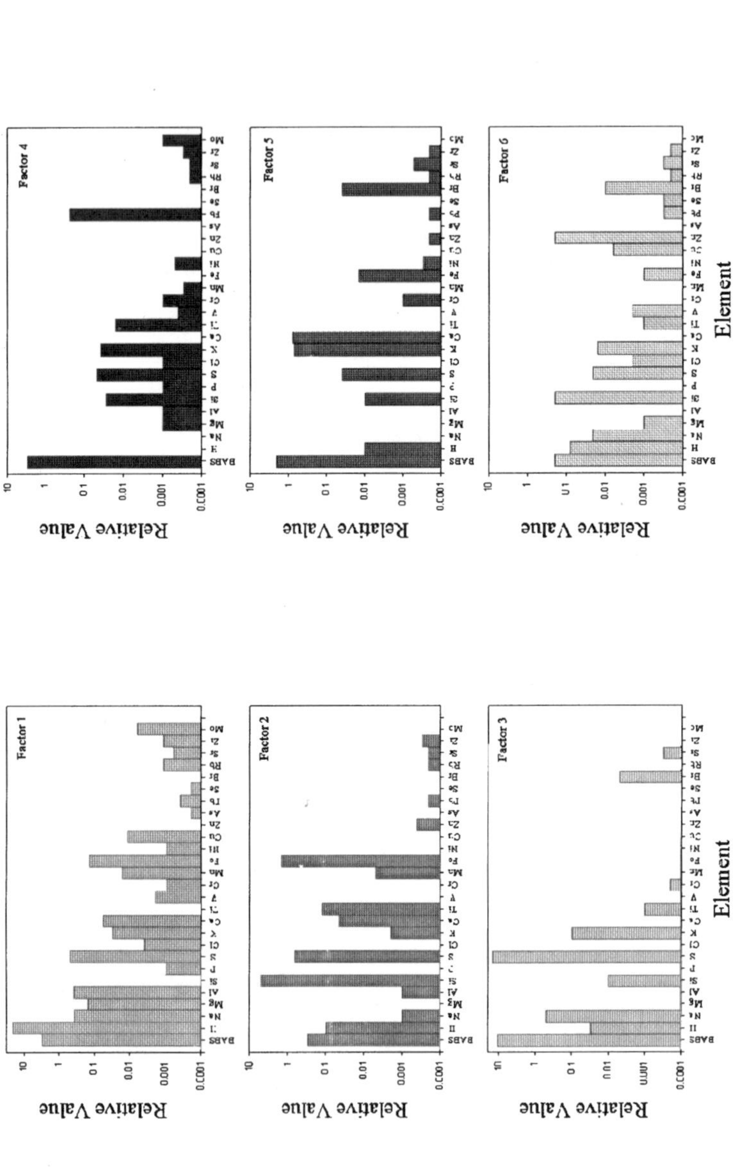

Figure 14. Source profiles as derived from PMF for the background samples in the NPS data set from Gates in the Arctic.

Results of these factor analyses show that the most likely factors with highest optical absorption coefficients and sulfur scores for original data sets, characterized two different types of aerosol at these Arctic locations: regional (wood smoke) and long range transported aerosol, respectively. The separation of the information for both locations provided data with a minimum amount of possible contributions of regional and local pollution sources.

CONCLUSION

There are a variety of methods that have been developed and applied to the problem of particle source identification and quantitative mass apportionment. For many problems, the methods described can provide information useful in the planning of air quality management strategies. However, there are limits to the ability of such methods to resolve sources particularly those with similar source compositions. In such cases additional information needs to be considered. For example, individual particle data can be obtained by automated electron microscopy (Hopke and Casuccio, 1991), and provide the ability to resolve quite similar composition sources. Alternatively, meteorological information can be included in the form of back trajectories which model the movement of the air carrying the particles to the site backward in time from the time the sample has been collected (Gao *et al.*, 1994). Source receptor modeling is a field in which developments are on-going. The ability to perform substantial computer-based analyses offers many additional possibilities for further improvements.

REFERENCES

Alpert, D.J. and P.K. Hopke, *Atmos. Environ.*, **15** (1981) 675.
Andrade, F., C. Orsini and W. Maenhaut, *Nucl. Instrum. Meth. in Phys. Res.*, **B75** (1993) 308.
Anttila, P., P. Paatero, U. Tapper and O. Jarvinen, *Atmos. Environ.*, **29** (1995) 1705.
Cahill, T.A., R.A. Eldred and P.J. Feeney, *Particulate Monitoring and Data Analysis for the National Park Service 1982–1985*, NPS contract number USDICX-3-0056 (1986).
Cartwright, H.M. and G. Mott, "Looking Around: Using Clues from the Data Space to Guide Genetic Algorithm Searches" in *Proceedings of the 3rd International Conference on Genetic Algorithms*, ed. by M. Kaufmann, San Mateo, CA (1991).
Cartwright, H.M. and S.P. Harris, *Atmos. Environ.*, **27A** (1993) 1783.
Cattell, R.B., *Handbook of Multivariate Experimental Psychology*, Rand McNally, Chicago (1966) pp. 174–243.
Cheng, M.D., P.K. Hopke and D. Jennings, *Chemom. Intell. Lab. Syst.*, **4** (1988) 239.
Cheng, M.D. and P.K. Hopke, *Atmos. Environ.*, **23** (1989) 1373.
Chow, J.C., J.G. Watson, L.C. Pritchett, W.R. Pierson, C.A. Frazier and R.G. Purcell, *Atmos. Environ.*, **27A** (1993) 1185.

Cooper, J.A., J.G. Watson and J.J. Huntzicker, *Atmos. Environ.*, **18** (1984) 1347.
Currie, L.A., R.W. Gerlach, C.W. Lewis, W.D. Bald, J.A. Cooper, S.L. Dattner, R.T. DeCesar, G.E. Gordon, S.L. Heisler, P.K. Hopke, J.J. Shah, G.D. Thurston and H.J. Williamson, *Atmos. Environ.*, **18** (1984) 1517.
Duewer, D.L., B.R. Kowalski and J.L. Fasching, *Anal. Chem.*, **48** (1976) 2002.
Dunker, A.M., *A Method for Analyzing Data on the Elemental Composition of Aerosols*, General Motors Research Laboratories Report MR-3074 ENV-67, Warren, Michigan (1979).
Dzubay, T.G., R.K. Stevens and L.W. Richards, *Atmos. Environ.*, **13** (1979) 653.
Dzubay T.G. and R.K. Stevens, "Sampling and Analysis Methods for Ambient PM-10 Aerosol", in *Receptor Modeling for Air Quality Management*, Hopke, P.K. (ed.), Elsevier Science Publishers, Amsterdam (1991) pp. 11–44.
Frank, I.E. and B.R. Kowalski, "Statistical Receptor Models Solved by Partial Least Squares", in *Environmental Applications of Chemometrics*, J.J. Breen and P.E. Robinson (eds.), Symposium Series No. 292, American Chemical Society, Washington, DC, 271–279 (1985).
Friedlander, S.K., *Environ. Sci Technol.*, **7** (1973) 235.
Gao, N., M.D. Cheng and P.K. Hopke, *Atmos. Environ.*, **23** (1994) 1447.
Glover, D.M., P.K. Hopke, S.J. Vermette, S. Landsberger and D.R. D'Auben, *J. Air and Waste Manage. Assoc.*, **41** (1991) 294.
Goldberg, D.E., *Genetic Algorithms in Search, Optimization, and Machine Learning*, Addison-Wesley, Reading, Massachusetts (1989).
Guttman, L., *Psychometrika*, **19** (1954) 149.
Harman, H.H., *Modern Factor Analysis*, 3rd Edition, University of Chicago Press, Chicago (1976).
Harrison, R.M. and W.T. Sturges, *Atmos. Environ.*, **17** (1983) 311.
Hebb, D.O., *The Organization of Behavior*, John Wiley, New York (1949).
Henry, R.C. and B.M. Kim, *Chemom. Intell. Lab. Syst.*, **8** (1989) 205.
Henry, R.C., Multivariate Receptor Models, in *Receptor Modeling for Air Quality Management*, P.K. Hopke (ed.), Elsevier Science Publishers, Amsterdam (1991) pp. 117–147.
Henry, R.V., C.W. Lewis and J.F. Collins, *Environ. Sci. Technol.*, **28** (1994) 823.
Holland, J.H., *Adaptation in Natural and Artificial Systems*, The University of Michigan Press, Ann Arbor, Michigan (1975).
Hopfield, J.J., *Proc. Nat. Acad. Sci.*, **79** (1982) 2554.
Hopke, P.K., E.S. Gladney, G.E. Gordon, W.H. Zoller and A.G. Jones, *Atmos. Environ.*, **10** (1976) 1015.
Hopke, P.K., R.E. Lamb and D.F.S. Natusch, *Environ. Sci. Technol.*, **14** (1980) 164.
Hopke, P.K., *Atmos. Environ.*, **16** (1982) 1279.
Hopke, P.K., *Receptor Modeling in Environmental Chemistry*, John Wiley, New York (1985).
Hopke, P.K. (ed.), *Receptor Modeling for Air Quality Management*, Elsevier Science Publishers, Amsterdam (1991).
Hopke, P.K. and G.S. Casuccio, "Scanning Electron Microscopy", in *Receptor Modeling for Air Quality Management*, P.K. Hopke (ed.), Elsevier Science Publishers, Amsterdam (1991) pp. 149–212.
Hopke, P.K., "The Mixture Resolution Problem Applied to Airborne Particle Source Apportionment", in *Chemometrics in Environmental Chemistry*, Springer-Verlag, Heidelberg (1995) pp. 47–86.
Houck, J.E., "Source Sampling for Receptor Modeling", in *Receptor Modeling for Air Quality Management*, Hopke, P.K. (ed.), Elsevier Science Publishers, Amsterdam (1991) pp. 45–82.
Kaiser, H.F. and S. Hunka, *Edu. Psych. Meas.*, **33** (1973) 99.
Kateman, G., Neural Ne2rks in Analytical Chemistry, Presentation at the Fifth International Meeting on Computer Applications in Analytical Chemistry, Jena, Germany (1992).
Kohln, T., *Self-Organization and Associative Memory*, 3rd Edition, Springer-Verlag, New York (1989).

Kowalczyk, G.S., C.E. Choquette and G.E. Gordon, *Atmos. Environ.*, **12** (1978) 1143.
Kowalczyk, G.S., G.E. Gordon and S.W. *Environ. Sci. Technol.*, **16** (1982) 79.
Leong, K.H., P.K. Hopke and J.J. Stukel, *J. Aerosol Sci.*, **14** (1983) 611.
Liu, C.K., B.A. Roscoe, K.G. Severin and P.K. Hopke, *Am. Ind. Hyg. Assoc. J.*, **43** (1982) 314.
Lucasius, C.B. and G. Kateman, Applications of Genetic Algorithms in Chemometrics, *Proceeding of the 3rd International Conference on Genetic Algorithms*, ed. by M. Kaufmann, Fairfax, Virginia (1989).
Maenhaut, W. and J. Cafmeyer, *J. Trace and Microprobe Techniques*, **5** (1987) 135.
Malm, W.C., J.F. Sisler, D. Huffman, R.A. Eldred and T.A. Cahill, *J. Geophys. Res.*, **99** (1994) 1347.
Martens, H. and T. Naes, *Multivariate Calibration*, John Wiley, Chicester (1989).
McCulloch, W.S. and W. Pitts, *Bull. Math. Biophys.*, **5** (1943) 115.
Miller, M.S., S.K. Friedlander and G.M. Hidy, *J. Colloid Interface Sci.*, **39** (1972) 65.
Paatero, P. and U. Tapper, *Chemom. Intell. Lab. Sys.*, **18** (1993) 183–194.
Pitts, W. and W.S. McCulloch, *Bull. Math. Biophys.*, **9** (1947) 127.
Pollisar, A.V., P.K. Hopke, W.C. Malm and J.F. Sisler, *Atmos. Environ.*, **30** (1996) 1147.
Prim, R.C., *Bell Syst. Tech. J.*, **36** (1958) 1389.
Rheingrover, S.W. and G.E. Gordon, *Aerosol Sci. Technol.*, **8** (1988) 29.
Roscoe, B.A., P.K. Hopke, S.L. Dattner and J.M. Jenks, *J. Air Pollut. Control Assoc.*, **32** (1982) 637.
Roscoe, B.A., C.Y. Chen and P.K. Hopke, *Anal. Chim. Acta*, **160** (1984) 121.
Severin, K.G., B.A. Roscoe and P.K. Hopke, *Particulate Sci. and Technol.*, **1** (1983) 183.
Thurston, G.D. and J.D. Spengler, *Atmos. Environ.*, **19** (1985) 9.
Vong, R., P. Geladi, S. Wold and K. Esbensen, *J. Chemom.*, **2** (1988) 281.
Vong, R., *Anal. Chim. Acta*, **277** (1993) 389.
Wang, S. and T. Larson, *Anal. Chim. Acta*, **272** (1993) 333.
Watson, J.G., *Chemical Element Balance Receptor Model Methodology for Assessing the Source of Fine and Total Suspended Particulate Matter in Portland, Oregon*, Ph.D. Thesis, (1979) Oregon Graduate Center, Beaverton, Oregon.
Watson, J.G., J.C. Chow and T.G. Pace, "Chemical Mass Balance", in *Receptor Modeling for Air Quality Management*, Hopke, P.K. (ed.), Elsevier Science Publishers, Amsterdam (1991) pp. 83–116.
Wienke, D. and P.K. Hopke, *Anal Chim. Acta*, **291** (1994a) 1.
Wienke, D. and P.K. Hopke, *Environ. Sci. Technol.*, **28** (1994b) 1015.
Wienke, D., N. Gao and P.K. Hopke, *Environ. Sci. Technol.*, **28** (1994) 1023.
Winchester J.W. and G.D. Nifong, *Water, Air, Soil Pollut.*, **1** (1971) 50–64.
Wolff, G.T., M.S. Ruthkosky, D.P. Stroup and P.E. Korsog, *Atmos. Environ.*, **25A** (1991) 2173.
Zupan, J. and J. Gasteiger, *Anal. Chim. Acta*, **248** (1991) 1.

INDEX